Practical Num

Methods with C#

Numerical Programming and Math Functions
for Real-World .NET Applications with C#

Practical Numerical Methods with C#

Numerical Programming and Math Functions for
Real-World .NET Applications with C#

Jack Xu, Ph.D

UniCAD Publishing

Practical Numerical Methods with C#

Copyright © 2008 by Jack Xu, Ph.D
Printed and bound in the United States of America 9 8 7 6 5 4 3 2 1UC

Editor: Betty Hsu

The author and publisher have made every effort in the preparation of this book to ensure the accuracy of the information, however this book is sold without warranty, either express or implied. No liability is assumed for incidental or consequential damages in connection with or arising out of the use of the information or programs contained in the book.

The publisher offers excellent discounts on this book when ordered in quantity for bulk purchases or special sales, which may include electronic versions and /or custom covers and content particular to your business, training goals, marketing focus, and branding interests. For more information, please contact:

sales@drxudotnet.com
Visit us on the website: www.drxudotnet.com

Published by UniCAD Publishing.
Phoenix, USA
ISBN-13: 978-0-9793725-3-7
ISBN-10: 0-9793725-3-4

Publisher's Cataloging-in-Publication Data

Xu, Jack
Practical Numerical Methods with C# – Numerical Programming and Math Functions for Real-World .NET Applications with C# / Jack Xu – 1st ed.
p.cm.
ISBN-13: 978-0-9793725-3-7
ISBN-10: 0-9793725-3-4

1. C# Programming. 2. Numerical Method. 3.Windows Presentation Foundation . 4. .NET Framework 5. Math Functions. 6. .NET Applications.
I. Title. II. Title III Title: Practical Numerical Methods with C#

For my wonderful family

Contents

Chapter 12 Numerical Integration 277

Chapter 13 Ordinary Differential Equations 301

Introduction

Overview

Welcome to *Practical Numerical Methods with C#*. This book is intended for scientists, engineers, and .NET developers who want to create scientific and engineering applications using C# and .NET Framework. For many years, FORTRAN has been the dominant language of scientific and engineering computation. As Microsoft's C# and .NET Framework gain popularity, you may find them also suitable for technical computing. This book presents C#-based procedures that perform fundamental mathematical and numerical computations critical to scientists and engineers.

The power of C# programming language, combined with the simplicity of implementing Windows Form, WPF desktop applications, and Silverlight Web applications based on the Visual Studio .NET framework, makes real-world .NET program development faster and easier than ever before. Visual C# is a versatile and flexible tool that allows users with even the most elementary programming abilities to not only perform complicated computations, but also display the calculated results in a variety of graphical representations. In this regard, C# is more powerful than FORTRAN, because it is hard to show results graphically using FORTRAN.

The main advantage of using FORTRAN in scientific and engineering computing is its rich math libraries. These libraries implement a complete collection of mathematical, statistical, and numerical algorithms, which have been evolving steadily for several decades. Each subroutine and algorithm in these libraries has undergone rigorous testing and quality assurance, providing users with more time to focus on their applications. On the other hand, the C# programming language is relatively new to the scientific and engineering community. The lack of C# math libraries prevents many researchers from using the C# programming language in their applications. In this book, I will show you that it's fairly easy to develop math libraries in C#, and that it is worth developing scientific and engineering applications using C# due to its computing power and graphical representations capability.

This book is aimed to provide scientists and engineers with a comprehensive explanation of scientific computing using C#. Much of the work in this book is original, based on my own programming experience in developing commercial Computer Aided Design (CAD) packages, which involve intensive scientific computations and sophisticated graphical representations. With FORTRAN, developing advanced graphics and chart applications is a difficult and time-consuming task. To add even simple charts or graphs to your applications, you have to waste effort in creating a chart program, or money in buying commercial graphics and chart add-on packages. Visual C# and its rich graphics features make it possible to easily implement both powerful math libraries and professional graphics using entirely managed C# codes.

Practical Numerical Methods with C# provides an in-depth introduction to performing complicated scientific computations using C# applications. In this book, I will begins with an overview of the C# and .NET Framework, and then present procedural descriptions of linear algebra, numerical solution of nonlinear and ordinary differential equations, optimization, parameter estimation, and special functions of mathematical physics. I will show you how to create useful C# mathematical and numerical libraries that you can use in real-world scientific and engineering problems. I will try my best to introduce the C# program to scientists and engineers in a simple way – simple enough for C# beginners to easily follow. From this book, you can learn how to perform complicated scientific computations and create your own math libraries based on C# and .NET Framework.

Practical Numerical Methods with C# is not simply a book, but a powerful C# math library. You may find that you can immediately use some of the examples in this book in your real-world problems, and that you can use others to give you inspiration on adding more advanced math libraries to your applications.

What This Book Includes

This book and its sample code listings, which are available for download from our website at www.drxudotnet.com, provide you with:

- A complete, in-depth instruction on practical scientific computing and programming using C#. After reading this book and running the example programs within, you will be able to create various math and numerical libraries in your own C# applications.

- Ready-to-run example programs that allow scientists and engineers to explore the numerical methods described in the book. You can use these examples to better understand how the mathematical model and algorithms work. You can also modify the code or add new features to them to form the basis of your own programs. Some of the example code listings provided in this book are already sophisticated math libraries; these can be used directly in your own real-world applications.

- Many C# classes in the sample code listings that you will find useful in your real-world scientific and engineering problems. These classes contain linear algebra, matrix manipulation, numerical approaches, and other useful utility classes. You can extract these classes and plug them into your own applications.

Is This Book for You?

You don't have to be an experienced C# developer or expert to use this book. I designed this book to be useful to scientists and engineers with all levels of C# programming experience. In fact, I believe that if you have some experience with programming languages other than C#, you will be able to sit down in front of your computer, start up Microsoft .NET Framework SDK or Visual Studio .NET, follow the examples that are provided with this book, and quickly become familiar with C# programming in scientific computing. For those of you who are already experienced C# developers, I believe this book has plenty to offer you as well. The information in this book about creating C# math libraries is not available in any other C# tutorial and reference book. In addition, most of the example programs provided with this book can be used directly in your real-world application development. This book will provide you with a level of detail, explanation, instruction, and sample program code that will enable you to do just about anything scientific and engineering computing related with visual C#.

This book is specifically designed for scientists and engineers. In fact, my own background is in theoretical physics, a field involving extensive numerical calculations as well as graphical representations of calculated data. I have been dedicated to this field for many years. My first computer experience was also with FORTRAN. Later on, I gained programming experience in Basic, C, C++, and MATLAB. I still remember how hard it was in those early days to present computational results graphically. I often spent hours creating a publication-quality chart by hand, using a ruler, graph paper, and rub-off lettering. During that time, I started to pay attention to various development tools that could be used to create integrated applications. I tried to find an ideal development tool that would allow me not only to easily generate data (computation capability) but also to easily represent data graphically (graphics and chart power). The C# and Microsoft Visual Studio .NET development environment made it possible to develop such integrated applications. Ever since Microsoft .NET 1.0 came out, I have been in love with the C# language, and have used this tool to successfully create powerful scientific and plotting applications, including commercial CAD packages.

The majority of the example programs in this book can be used routinely by scientists and engineers. Throughout this book, I will emphasize the *usefulness* of C# programming in real-world scientific and engineering problems. If you follow this book closely, you will be able to easily develop various math and numerical libraries. At the same time, I will not spend too much time discussing program style, execution speed, and code optimization, because there is a plethora of books out there that already deal with those topics. Most of the example programs in this book omit error handlings. This makes the code easier to understand by focusing on the key concepts.

Note that this book focuses on numerical computing methods and math library development using C#. It will not address the graphical representations of your calculation results. In fact, the real power of the .NET Framework is its ability to create graphics and user interfaces. If you are interested in graphics and user interface programming in C#, you can read my other books:

Practical C# Charts and Graphics – This book is a perfect guide to learning all the basics for creating advanced chart and graphics applications in C#, GDI+, and Windows Form. It clearly explains practical chart and graphics methods and their underlying algorithms. The 2D and 3D chart packages contained in the book can be directly used in your C# applications.

Practical WPF Programming – This book provides all the information you need to add advanced graphics to your .NET applications using C# and Windows Presentation Foundation (WPF), which comes with the new version (3.0 or later) of .NET framework. From 2D shapes and charts to complex interactive 3D models, this book uses code examples to explain every step it takes to build a variety of WPF graphics applications.

Practical Silverlight Programming – This book shows you how to develop rich interactive applications (RIAs) for Web using C# and Silverlight. Silverlight is a subset of WPF and enables you to create advanced graphics and user interfaces for Web applications. You will learn from this book how to display your computation results graphically and interactively over the internet.

What Do You Need to Use This Book?

You'll need no special equipment to make the best use of this book and understand the algorithms. To run and modify the sample programs, you need a computer capable of running either the Windows Vista or XP operating systems. The software installed on your computer should include .NET Framwork SDK 2.0 or later, which is available at Microsoft Website for free download. It will be better if you have Visual Studio 2005 or 2008 standard edition or higher. Please note that all of the example programs and math libraries were created and tested on Visual Studio 2008. However, the example code should be independent of which platform you use.

How the Book Is Organized

This book is organized into fifteen chapters, each of which covers a different topic of numerical computating. The following summaries of each chapter should give you an overview of the book's contents:

Chapter 1, *Overview of C# Programming*

This chapter introduces the basics of C# programming, including the basic types, properties, methods, mathematical operations, and how to create branches and loops.

Chapter 2, *Complex Numbers and Functions*

This chapter demonstrates how to implement a complex structure, which contains the definition of complex numbers, complex operators, and commonly used complex functions. This structure allows you to perform various computations using complex numbers and functions.

Chapter 3, *Vectors and Matrices*

This chapter introduces a more general n-dimensional vector class and a general matrix class with $n \times m$ dimension, which can be used in many scientific and engineering computations involving the solution of linear equations with multiple variables. Matrix analysis is a basic theory of these linear operations.

Chapter 4, *LinearAlgebraic Equations*

This chapter introduces various numerical methods for solving linear equations with an arbitrary number of unknowns. Solving linear equations is one of the most commonly used operations in numerical analysis and scientific and engineering applications.

Chapter 5, *Nonlinear Equations*

This chapter describes several numerical methods for solving nonlinear equations. These numerical methods are all iterative in nature, and may be used for equations that contain one or several variables.

Chapter 6, *Special Functions*

This chapter discusses a special function class, which contains popular special functions such as the gamma function, beta function, error function, elliptic intergral, Laguerre function, Hermit function, Chebyshev function, Legendre function, and Bessel function, etc.

Chapter 7, *Random Numbers and Distribution Functions*

This chapter covers a variety of random number generators and different probability distribution functions, which can be used to simulate the different chaotic circumstances that can be found in the real world.

Chapter 8, *Interpolation*

This chapter explains the implementation of several interpolation methods, which can be used to construct new data points within the range of a discrete set of known data points. The interpolation is usually called curve fitting or regression analysis, and can be regarded as a special case of curve fitting, in which the function must go exactly through the data points.

Chapter 9, *Curve Fitting*

This chapter explains a variety of curve fitting approaches that can be applied to data containing noise, usually due to measurement errors. Curve fitting tries to find the best fit to a set of given data. Thus, the curve does not necessarily pass through all of the given data points.

Chapter 10, *Optimization*

This chapter covers several popular methods for optimizing functions with multiple variable, including the golden search, Newton, simplex, simulated annealing, and differential evolution techniques. In particular, simulated annealing and differential evolution can deal with highly nonlinear models, chaotic, noisy data, and constraints.

Chapter 11, *Numerical Differentiation*

This chapter discusses several methods of numerical differentiation, such as forward and backward difference, central difference, extended central difference, Richardson extrapolation, and derivatives by interpolation. These methods provide you with different tools for estimating the derivative of a function.

Chapter 12, *Numerical Intergration*

This chapter covers a variety of methods for numerical integration, including methods based on Newton-Cotes formulas, Romberg integration, and Gaussian quadrature methods. These methods can be used to estimate the finite and infinite integrals of functions.

Chapter 13, *Ordinary Differential Equations*

This chapter focuses on solving ordinary differential equations numerically. It presents several popular methods including the Euler method, second- and fourth-order Runge-Kutta methods, Adaptive Runge-Kutta method, and the Runge-Kutta methods that can be used for solving a system of ordinary differential equations.

Chapter 14, *Boundary Value Problems*

This chapter discusses two methods for solving boundary value problems: solution by the shooting method and by finite differences. The shooting method involves guessing the missing values and the resulting solution is very unlikely to satisfy boundary conditions at the other end. The finite difference method involves approximating the differential equations by finite differences at evenly spaced mesh points.

Chapter 15, *Eigenvalue Problems*

This chapter presents several popular methods for solving eigenvalue problems, including the Jacobi method, power iteration, Rayleigh method, Rayleigh-quotient method, and matrix tridiagonalization method. These methods offer you nontrivial tools for calculating eigenvalues and eigenvectors of a real symmetric matrix system.

Using Code Examples

You may use the code in this book in your applications and documentations. You don't need to contact the author for permission unless you are reproducing a significant portion of the code. For example, writing a program that uses several chunks of code from this book doesn't require permission. Selling or distributing the example code listings does require permission. Incorporating a significant amount of example code from this book into your applications and documentation also requires permission. Integrating the example code from this book into commercial products isn't allowed except with the author's written permission.

Customer Support

I am always interested in hearing from readers, and would like to know your thoughts on this book. You can send me comments by e-mail to jxu@drxudotnet.com. I also provide updates, bug fixes, and ongoing support via my website: www.drxudotnet.com.

You can also obtain the complete source code for all of examples in this book from the above website.

Note that this website was developed using Silveright. In order to view the site, you'll be prompted to install Silverlight runtime to your browser.

Chapter 1
Overview of C# Programming

The Microsoft .NET framework is a programming environment that contains a large class library, a programming language – C# with which one writes applications manipulating these classes, and a Common Language Runtime (CLR). With this framework, you are not restricted to using C# to access the .NET class library. This library can be manipulated using a host of different programming languages, meaning that developers can choose the programming language they're most familiar with for .NET application development.

The most interesting part of .NET, however, is the CLR, which is a kind of virtual machine that executes Microsoft Interpreted Language code. The CLR manages the execution of code compiled for the .NET platform. It has two interesting features: firstly, its specification has been opened up so that it can be ported to non-Windows platforms; secondly, any number of different languages can be used to manipulate the .NET framework classes, and the CLR will support them.

Note that not all of the supported languages fit entirely into the .NET framework. The one language that is guaranteed to fit in perfectly is C#. This new language, a successor to C++, has been released in conjunction with the .NET framework, and is likely to be the language of choice for many developers working on .NET applications. C# is the flagship language of the .NET framework. Unlike most other languages, C# can access every feature of .NET framework.

Visual C# and .NET framework are powerful and useful programming tools for developing applications. In this chapter, I will show you how to create a simple C# program and introduce you to the class concept. I will also overview basic types, mathematical operations, properties, methods, and how to create branches and loops.

Your First C# Program

As I mentioned in the Introduction section, in order to use C# and the .NET framework classes, you will first need to install either the .NET framework SDK, or Visual Studio .NET. In this book, I will create all of the examples and math libraries using Visual Studio 2008.

As with the C programming language, your first C# program is also a "Hello World!" program, which simply displays a welcome message on your computer screen. You can perform this simple task using the Console application.

Start a new project with Visual Studio 2008 to bring up the New Project dialog window, as shown in Figure 1-1. Select Visual C# from Project types, select Console Application from Templates, and enter the project name: *FirstCSharpProgram*. Click the OK button.

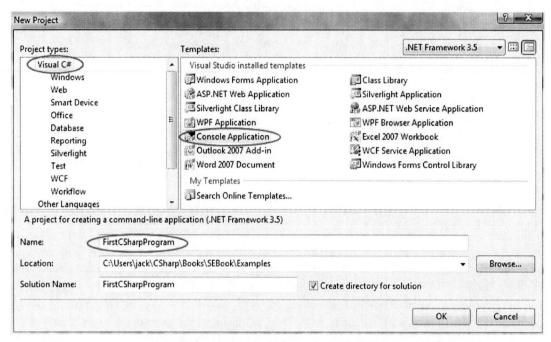

Figure 1-1 Start a new project in Visual Studio.

Visual Studio then generates the project and opens the *Program.cs* file on your computer screen.

```
using System;
using System.Collections.Generic;
using System.Linq;
using System.Text;

namespace FirstCSharpProgram
{
    class Program
    {
        static void Main(string[] args)
        {
        }
    }
}
```

This file is always automatically generated by Visual Studio whenever you start a new C# Console Application. There are several points about C# that need to be addressed:

• C# is a case-sensitive programming language. You will get compiler errors if , for instance, you write "system" instead of "System".

• Every statement in C# must finish with a semicolon (;) or else take a code block curly braces.

- Because C# is an object-oriented language, C# programs must be placed in classes. In this example, the default class name, *Program*, is used.

- The *using* statements at the beginning of the program declare different namespaces, which will save your time typing.

Now, we need to add the following code snippet to the *Main* method:

```
Console.WriteLine("Hello, World!");
```

You can then compile and run the project by pressing F5. When you run the program, you should get a simple output on your command (or DOS) prompt screen:

```
Hello, World!
```

There is a problem associated with your command prompt screen. You may find that the screen disappears very quickly even before you see the result. To fix this, you can add another line of code to the *Main* method:

```
Console.ReadLine();
```

Using the *Console.Readline()* method here means that the command prompt launched by Visual Studio remains visible during the application session until the ENTER key is pressed by the user.

Now, rerunning the application by pressing F5 will produce the result shown in Figure 1-2.

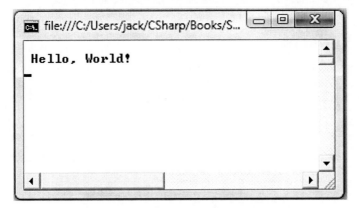

Figure 1-2 Output from Visual Studio.

As mentioned previously, C# is an object-oriented programming language, and represents its programs as objects. The idea is to separate your programs into nouns and verbs, where every noun can be represented as an object. For example, if you create a mathematical model that involves matrices, you can think of the matrices as objects.

A class in a C# program describes an object; it tells your computer what kind of data your objects will have and what kind of actions can be done to them. A matrix class might tell the computer how an identity matrix is defined, and how can you perform matrix operations such as multiplication, inversion, etc. In C#, your entire program is actually a class. In the above C# example, you have the *Program* class.

In C#, every program should have an entry point, which is the place in the code where the computer will start execution. The entry point for every C# program is a static method called *Main* inside a class, like the one you saw defined in the above example.

Every C# program must also have a class with a static *Main* method; if it doesn't, then the computer won't know where to start running the program. Furthermore, you can only have one *Main* method defined in your program; if you have more than one, then the computer won't know which one to start with.

Almost every programming language has common properties. For one thing, programming languages generally know how to store data. They must also operate on that data by moving it around and performing calculations on it.

In the following sections, we will review some C# programming basics.

Data Types

C# is a type-safe programming language. Every variable in C# must be declared as being of a particular data type. Variables in C# can hold either value types or reference types. A value-type variable directly contains an object with some value. This object cannot be contained by the other variables. On the other hand, if a variable is the reference type, it only contains a reference to an object. Other reference-type variables can also contain references to the same object.

Value-Type Variables

Like most programming language, C# has a large number of built-in data types, mostly representing numbers of various formats. Table 1-1 represents the built-in value types. All of the apparently fundamental value types in C# are in fact built up from the object type.

Table 1-1. Built-in value types in C#

C# Type	.NET System Type	Size	Values
bool	System.Boolean	1	True or False
byte	System.Byte	1	0 to 255
sbyte	System.Sbyte	1	-128 to 127
char	System.Char	2	Any Unicode character
short	System.Int16	2	-32768 to 32767
ushort	System.Uint16	2	0 to 65535
int	System.Int32	4	-2147483648 to 2147483647
uint	System.Uint32	4	0 to 4294967295
float	System.Single	4	$\pm 1.5 \times 10^{-45}$ to $\pm 3.4 \times 10^{38}$
long	System.Int64	8	-9223372036854775808 to 9223372036854775807
ulong	System.Uint64	8	0 to 18446744073709551615

| double | System.Double | 8 | $\pm 5.0 \times 10^{-324}$ to $\pm 1.7 \times 10^{308}$ |
| decimal | System.Decimal | 12 | $\pm 1.0 \times 10^{-28}$ to $\pm 7.9 \times 10^{28}$ |

The integer-based types (byte, short, int, long, and so on) can only store whole numbers, such as 0, 1, 2, and so on; they cannot hold decimal numbers, such as 1.5 or 3.1415926. In order to hold decimal numbers, you need to switch to either a floating point or a fixed-point format. The exact details of how these kinds of numbers are stored are beyond the scope of this book.

In C#, you can create instances of the basic data types, called variables, and perform mathematical operations on them. It is easy to declare a variable. All you need to do is to put in the name of the data type, followed by the name of the variable, and then (optionally) initialize the data with a value. For example:

```
int x = 5;
float y = 1.5f;
double z;
```

After the variables are created, you can use them to perform various mathematical operations.

C# also allows you to define your own value types by declaring enumerations or structs, which will be discussed later. These user-defined types are treated in exactly the same way as C#'s built-in value types, although compilers are optimized for the latter.

Reference-Type Variables

The built-in reference types in C# are object and string, where object is the base class of all other types. You can define new reference type variables using class, interface, and delegate declarations.

Reference-type variables hold the value of a memory address occupied by the object they reference. For example, consider the following code snippet, in which two variables reference to the same object named *MyObject*, which contains a numeric property called *myValue*:

```
MyObject object1 = new MyObject();
object1.myValue = 5;
MyObject object2 = object1;
object2.myValue = 10;
```

After executing the last statement, both *object1.myValue* and *object2.myValue* have the same value of 10. The above code snippet demonstrates that when you change a property of an object using a particular reference to it, this change will be reflected in all other references to that property value.

C# allows you to convert any value type into a corresponding reference type, and to convert the resultant "boxed" type back again. Let's take a look at the following code snippet:

```
double x = 1.5;
object o = x;
if (o is double)
{
    Console.WriteLine(o.ToString());
}
```

When the second line executes, an object is initiated as the value of the object (reference type), and the value held by *x* is copied to this object. Note that the runtime type of the object is returned as the boxed value type. The "is" operator returns the type of the object as "double".

Math Operators and Functions

C# has implemented some default math operators, logical operators, and math functions, which allow you to perform basic math and logical operations.

Math Operators

Math operators are symbols that appear in a computer language; they tell the computer to perform certain calculations on data. Operators are commonly used in math equations, so I'm sure this concept will be very familiar to you.

The C# language has several built-in operators. If you've ever used C, C++, or FORTRAN, then you probably already know most of them. Some of the C# math operators are listed in Table 1-2:

Table 1-2. Basic math operators in C#

Operator	Symbol	Example	Result
Addition	+	2+3	5
Subtraction	-	4-3	1
Multiplication	*	3*4	12
Division	/	9/3	3
Modulus	%	8%2	0
Increment	++	2++	3
Decrement	--	2--	1

The first four operators are conventional mathematical operations. The fifth operator, modulus, may be new to you if you haven't done a lot of programming before. Modulus is sometimes known as the remainder operator. Basically, the result from a modulus operator is the same as the remainder if you took the first number and divided by the second. In the example given in Table 1-2, 8 is divided by 2 evenly, so the remainder is 0. If you took 7%3, the result would be 1 because the remainder of 7/3 is 1.

Please note that all the operators have alternative versions that allow you to directly modify a variable. For example, if you want to add 5 to x, you can do this:

```
x = x + 5;
```

But that's somewhat redundant. Instead, you can write this:

```
x += 5;
```

All of the other math operators have similar versions:

```
x *= 5;          // multiply by 5
x /= 5;          // divide by 5
x -= 5           // subtract 5
x %= 5;          // modulus by 5;
```

Logical Operators

There are a few common logical operators that perform comparisons on things and return the Boolean values of *true* or *false*, depending on the outcome. Table 1-3 lists the logical operators.

Table 1-3. Basic logical operators in C#

Operation	Symbol	Example	Result
Equals	==	2 == 3	false
Not equal	!=	2 != 3	true
Less than	<	2 < 3	true
Greater than	>	2 > 3	false
Less than or equal to	<=	2 <= 3	true
Greater or equal to	>=	2 >= 3	false
Logical and	&&	true && false	false
Logical or	||	true || false	true
Logical not	!	!false	true

In addition to logical operators, there are also bitwise and shifting operators. The bitwise operators perform binary math operations on numbers, while the shifting operators (>> and <<) shift the bits in a number up or down.

Built-In Math Functions

C# has a built-in *System.Math* namespace, which contains the base classes and standard math functions. It provides constants and static methods for trigonometric, logarithmic, and other common mathematical functions. You can perform many types of calculations simply by calling the appropriate methods. Table 1-4 lists some commonly used methods in the math class, and their description.

Table 1-4. Static methods in C# Math class

Method	Description
Abs	Returns the absolute value of a specified number.
Acos	Returns the arc cosine of a value.
Asin	Returns the arc sine of a value.
Atan	Returns the arc tangent of a value.
Ceiling	Returns the smallest integer greater than or equal to the specified number.
Cos	Returns the cosine of an angle.
Cosh	Returns the hyperbolic cosine of an angle.
Exp	Returns *e* raised to a specified power.
Floor	Returns the largest integer less than or equal to the specified number.

IEEERemainder	Returns the remainder resulting from the division of a specified number by another specified number.
Log	Returns the logarithm of a specified number.
Log10	Returns the base 10 logarithm of a specified number.
Max	Returns the larger of two specified numbers.
Min	Returns the smaller of two specified numbers.
Pow	Returns a specified number raised to the specified power.
Round	Rounds a value to the nearest integer or specified number of decimal places.
Sign	Returns a value indicating the sign of a number.
Sin	Returns the sine of an angle.
Sinh	Returns the hyperbolic sine of the specified angle.
Sqrt	Returns the square root of a value.
Tan	Returns the tangent of an angle.
Tanh	Returns the hyperbolic tangent of an angle.

The *Math* class is sealed, meaning it cannot be used for inheritance. Additionally, all the classes and data members in the *System.Math* namespace are static, so you cannot create an object of type *Math*. Instead, you use the members and methods with the class name. The *Math* class also includes two constants: *PI* and *E*. *PI* returns the value of 3.14159265358979323846 and the *E* data member returns the value of 2.7182818284590452354.

Most of the math methods in the above table are standard math functions and easy to understand.

Math Operation Example

You can easily perform basic math operations in C#. Let's consider an example. Start with a new Console application with Visual Studio, and name the project *BasicMathOperations*. Here is the code listing of the *Program.cs* file:

```
using System;

namespace BasicMathOperations
{
    class Program
    {
        static void Main(string[] args)
        {
            double x;
            x = 5 + 3;
            Console.WriteLine("\n Addition: \n 4 + 5 = {0}\n", x);
            x = 5 - 3;
            Console.WriteLine(" Subtraction: \n 5 - 3 = {0}\n", x);
            x = 5 * 3; ;
            Console.WriteLine(" Multiplication: \n 5 * 3 = {0}\n", x);
            x = 5 / 3.0;
            Console.WriteLine(" Division: \n 5 / 3 = {0}", x);
```

```
            Console.ReadLine();
        }
    }
}
```

Here, we implement the basic math operations – addition, subtraction, multiplication, and division. The results will displayed on the command prompt screen.

Running this project by pressing F5 generates results shown in Figure 1-3.

```
Addition:
4 + 5 = 8

Subtraction:
5 - 3 = 2

Multiplication:
5 * 3 = 15

Division:
5 / 3 = 1.66666666666667
```

Figure 1-3 Basic math operations in C#.

Selection Statements

In the previous examples, the programs contain a limited amount of sequential steps and then stop. You do not need to make any decisions with the input. In this section, we will overview the selection statements in C#, which allow you to branch into separate logical sequences based on decisions you make. The selections in C# consist of several statements, including *if* statement, *switch* statement, and *goto* statement.

If Statement

The *if* statement allows you to take different paths of logic, depending on a specified condition. When the condition is true, a block of code for that true condition will execute. C# provides you the option of a single *if* statement, multiple *else if* statements, and an optional *else* statement.

Let's use an example to show how each of these types of *if* statements work. Start with a C# Console application and name it *IfStatement*. Here is the code list of this example:

```
using System;

namespace IfStatement
{
    class Program
    {
        static void Main(string[] args)
        {
            string myString;
            int stringLength;

            Console.Write("Please enter a string: ");
            myString = Console.ReadLine();
```

```csharp
        stringLength = myString.Length;

        // Single decision:
        if (stringLength < 1)
        {
            Console.WriteLine("The string length (= {0}) is less than one.",
                            stringLength);
        }

        // Either-or decision:
        if (stringLength != 1)
        {
            Console.WriteLine("The string length (= {0}) is not equal to
                            one.", stringLength);
        }
        else
        {
            Console.WriteLine("The string length (= {0}) is equal to one.",
                            stringLength);
        }

        // Multiple-case decision:
        if (stringLength < 1 || stringLength == 1)
        {
            Console.WriteLine("The string length (= {0}) is less than or
                            equal to one.", stringLength);
        }
        else if (stringLength > 1 && stringLength <= 5)
        {
            Console.WriteLine("The string length (= {0}) is in the range
                            from 2 to 5.", stringLength);
        }
        else if (stringLength > 5 && stringLength <= 10)
        {
            Console.WriteLine("The string length (= {0}) is in the range
                            from 5 to 10.", stringLength);
        }
        else if (stringLength > 10 && stringLength <= 15)
        {
            Console.WriteLine("The string length (= {0}) is in the range
                            from 11 to 15.", stringLength);
        }
        else
        {
            Console.WriteLine("The string length (= {0}) is greater than
                            15.", stringLength);
        }

        Console.ReadLine();
    }
  }
}
```

The statements in the above code listing calculate the *stringLength* variable from the input string (*myString*). The first *if* statement is of the form:

```
if (Boolean expression)
{
    statements
}
```

Here, you must begin with the keyword *if*. Next is the Boolean expression between parentheses, which must evaluate the true or false value. In this case, we check the user's input to see if the string length is greater than one. If this expression evaluates to true, the statements within the curly braces are executed. On the other hand, if the Boolean expression evaluates to false, we ignore the statements inside the block and continue executing the program with the next statement after the block.

Sometimes, you may want to use the *if-else* statement, which allows you to make an either/or decision. The second *if* statement in the above listing presents this idea. In this case, when the Boolean expression evaluates to true, the statements in the block immediately following the *if* statement are executed. However, when the Boolean expression evaluates to false, the statements in the block following the *else* keyword are executed.

When you need to evaluate multiple expressions, you can use the *if/else* form of the *if* statement. The above listing shows this form in the third *if* statement. In this case, the program executes the following block if the Boolean expression evaluates to true. However, here you can evaluate multiple subsequent conditions using the *else if* key word combination. The *else if* statement also takes a Boolean expression, just like the *if* statement. The rules are the same: when the Boolean expression for *the else if* statement evaluates to true, the block immediately following the Boolean expression is executed. When none of the either *if* or *else if* Boolean expressions evaluate to true, the block following the *else* keyword will be executed. Note that only one section of an *if/else* statement will be executed.

Switch Statement

Another form of selection statement is the switch statement, which handles multiple selections by passing control to one of the *case* statements within its body. The types of the value a switch statement can operate on are booleans, enums, integral types, and string. The switch statement takes the following form:

```
Switch (expression)
{
    case constant-expression:
        statement
        jump-statement
    [default:
        statement
        jump-statement]
}
```

where *expression* is an integer or string type expression; *statement* is the embedded statement(s) to be executed if control is transferred to the *case* or the *default*; *jump-statement* is used to transfer control out of the *case* body; and the value of the *constant-expression* is used to transfer control to a specific *case*.

Let's use an example to show how to apply the switch statement. Start with a new C# Console Application and name it *SwitchStatement*. Here is the code listing of this example:

```csharp
using System;

namespace SwitchStatement
{
    class Program
    {
        static void Main(string[] args)
        {
            Console.WriteLine("Which major do you like: 1 = Math,
                              2 = English, 3 = Physics");
            Console.Write("Please enter your selection: ");
            int selection = int.Parse(Console.ReadLine());
            string mySelection;
            switch (selection)
            {
                case 1:
                    mySelection = "I like math.";
                    break;
                case 2:
                    mySelection="I like English.";
                    break;
                case 3:
                    mySelection = "I like Physics.";
                    break;
                default:
                    mySelection = "I like none of the above.";
                    break;
            }
            Console.WriteLine(mySelection);
            Console.ReadLine();
        }
    }
}
```

You can see that the *switch* block follows the *switch* expression, where one or more choices are evaluated for a possible match with the *switch* expression. When the result evaluated in the *switch* expression matches one of these choices, the statements immediately following the matching choice are executed, up to and including a jump statement. In this example, the jump statement, *break*, is used, which leaves the switch block.

In addition to the *break* statement, C# supports several different jump statements, including *continue*, *goto*, *return*, and *throw*. The following listing describes the various jump statements:

- *break* – Leaves the switch block.

- *continue* – Leaves the switch block, skips remaining logic in enclosing loop, and goes back to loop condition to determine if loop should be executed again from the beginning.

- *goto* – Leaves the switch block and jumps directly to a label of the form "*<labelname>*".

- *return* – Leaves the current method.

- *throw* – Throws an exception.

Note that in the above example, we also include a *default* choice following all other choices. If none of the other choices match, then the default choice is executed. Although use of the *default* label is optional, I highly suggest that you always include it. This will help you catch unforeseen circumstances and make your programs more reliable.

Loop Statements

In this section, we discusss various C# looping statements, including the *for* loop, *foreach* loop, *while* loop, and *do* loop.

For Loop

The *for* loop excecutes a statement or a block of code repeatedly until a specified expression evaluates to false. The *for* loop is useful for iterating over arrays and for sequential processing. It is usually used when you know exactly how many times you want to perform the statements within the loop.

The contents within a *for* loop parentheses holds three sections that are separated by semicolons:

```
for (<initializer list>; <boolean expression>; <iterator list>)
{
    <statements>
}
```

The initializer expressions are evaluated only once during the lifetime of the *for* loop. This is a one-time operation before loop execution, commonly used to initialize an integer to be used as a counter.

Once the initializer list has been evaluated, the *for* loop processes the second section, the Boolean expression. There is only one Boolean expression, which is commonly used to verify the status of a counter variable.

When the Boolean expression evaluates to *true*, the statements within the curly braces of the *for* loop are executed. After executing the *for* loop statements, the program moves to the top of the loop and executes the iterator list, which is normally used to increment or decrement a counter.

The following code listing shows how to implement a typical *for* loop:

```
using System;

namespace ForLoop
{
    class Program
    {
        static void Main(string[] args)
        {
            for (int i = 0; i < 11; i++)
            {
                Console.WriteLine(2 * i);
            }
            Console.ReadLine();
        }
    }
}
```

The output of the example will be 0, 2, 4,…, 20.

In this listing, the initial value of the variable *i* is evaluated, and then – as long as the value of *i* is less than 11 – the condition evaluates to true, the *Console.WriteLine* statement is executed and *i* is reevaluated. When *i* is greater than 11, the condition becomes false and control is transferred outside the loop.

Since the test of the Boolean expression occurs before the execution of the loop, a *for* loop executes zero or more times.

Note that all of the expressions of the *for* loop are optional. For example, the following statement is used to create an infinite loop:

```
for (; ;)
{
    <statements>
}
```

Foreach Loop

The *foreach* statement repeats a group of embedded statements for each element in an array or an object collection such as *ArrayList*. The *foreach* statement is used to iterate through the collection to get the information that you need, but cannot be used to add or remove items from the source collection to avoid unpredictable side effects.

The syntax of a *foreach* loop has the form:

```
foreach (<type> <item name> in <collection>
{
    <statements>
}
```

Here, "type" is the type of item contained in the collection. For example, if the type of the array or collection is *double[]* then the type would be *double*. The item name is an identifier that you specify, which could be anything but should be meaningful. The *in* keyword is required in the *foreach* loop.

On each iteration through a *foreach* loop, the array or collection is queried for a new value. As long as the collection can return a value, this value will be put into the item name variable, causing the statements in the *foreach* block to be executed. When the collection has been fully iterated, control will transfer to the first executable statement following the end of the *foreach* loop.

The following code listing shows how to use the *foreach* loop:

```
using System;

namespace ForeachLoop
{
    class Program
    {
        static void Main(string[] args)
        {
            string[] students = { "Anna", "Betty", "Tyler" };
            foreach (string student in students)
            {
                Console.WriteLine(student);
            }
            Console.ReadLine();
```

```
            }
        }
    }
```

Here, we first create a string array that contains three strudent names, which is used in the *foreach* loop. Inside the *foreach* loop, we use string variable, *student*, as the item name to hold each element of the *students* array. As long as there are *students* in the array that have not been returned, the *Console.WriteLine* method will print each value of the *student* variable on the screen.

While Loop

A *while* loop will check the condition and then continues to execute a block of code as long as the condition evaluates to true. The syntax of the *while* loop has the following form:

```
while (<boolean expression>)
{
    <statements>
}
```

The statements can be any valid C# statements. The Boolean expression is evaluated before any code in the following block has executed. When the Boolean expression evaluates to true, the statements will execute. Once the statements have executed, control returns to the beginning of the *while* loop to check the condition again.

When the condition evaluates to false, the *while* loop statements are skipped and execution begins after the *while* loop. Before starting the *while* loop, ensure that the variables evaluated in the loop condition are initialized. During execution, make sure that you update variables associated with the condition expression so that the loop will be terminated when you want it to.

The following code listing shows how to implement a *while* loop:

```
using System;

namespace WhileLoop
{
    class Program
    {
        static void Main(string[] args)
        {
            int n = 0;
            while (n < 10)
            {
                Console.WriteLine(n);
                n++;
            }
            Console.ReadLine();
        }
    }
}
```

You can see that the *while* loop begins with the keyword *while*, followed by a condition expression. All control statements use Boolean expressions as their condition for entering the loop. This means that the expression must evaluate to either a true or false value. In this example, we check the integer *n* variable to see if it is less than 10. Since *n* was initialized to 0, the condition will return true the first time is evaluated. When the condition evaluates to true, the block immediately following the condition expression will be executed.

Within the *while* loop block, we print the number to the console screen. Then we increment (*n++*) to the next integer. Once the statements in the *while* block have been executed, the condition expression is evaluated again. This iteration will continue until the condition evaluates to false. Once the condition becomes false, control will jump to the first statement following the *while* block.

A *while* loop can be also terminated when a *break*, *goto*, *return*, or *throw* statement transfers control outside the loop. To pass control to the next iteration without exiting the loop, you can use the *continue* statement.

Do Loop

A *do* loop is similar to the *while* loop, except that it checks its condition at the end of the loop. This means that the *do* loop is guaranteed to execute at least one time. On the other hand, a *while* loop evaluates its condition at the beginning and there is generally no guarantee that the statements inside the loop will be executed.

The syntax of the *do* loop has the form:

```
do
{
    <statements>
}
while (<condition expression>);
```

The statements can be any valid C# statements. The condition expression is the same as in the *while* loop.

In the following example, the *do* loop statements execute as long as the variable *n* is less that 10:

```
using System;

namespace DoLoop
{
    class Program
    {
        static void Main(string[] args)
        {
            int n = 0;
            do
            {
                Console.WriteLine(n);
                n++;
            } while (n < 10);
            Console.ReadLine();
        }
    }
}
```

Methods

In C#, a method is simply a discrete piece of code that can be called by other functions. When the method is called, control of the program is passed to the method, which performs its activities before returning control back to the command following the calling statement.

In previous examples, the functionality of each program was implemented in the *Main* method, which may be adequate for simple programs. However, for a more complicated application, we need a better way to organize the code using methods. C# methods can help you separate your code into modules that perform a specific task.

The use of the methods provides several benefits. The key advantage is that code that would otherwise be repeated can be held in a single location. If the code for a particular function exists in a method, modifications and bug fixes are much simpler than if the same code has been duplicated in many locations. The other benefit is that when a method is given a descriptive name, it can make your code easier to understand.

Creating a Method

The method in C# is very useful because it allow you to separate your logic into different units. You can pass information to a method, have it perform one or more statements, and retrieve a return value. The capability to pass parameters and return values is optional, depending on what you want the method to do. Here is the syntax required for creating a method:

```
<attributes> <modifiers> <return type> <method name>(<parameters>)
{
    <statements>
}
```

Note that methods in C# are always defined within the bounds of a class or struct. They can have various attributes. For example, methods can be *instance* (called as an instance of the type within which the method is defined) or *static*, where the method is associated with the type itself. They can also have attributes such as *virtual, abstract , or sealed*. Methods can be *overloaded, overridden* and *hidden*.

Access modifiers of a method are specified as part of the method declaration syntax and can be:

- *public* – indicates the method is freely accessible inside and outside of the class in which it is defined.

- *internal* – indicates the method is only accessible to types defined in the same assembly.

- *protected* – means the method is accessible in the type in which it is defined, and in derived types of that type. This is used to give derived classes access to the methods in their base class.

- *protected internal* – means the method is accessible to types defined in the same assembly or to types in a derived assembly.

- *private* – indicates the method is only accessible in the class in which it is defined.

If no modifier is specified, the method is given *private* access. The *return type* can be any C# type. It can be assigned to a variable for use later in the program. The method *name* is a unique identifier, which should be meaningful and associated with the task the method performs. *Parameters* allow you to pass information to and from a method. They are surrounded by parentheses. *Statements* within the method carry out the functionality.

Here, we will create a simple method that obtains the value of the square root of a input number. The result will be displayed to the Console. Although the functionality is very simple, it serves as a good demonstration of the structure of a method. Here is the code snippet of this method:

```
void DisplaySquareRoot(double x)
{
    Console.WriteLine("Square root of x = {0}", Math.Sqrt(x));
}
```

This method simply contains one statement. After this statement, the method has no further code to execute, so control naturally returns the point after where the method was originally called. However, it is also possible to instruct the program to return from a method at any point by adding a *return* statement. The following code shows this by returning without displaying the result to the Console if the input parameter $x < 0$:

```
void DisplaySquareRoot(double x)
{
    if(x < 0)
        return;
    Console.WriteLine("Square root of x = {0}", Math.Sqrt(x));
}
```

The *return* command may be used several times within a method to exit at different points according to your requirements. However, the number of return points should be kept to a minimum in order to keep the code as easy to understand as possible. In ideal conditions there should be only one return statement or the code should end naturally when control reaches the closing brace character.

Calling a Method

Now that the method is complete it can be called. The method belongs to a class. In a standard Console application, this class is called *Program* and will now contain a *Main* method as well as *DisplaySquareRoot*. To use the new method, you must first create a new *Program* object before you can use the *DisplaySquareRoot* method. This is because of the way the method was declared. Since we did not specify a *static* modifier, as for *Main* method, The *DisplaySquareRoot* method becomes an instance method. The difference between *instance* methods and *static* methods is that multiple instances of a class can be created (or instantiated) and each instance has its own separate *DisplaySquareRoot* method. However, when a method is *static*, there are no instances of that method, and you can invoke only that one definition of the *static* method.

The *DisplaySquareRoot* method can be called within the *Main* method, making the final code as follows:

```
using System;

namespace MethodDemo
{
    class Program
    {
        static void Main(string[] args)
        {
            Program p = new Program();
```

```
        p.DisplaySquareRoot(3);

        Console.ReadLine();
    }

    void DisplaySquareRoot(double x)
    {
        if (x < 0)
            return;
        Console.WriteLine("Square root of x = {0}", Math.Sqrt(Math.Abs(x)));
    }
  }
}
```

Running the above example produces the results:

```
Square root of x = 1.73205080756888.
```

Properties

Properties are an integral part of C# language, which allow you to control the accessibility of the variables in a class. This is the recommended way to access variables from the outside in C#. A property in C# is much like a combination of a variable and a method – it cannot take any parameters, but it allows you to process the value before it is assigned to your return.

The advantage of properties over fields is that you can change their internal implementation over time. With a public field, the underlying data type must always be the same. However, with a public property, you can change the implementation.

Field Members

In many cases, the information made public through a property is held directly within the object as a field member. This field number has a scope that makes it visible to the entire class. To define a class-level field number, the declaration is made within the class' code block but outside of any methods or properties. For example, we can add a field, *side*, to a new *Square* class to hold the side length of a square object:

```
using System;
namespace Property
{
    class Program
    {
        static void Main(string[] args)
        {

        }
    }

    class Square
    {
        private double side;
    }
}
```

Exposing a Property

Adding a property to a class is similar to adding a field. In fact, it is possible to create a public field by simply declaring it as *public* instead of *private*. However, this is inadvisable as it means that the class will have no control over the values that the properties are set to and some standard .NET functionality will be lost. To avoid these problems, the property should be declared with a code block as follows:

```
public double Side
{
}
```

Once the property is declared, functionality must be added to it. This can be achieved by using *get* and *set* statements. These create two code blocks that control how the property's value is retrieved via the *get* statement and how the property's value is validated, processed, and stored via the *set* statement:

```
public double Side
{
    get {}
    set {}
}
```

For the *Side* property, we will validate the specified value before storing it. The value of the side length of a square must be positive. If the value is negative, an exception will be throw and the property will remain unchanged. This is possible by using the *get* and *set* statements correctly. Note that this level control is impossible if a simple public field is used. Here is the code listing of the *Side* property:

```
class Square
{
    private double side;

    public double Side
    {
        get { return side; }
        set
        {
            if (value < 0)
                throw new OverflowException();
            side = value;
        }
    }
}
```

Using Properties

Properties of an object can be accessed by using the object name followed by the member access operator (.) and the property name. The property can be read from and written to using similar syntax as for a standard variable. To demonstrate this, consider the following example code added to the *Main* method of the program. It creates three objects based upon the *Square* class and assigns and reads their properties individually. When trying to assign an invalid side length to a sqaure, an exception will be thrown by the validation code. Here is the code snippet:

```
static void Main(string[] args)
{
```

```
Square square1 = new Square();
square1.Side = 10.5;

Square square2 = new Square();
square2.Side = 22.5;

Console.WriteLine(square1.Side);   // Output = 10.5
Console.WriteLine(square2.Side);   // Output = 22.5

Square square3 = new Square();
square3.Side = -5.5;                // Throws an overflow exception.
Console.ReadLine();
}
```

Here, the state of the three squares is held internally and the implementation details of the side length property are hidden from the program, which achieves the encapsulation.

Read-Only and Write-Only Properties

The *Side* property of a square is an example of the read-write property, because the information held within it can be both read and written. Sometimes you may want read-only or write-only property, which can be created easily simply by not specifying the accessors that are not required.

For example, we can now add a read-only property for the area of the square without *set* statements. This property is calculated from the existing field member "side", so it requires no direct internal storage. Here is the code listing:

```
public double Area
{
    get { return side * side; }
}
```

The new read-only property can be read using the same syntax as the read-write property. You can test it by modifying the program's *Main* method:

```
static void Main(string[] args)
{

    Square square = new Square();
    square.Side = 10;
    Console.WriteLine(square.Area);   // Output = 100

    Console.ReadLine();
}
```

We can also add a write-only property for the perimeter of the square, without *get* statements:

```
private double perimeter;
public double Perimeter
{
    set
    {
        perimeter = value;
        Console.WriteLine("The perimeter of the square = {0}", perimeter);
    }
}
```

Here you can see how to implement a write-only property, which has no *get* accessor. The *set* accessor is varied a little by printing the value of the *perimeter* after it is assigned. You can test this write-only property by modifying the program's *Main* method:

```
static void Main(string[] args)
{
    Square s = new Square();
    s.Side = 10;
    s.Perimeter = 4 * s.Side;          // Output = 40

    Console.ReadLine();
}
```

Auto-Implemented Properties

C# 3.0 (or later) introduces a new feature, auto-implemented properties, which allow you to create properties without *get* and *set* accessor implementations. This feature makes property-declaration more concise when no additional logic is required in the property accessors. When you create auto-implemented properties in an application, the compiler creates a private, anonymous backing field that can only be accessed through the property's *set* and *get* accessors.

Assume we have a class, *Customer*, with three properties, *ID*, N*ame*, and *Purchases*:

```
class Customer
{
    public int ID { get; set; }
    public string Name { get; set; }
    public double Purchases { get; set; }
}
```

Following is the usual method to set values to these properties:

```
Customer customer = new Customer();
customer.ID = 1;
customer.Name = "Tyler";
customer.Purchases = 10.5;
```

However, in C# 3.0 or later, you can also use a different type of initialization:

```
Customer customer = new Customer { ID = 1, Name = "Tyler", Purchases = 10.50 };
Console.WriteLine("Customer ID = {0}, Name = {1}, Puerchases = {2}",
                    customer.ID, customer.Name, customer.Purchases);
```

You can set values in this way for properties and public variables. These features make the programming much easier.

Chapter 2

Complex Numbers and Functions

Complex numbers, complex functions, and complex analysis are part of an important branch of mathematics. They find wide application in solving real scientific and engineering problems. However, there is no built-in class for complex numbers and complex functions in C#. In this chapter, I will show you how to implement a complex class, which contains the definition of complex numbers, complex operators, and commonly used complex functions. Based on this class, you'll be able to perform various computations using complex numbers and functions, as well as add your own complex functions to this class.

Complex Numbers and Operators

In mathematics, a complex number is a number of the form $z = x + iy$, where z is the complex variable; x, and y are real numbers, and i is the imaginary unit, with the property $i^2 = -1$. The real number x is called the real part of the complex number, and the real number y is the imaginary part. Real numbers may be considered to be complex numbers with imaginary part set to zero; that is, the real number x is equivalent to the complex number $x + 0i$.

You can perform mathematical operations on complex numbers. These complex numbers can be added, subtracted, multiplied, and divided just like real numbers. In addition, you can also define standard math functions using complex variables. In this section, I'll show you how to define complex numbers, complex functions, and their basic operations in C#.

Complex Numbers in C#

Start with a new C# Console application and name it *ComplexNumberTest*. Now, in the Solution Explorer of the project, right click the *ComplexNumberTest* project, select Add, then select Class…, and name it *Complex*. This way, you add a new *Complex* class to the project. Then change its namespace to *XuMath*. This class will implement complex numbers and complex functions.

Next, we'll change the class to a structure, *struct*. The structures in C# seem similar to classes, but they are two entirely different aspects of the platform. The classes are reference types, while a structure is a value type. The objects of class types are always created on heal while the objects of structure types are always created on the stack. However, C# structures are useful for small data systems that have value semantics. Complex numbers are good examples of structure types. Of course, you can perform the same task using classes.

The simplest approach to represent complex numbers is to declare a structure with two floating point field members of double type representing real and imaginary parts. We can then define the complex numbers in this structure:

```
using System;
using System.Collections.Generic;
using System.Text;

namespace XuMath
{
    public struct Complex
    {
        public double Real;
        public double Imaginary;

        public Complex(double real, double imaginary)
        {
            this.Real = real;
            this.Imaginary = imaginary;
        }
    }
}
```

You can see that the original class name is changed to a structure with the following code snippet:

```
public struct Complex
```

We then define two field members: one is for the real part and the other for imaginary part of a complex number. We also create a constructor with two parameters. Using this simple structure, we can easily create complex numbers. To see this more clearly, we add a utility method to this structure:

```
public override string ToString()
{
    string s = "";
    if (Imaginary >= 0)
        s = String.Format("{0} + {1} i", Real, Imaginary);
    else
        s = String.Format("{0} - {1} i", Real, Math.Abs(Imaginary));
        return s;
}
```

Here, the *ToString* method is overrided, which allows you to display complex numbers easily.

Common Public Properties

In the *Complex* structure, we first define a static property of I with $I^2 = -1$:

```
public static Complex I
{
    get { return new Complex(0.0, 1.0); }
}
```

Since I represents the unit imaginary number, it is a static constant.

For a given complex number, we can easily obtain its complex conjugate by simply changing the sign of the imaginary part. This can be achieved by adding a *Conjugate* property to the *Complex* structure using the *get* and *set* accessors:

```
public Complex Conjugate
{
    get { return new Complex(Real, -Imaginary); }
    set { this = value.Conjugate; }
}
```

The modulus, or absolute value, of a complex number can be calculated using the following *Modulus* property:

```
public double Modulus
{
    get { return Math.Sqrt(Real * Real + Imaginary * Imaginary); }
    set
    {
        if (Real == 0.0 && Imaginary == 0.0)
        {
            Real = value;
            Imaginary = 0;
        }
        else
        {
            double x1 = value / Math.Sqrt(Real * Real +
                                Imaginary * Imaginary);
            Real *= x1;
            Imaginary *= x1;
        }
    }
}
```

The argument angle of a complex number, $z = x + iy$, is given by the following formula:

$$\tan \varphi = \frac{y}{x}$$

We can add the following property to the *Complex* structure, which can be used to calculate the argument angle of a complex number:

```
public double Angle
{
    get
    {
        if (Imaginary == 0 && Real < 0)
        {
            return Math.PI;
        }
        if (Imaginary == 0 && Real >= 0)
        {
            return 0;
        }
        return Math.Atan2(Imaginary, Real);
    }
    set
    {
```

```
            double modulus = Modulus;
            Real = Math.Cos(value) * modulus;
            Imaginary = Math.Sin(value) * modulus;
        }
    }
```

Test Public Properties

We can now test the definition of complex numbers and public properties. First, we need to add the *using* statement at the top of the *Program.cs* file:

```
using XuMath;
```

This *using* statement allows you to use the math functions defined in the *XuMath* namespace. The following is the code of the *Program.cs* file:

```
using System;
using XuMath;

namespace ComplexNumberText
{
    class Program
    {
        static void Main(string[] args)
        {
            TestProperties();

            Console.ReadLine();
        }

        static void TestProperties()
        {
            // Test public properties:
            Complex c = new Complex(2, 3);
            Console.WriteLine("\n Complex nember = {0}\n", + c);
            Console.WriteLine(" Test public properties:");
            Console.WriteLine(" Real part = {0}", c.Real);
            Console.WriteLine(" Imaginary part = {0}", c.Imaginary);
            Console.WriteLine(" Conjugate = {0}", c.Conjugate);
            Console.WriteLine(" Modulus = {0}", c.Modulus);
            Console.WriteLine(" Angle = {0}\n", c.Angle);
        }
    }
}
```

This file creates a new complex number by calling the name of the public *Complex* structure, using the following code snippet:

```
Complex c = new Complex(1, 2);
```

Then various public properties are printed on your command prompt screen.

Pressing F5 will produce the results shown in Figure 2-1.

```
Complex nember = 2 + 3 i

Test public properties:
Real part = 2
Imaginary part = 3
Conjugate = 2 - 3 i
Modulus = 3.60555127546399
Angle = 0.982793723247329
```

Figure 2-1 Complex properties in C#.

Here, we use public properties to create the complex conjugate, modulus, and argument angle for complex numbers, which allow you to access the corresponding values through the access (.) operator:

```
Console.WriteLine("\n Complex nember = {0}\n", + c);
Console.WriteLine(" Test public properties:");
Console.WriteLine(" Real part = {0}", c.Real);
Console.WriteLine(" Imaginary part = {0}", c.Imaginary);
......
```

However, these quanlities can also be rewritten in terms of public methods. The advantage of using public methods instead of properties is that you can use notations resembling ordinary mathematical representations. In mathematics, we would write the modulus and argument angle of the complex number c as $Mod(c)$ and $Arg(c)$, respectively, rather than $c.Modulus$ and $c.Angle$. To do that, we can simply add corresponding public methods to the *Complex* structure. In the following sections, we will use public methods to implement complex functions.

Common Public Methods

We can easily convert the public properties discussed in the previous section into public methods. Add the following static methods to the *Complex* structure:

```
public static Complex Conj(Complex c)
{
    return c.Conjugate;
}

public static double Re(Complex c)
{
    return c.Real;
}

public static double Im(Complex c)
{
    return c.Imaginary;
}

public static double Mod(Complex c)
{
    return c.Modulus;
}

public static double Arg(Complex c)
{
    return c.Angle;
```

```
    }
```

These methods can be examined using the previous *ComplexNumbersTest* example by modifying the *Program.cs* file according to the following listing:

```csharp
using System;
using XuMath;

namespace ComplexNumberText
{
    class Program
    {
        static void Main(string[] args)
        {
            TestProperties();
            TestMethods();
            Console.ReadLine();
        }

        static void TestProperties()
        {
            // Test public properties:
            Complex c = new Complex(2, 3);
            Console.WriteLine("\n Complex nember = {0}\n", + c);
            Console.WriteLine(" Test public properties:");
            Console.WriteLine(" Real part = {0}", c.Real);
            Console.WriteLine(" Imaginary part = {0}", c.Imaginary);
            Console.WriteLine(" Conjugate = {0}", c.Conjugate);
            Console.WriteLine(" Modulus = {0}", c.Modulus);
            Console.WriteLine(" Angle = {0}\n", c.Angle);
        }

        static void TestMethods()
        {
            //Test public methods:
            Complex c = new Complex(2, 3);
            Console.WriteLine(" Test public methods:");
            Console.WriteLine(" Real part = {0}", Complex.Re(c));
            Console.WriteLine(" Imaginary part = {0}", Complex.Im(c));
            Console.WriteLine(" Conjugate = {0}", Complex.Conj(c));
            Console.WriteLine(" Modulus = {0}", Complex.Mod(c));
            Console.WriteLine(" Angle = {0}\n", Complex.Arg(c));
        }
    }
}
```

Here we display a complex number's real part, imaginary part, comjugate, modulus, and argument angle using both public properties and public methods.

Pressing F5 generates the same results shown in Figure 2-2.

```
Complex nember = 2 + 3 i

Test public properties:
Real part = 2
Imaginary part = 3
Conjugate = 2 - 3 i
Modulus = 3.60555127546399
Angle = 0.982793723247329

Test public methods:
Real part = 2
Imaginary part = 3
Conjugate = 2 - 3 i
Modulus = 3.60555127546399
Angle = 0.982793723247329
```

Figure 2-2 Results from public properties and methods.

Equality and Hashing

In condition statements, we need to know if two complex variables are equal or not. The *System.Object* type offers a virtual method, named *Equals*. The purpose of this method is to return true if two objects have the same value. Because *Equals* is defined by *Object* and every type is ultimately derived from *Object*, every instance of any type offers the *Equals* method. It will be useful if any instance of any object could be placed into a hash table collection. To this end, *System.Object* provides a virtual *GetHashCode* method so that an *int* hash code can be obtained for any and all objects.

In order to compare two complex numbers, we need to implement the *Equals* method in the *Complex* structure using the following code snippet:

```
public override bool Equals(object obj)
{
    return (obj is Complex) && this.Equals((Complex)obj);
}

public bool Equals(Complex c)
{
    return Real == c.Real && Imaginary == c.Imaginary;
}

public override int GetHashCode()
{
    return Real.GetHashCode() ^ Imaginary.GetHashCode();
}

public static bool operator ==(Complex c1, Complex c2)
{
    return c1.Equals(c2);
}

public static bool operator !=(Complex c1, Complex c2)
{
    return !c1.Equals(c2);
}
```

You can examine the above *Equals* method by adding the following method to the *Program.cs* file:

```
private void TestEquals()
{
    Complex c1 = new Complex(2, 3);
    Complex c2 = new Complex(3, 4);
    bool b = c1 == c2;
    Console.WriteLine(b.ToString());
}
```

Calling this method from the *Main* method will give a *False* output. On the other hand, if you set:

```
bool b = c1 != c2;
```

you'll get a *True* output.

Complex Operators

In this section, we will implement the standard math operators for complex numbers. Structures or classes in C# can contain all the common operations for combining objects together into expressions. In this simplified exposition of the structure *Complex*, we will confine ourselves to the operators for addition, subtraction, multiplication, and division, but similar considerations apply to all the other math operators. You can easily add more math operators to this *Complex* structure according to the requirements of your applications.

Unary Operators

In C#, the operator methods are denoted by the keyword *operator*. Here, we will first implement two unary operators, unary + and unary −, in the *Complex* structure. Although the use of the unary + can often be avoided, it is helpful to add it for mathematical convenience. The unary − is often used to change the sign of a complex number. Add the following two methods to the *Complex* struct:

```
public static Complex operator +(Complex c)
{
    return c;
}

public static Complex operator -(Complex c)
{
    return new Complex(-c.Real, -c.Imaginary);
}
```

Note that you can change sign of a complex number using the unary − operator.

Addition Operator

We'll implement an addition operator with three overloaded methods: complex + complex, complex + double, and double + complex:

```
public static Complex operator +(Complex c1, Complex c2)
{
    return new Complex(c1.Real + c2.Real, c1.Imaginary + c2.Imaginary);
}

public static Complex operator +(Complex c1, double c2)
```

```
{
    return new Complex(c1.Real + c2, c1.Imaginary);
}

public static Complex operator +(double c1, Complex c2)
{
    return new Complex(c1 + c2.Real, c2.Imaginary);
}
```

Using the above methods, you can test the addition of two complex numbers. Add the following static method to the *Program.cs* file:

```
static void TestAddition()
{
    Complex c1 = new Complex(2, 3);
    Complex c2 = new Complex(3, 4);
    double d = 5.0;
    Complex c3 = c1 + c2;        // result: 5 + 7i
    Complex c4 = c1 + d;         // result: 7 + 3i
    Complex c5 = d + c2;         // result: 8 + 4i

    Console.WriteLine("\n c3 = {0}\n c4 = {1} \n c5 = {2}", c3, c4, c5);
}
```

Calling this method from the *Main* and running this application will give you the results shown in Figure 2-3, as expected.

```
c3 = 5 + 7 i
c4 = 7 + 3 i
c5 = 8 + 4 i
```

Figure 2-3 Addition of two complex numbers.

Subtraction Operator

Similar to the case of the addition operator, the subtraction operator also has three overloaded methods:

```
public static Complex operator -(Complex c1, Complex c2)
{
return new Complex(c1.Real - c2.Real, c1.Imaginary - c2.Imaginary);
}

public static Complex operator -(Complex c1, double c2)
{
    return new Complex(c1.Real - c2, c1.Imaginary);
}

public static Complex operator -(double c1, Complex c2)
{
    return new Complex(c1 - c2.Real, - c2.Imaginary);
}
```

Multiplication Operator

Similar to the case of the addition operator, the multiplication operator also has three overloaded methods:

```
public static Complex operator *(Complex c1, Complex c2)
{
return new Complex(c1.Real * c2.Real -  c1.Imaginary * c2.Imaginary,
                        c1.Real * c2.Imaginary + c1.Imaginary * c2.Real);
}

public static Complex operator *(Complex c1, double c2)
{
    return new Complex(c1.Real * c2, c1.Imaginary * c2);
}

public static Complex operator *(double c1, Complex c2)
{
    return new Complex(c1 * c2.Real, c1 * c2.Imaginary);
}
```

Division Operator

Similar to the case of the addition operator, the division operator also has three overloaded methods:

```
public static Complex operator /(Complex c1, Complex c2)
{
    if (c2.Real == 0.0 && c2.Imaginary == 0)
{
        return (new Complex(1.0e300, 1.0e300));
    }
    double c2m2 = c2.Modulus * c2.Modulus;
    return new Complex((c1.Real * c2.Real + c1.Imaginary * c2.Imaginary) / c2m2,
                    (c1.Imaginary * c2.real - c1.Real * c2.Imaginary) / c2m2);
}

public static Complex operator /(Complex c1, double c2)
{
    if (c2 == 0.0)
    {
        return (new Complex(1.0e300, 1.0e300));
    }
    return new Complex(c1.Real / c2, c1.Imaginary / c2);
}

public static Complex operator /(double c1, Complex c2)
{
    if (c2.Real == 0.0 && c2.Imaginary == 0)
    {
        return (new Complex(1.0e300, 1.0e300));
    }
    double c2m2 = c2.Modulus * c2.Modulus;
    return new Complex(c1* c2.Real / c2m2, - c1 * c2.Imaginary / c2m2);
}
```

Testing Math Operators

We can examine the math operations of the complex variables. Add a new static method, *TestOperators*, to the *Program.cs* file:

```
static void TestOperators()
{
    Complex c1 = new Complex(4, 5);
    Complex c2 = new Complex(2, 3);

    Console.WriteLine("\n c1 = {0}", c1);
    Console.WriteLine(" c2 = {0}", c2);
    Console.WriteLine(" c1 + c2 = {0}", c1 + c2);
    Console.WriteLine(" c1 - c2 = {0}", c1 - c2);
    Console.WriteLine(" c1 * c2 = {0}", c1 * c2);
    Console.WriteLine(" c1 / c2 = {0}", c1 / c2);
}
```

Calling this method from *Main* and running this project generate the results shown in Figure 2-4.

```
c1 = 4 + 5 i
c2 = 2 + 3 i
c1 + c2 = 6 + 8 i
c1 - c2 = 2 + 2 i
c1 * c2 = -7 + 22 i
c1 / c2 = 1.76923076923077 - 0.153846153846154 i
```

Figure 2-4 Math operations for complex numbers in C#.

Complex Functions

In this section, we will implement various complex functions in the *Complex* structure, including basic math functions, trigonometric functions, inverse trigonometric functions, trigonometric hyperbolic functions, and inverse trigonometric hyperbolic functions.

Basic Math Functions

Similar to the basic math functions for real variables, we can also define basic math functions for complex variables, including square root, exponential, power, and logarithm functions.

Square Root Function

The square root of a complex number *z* can easily be calculated using the polar form:

$$\sqrt{z} = \sqrt{x + iy} = \sqrt{r}e^{i\varphi/2}$$

where

$$r = \sqrt{x^2 + y^2}$$

is the modulus of *z*. We can easily implement the square root function of a complex number using the above equations. Add the following public method to the *Complex* structure:

```
public static Complex Sqrt(Complex c)
{
    return new Complex(Math.Sqrt(c.Modulus) *  Math.Cos(c.Angle / 2.0),
                       Math.Sqrt(c.Modulus) *  Math.Sin(c.Angle / 2.0));
}
```

Exponential Function

We now consider the exponential function for a complex variable z:

$$e^z = e^{x+iy} = e^x e^{iy} = e^x (\cos y + i \sin y)$$

Using this relationship, we can easily implement the complex exponential function in the *Complex* structure:

```
public static Complex Exp(Complex c)
{
return new Complex(Math.Exp(c.Real) *  Math.Cos(c.Imaginary),
                   Math.Exp(c.Real) *  Math.Sin(c.Imaginary));
}
```

Pow Function

Here, we want to calculate z^c where both z and c are complex. If $z = x + iy$ and $c = a + ib$, we can first rewite z in polar coordinates:

$$z^c = (re^{i\varphi})^{a+ib} = e^{a \log r - b\varphi + i(a\varphi + b \log r)}$$

$$= e^{a \log r - b\varphi}\left[\cos(a\varphi + b \log r) + i\sin(a\varphi + b \log r)\right]$$

Then, we can use this equation to create the complex *Pow* function in the *Complex* structure:

```
public static Complex Pow(Complex c1, Complex c2)
{
    double x1 = Math.Exp(c2.Real * Math.Log(c1.Modulus) -
             c2.Imaginary * c1.Angle);
    double x2 = Math.Cos(c2.Real * c1.Angle +
             c2.Imaginary * Math.Log(c1.Modulus));
    double x3 = Math.Sin(c2.Real* c1.Angle +
             c2.Imaginary * Math.Log(c1.Modulus));
    return (new Complex(x1 * x2, x1 * x3));
}
```

Logarithm Function

The complex log function has a simple relation in polar coordinates:

$$\log z = \log(re^{i\varphi}) = \log r + i\varphi$$

The corresponding public method for computing the complex log function is given below:

```
public static Complex Log(Complex c)
{
    return new Complex(Math.Log(c.Modulus), c.Angle);
}
```

Testing Math Functions

We can examine math functions of the complex variables defined in the previous sections. Add a static method, *TestMathFunctions*, to the *Program.cs* file:

```
static void TestMathFunctions()
{
    Complex c1 = new Complex(4, 5);
    Complex c2 = new Complex(2, 3);
    Console.WriteLine("Test Math Functions:");
    Console.WriteLine("c1 = " + c1.ToString());
    Console.WriteLine("c2 = " + c2.ToString());
    Console.WriteLine("Sqrt(c1) = " + Complex.Sqrt(c1).ToString());
    Console.WriteLine("Pow(c1, c2) = " + Complex.Pow(c1, c2).ToString());
    Console.WriteLine("Exp(c1) = " + Complex.Exp(c1).ToString());
    Console.WriteLine("Log(c1) = " + Complex.Log(c1).ToString());
}
```

Calling this method from the *Main* method produces the output shown in Figure 2-5.

```
Test Math Functions:
c1 = 4 + 5 i
c2 = 2 + 3 i
Sqrt(c1) = 2.2806933416653 + 1.09615788950152 i
Pow(c1, c2) = 1.31593192903841 + 2.45815642480931 i
Exp(c1) = 15.4874305606508 - 52.355491418482 i
Log(c1) = 1.85678603335215 + 0.896055384571344 i
```

Figure 2-5 Complex math functions in C#

Trigonometric Functions

In this section, we will implement various complex trigonometric functions, including *Sine*, *Cosine*, and *Tangent*.

Sine Function

A complex sine function can be calculated using complex exponential functions:

$$\sin z = \frac{1}{2i}(e^{iz} - e^{-iz}) = \frac{1}{2i}\left[e^x(\cos y + i\sin y) - e^{-x}(\cos y - i\sin y)\right]$$

$$= \sin x \cosh y + i \cos x \sinh y$$

So we can easily add the complex *Sine* method to the *Complex* structure:

```
public static Complex Sin(Complex c)
{
return new Complex(Math.Cosh(c.Imaginary) * Math.Sin(c.Real),
                   Math.Sinh(c.Imaginary) * Math.Cos(c.Real));
}
```

Cosine Functions

Similar to the case of the complex sine function, a complex cosine function can be also calculated using complex exponential functions:

$$\cos z = \frac{1}{2i}(e^{iz} + e^{-iz}) = \frac{1}{2i}\left[e^x(\cos y + i\sin y) + e^{-x}(\cos y - i\sin y)\right]$$

$$= \cos x \cosh y - i\sin x \sinh y$$

Using the above equation, add the following method to the Complex structure:

```
public static Complex Cos(Complex c)
{
return new Complex(Math.Cosh(c.imaginary) * Math.Cos(c.real),
                   -Math.Sinh(c.imaginary) * Math.Sin(c.real));
}
```

Tangent Function

A complex tangent function is computed using complex sine and cosine functions:

$$\tan z = \sin z / \cos z$$

Add the following complex *Tangent* method to the *Complex* structure:

```
public static Complex Tan(Complex c)
{
    return Sin(c) / Cos(c);
}
```

Testing Trigonometric Functions

We can examine complex trigonometric functions defined in the previous sections. Add a new static method, *TestTrigonometricFunctions*, to the *Program.cs* file:

```
static void TestTrigonometricFunctions()
{
    Console.WriteLine("\n Test complex trigonometric functions:");
    Complex c = new Complex(2, 3);
    Console.WriteLine(" c = {0}", c);
    Console.WriteLine(" Sin(c) = {0}", Complex.Sin(c));
    Console.WriteLine(" Cos(c) = {0}", Complex.Cos(c));
    Console.WriteLine(" Tan(c) = {0}", Complex.Tan(c));
}
```

Calling this method produces the results shown in Figure 2-6.

```
Test complex trigonometric functions:
c = 2 + 3 i
Sin(c) = 9.15449914691143 - 4.16890695996656 i
Cos(c) = -4.18962569096881 - 9.10922789375534 i
Tan(c) = -0.00376402564150415 + 1.00323862735361 i
```

Figure 2-6 Complex trigonometric functions in C#.

Inverse Trigonometric Functions

In this section, we'll implement inverse trigonometric functions of complex variables. The complex inverse trigonometric functions can be expressed in terms of the complex square root and logarithmic functions.

Inverse Sine Function

The complex inverse sine function can be expressed in the following equation:

$$\arcsin z = -i\log\left(iz + \sqrt{1-z^2} \right)$$

Using this relationship, we can add the following method to the *Complex* structure:

```
public static Complex Asin(Complex c)
{
    return -I * Log(Sqrt(1 - c * c) + I * c);
}
```

Inverse Cosine Function

The complex inverse cosine function can be represented by

$$\arccos z = -i\log\left(z + i\sqrt{1-z^2} \right)$$

Using the above equation, we can add the following method to the *Complex* structure:

```
public static Complex Acos(Complex c)
{
    return -I * Log(I * Sqrt(1 - c * c) + c);
}
```

Inverse Tangent Function

The complex inverse tangent function has the following relation:

$$\arctan z = \frac{i}{2}\left[\log(1-iz)-\log(1+iz)\right]$$

We can easily add the complex inverse *Tangent* method to the *Complex* structure:

```
public static Complex Atan(Complex c)
{
    return 0.5 * I * (Log(1 - I * c) - Log(1 + I * c));
}
```

Testing Inverse Trigonometric Functions

We can examine complex inverse trigonometric functions defined in the previous sections. Add a new static method, *TestInverseTrigonometricFunctions*, to the *Program.cs* file:

```
static void TestInverseTrigonometricFunctions()
{
    Console.WriteLine("\n Test complex inverse trigonometric functions:");
    Complex c = new Complex(2, 3);
    Console.WriteLine(" c = {0}", c);
    Console.WriteLine(" Asin(c) = {0}", Complex.Asin(c));
    Console.WriteLine(" Acos(c) = {0}", Complex.Acos(c));
    Console.WriteLine(" Atan(c) = {0}", Complex.Atan(c));
}
```

Calling this method will create the result shown in Figure 2-7.

```
Test complex inverse trigonometric functions:
c = 2 + 3 i
Asin(c) = 0.570652784321097 + 1.98338702991653 i
Acos(c) = 1.0001435424738 - 1.98338702991654 i
Atan(c) = 1.40992104959658 + 0.229072682968539 i
```

Figure 2-7 Complex inverse trigonometric functions in C#.

Hyperbolic Trigonometric Functions

In this section, we'll implement various complex hyperbolic trigonometric functions, including *Sinh*, *Cosh*, and *Tanh* functions.

Hyperbolic Sine Function

A complex *Sinh* function can be calculated using complex exponential functions:

$$\sinh z = \frac{1}{2}(e^z - e^{-z}) = \frac{1}{2}\left[e^x(\cos y + i\sin y) - e^{-x}(\cos y - i\sin y)\right]$$

$$= \sinh x \cos y + i\cosh x \sin y$$

So we can easily add the complex *Sinh* method to the *Complex* structure:

```
public static Complex Sinh(Complex c)
{
return new Complex(Math.Sinh(c.Real) * Math.Cos(c.Imaginary),
                Math.Cosh(c.Real) * Math.Sin(c.Imaginary));
}
```

Hyperbolic Cosine Functions

Similar to the case of the complex *Sinh* function, a complex *Cosh* function can be calculated using complex exponential functions:

$$\cosh z = \frac{1}{2}(e^z + e^{-z}) = \frac{1}{2}\left[e^x(\cos y + i\sin y) + e^{-x}(\cos y - i\sin y)\right]$$

$$= \cosh x \cos y + i\sinh x \sin y$$

Using the above equation, add the following method to the *Complex* structure:

```
public static Complex Cosh(Complex c)
{
    return new Complex(Math.Cosh(c.Real) * Math.Cos(c.Imaginary),
                       Math.Sinh(c.Real) * Math.Sin(c.Imaginary));
}
```

Hyperbolic Tangent Function

A complex *tanh* function can be computed using complex *Sinh* and *Cosh* functions:

$$\tanh z = \sinh z / \cosh z$$

Add the following complex *Tanh* method to the *Complex* structure:

```
public static Complex Tanh(Complex c)
{
    return Sinh(c) / Cosh(c);
}
```

Testing Hyperbolic Trigonometric Functions

We can examine complex hyperbolic trigonometric functions defined in the previous sections. Add a new static method, *TestHyperbolicTrigonometricFunctions*, to the *Program.cs* file:

```
static void TestHyperbolicTrigonometricFunctions()
{
    Console.WriteLine("\n Test complex hyperbolic trigonometric functions:");
    Complex c = new Complex(2, 3);
    Console.WriteLine(" c = {0}", c);
    Console.WriteLine(" Sinh(c) = {0}", Complex.Sinh(c));
    Console.WriteLine(" Cosh(c) = {0}", Complex.Cosh(c));
    Console.WriteLine(" Tanh(c) = {0}", Complex.Tanh(c));
}
```

This method generates the result shown in Figure 2-8.

```
Test complex hyperbolic trigonometric functions:
c = 2 + 3 i
Sinh(c) = -3.59056458998578 + 0.53092108624852 i
Cosh(c) = -3.72454550491532 + 0.511822569987385 i
Tanh(c) = 0.965385879022133 - 0.00988437503832252 i
```

Figure 2-8 Complex hyperbolic trigonometric functions in C#.

Inverse Trigonometric Hyperbolic Functions

In this section, we will implement inverse hyperbolic trigonometric functions of complex variables. The complex inverse hyperbolic trigonometric functions can be expressed in terms of the complex square root and logarithmic function.

Inverse Hyperbolic Sine Function

The complex inverse *Sinh* function can be expressed in the following equation:

$$\arcsin h(z) = \log\left(z + \sqrt{1 + z^2}\right)$$

Using this relationship, add the following method to the *Complex* structure:

```
public static Complex Asinh(Complex c)
{
    return Log(c + Sqrt(1.0 + c * c));
}
```

Inverse Hyperbolic Cosine Functions

The complex inverse *Cosh* function can be represented by

$$\arccos h(z) = \log\left(z + \sqrt{z^2 - 1}\right)$$

Using the above equation, add the following method to the *Complex* structure:

```
public static Complex Acosh(Complex c)
{
    return Log(c + Sqrt(c * c - 1.0));
}
```

Inverse Hyperbolic Tangent Function

The complex inverse *Tanh* function has the following relation:

$$\arctan h(z) = \frac{1}{2}\left[\log(1 + z) - \log(1 - z)\right]$$

We can easily add the complex inverse *Tanh* method:

```
public static Complex Atanh(Complex c)
{
    return 0.5 * (Log(1 + c) - Log(1 - c));
}
```

Testing Inverse Hyperbolic Trigonometric Functions

We can examine complex inverse hyperbolic trigonometric functions defined in previous sections. Add a new static method, *TestInverseHyperbolicTrigonometricFunctions*, to the *Program.cs* file:

```
static void TestInverseHyperbolicTrigonometricFunctions()
{
    Console.WriteLine("\n Test complex inverse hyperbolic trigonometric
                        functions:");
    Complex c = new Complex(2, 3);
    Console.WriteLine(" c = {0}", c);
    Console.WriteLine(" Asinh(c) = {0}", Complex.Asinh(c));
    Console.WriteLine(" Acosh(c) = {0}", Complex.Acosh(c));
    Console.WriteLine(" Atanh(c) = {0}", Complex.Atanh(c));
}
```

This method creates the result shown in Figure 2-9.

```
Test complex inverse hyperbolic trigonometric functions:
c = 2 + 3 i
Asinh(c) = 1.9686379257931 + 0.964658504407603 i
Acosh(c) = 1.98338702991654 + 1.0001435424738 i
Atanh(c) = 0.14694666622553 + 1.33897252229449 i
```

Figure 2-9 Complex inverse hyperbolic trigonometric functions in C#.

Chapter 3
Vectors and Matrices

In science and engineering, we often encounter many physical quantities that are vectors, including velocity, acceleration, force, and momentum. Many scientific and engineering computations involve solving linear equations with multiple variables. Matrix analysis is a basic theory of these linear operations. In this chapter, we will implement vector and matrix classes in C#. The vector will be defined using a double or complex array, and a matrix will be defined using an array of vectors.

If you have ever programmed with WPF, you probably know that WPF implements a *Vector* and a *Matrix* structure in homogeneous coordinates in 2D space. It also includes corresponding structures, *Vector3D* and *Matrix3D*, for 3D applications. WPF uses a convention of pre-multiplying matrices by row vectors. If you want to know more about the vector and matrix structures in WPF and their applications, please refer to my other book – *Practical WPF Graphics Programming*.

Note that the vector and matrix structures contained in WPF are designed specifically for transformation operations on various framework and user interface elements. They are not general enough for applications such as the solution of linear equation systems. For example, if a linear system consists of 10 equations and 10 unknowns, the vector and matrix involved in this system will be 10 dimensional. The 2D *Vector* and *Matrix* structrures or 3D *Vector3D* and *Matrix3D* structures in WPF cannot be applied to this system. Thus, we need to create general n-dimensional vectors and n by n matrix structures.

Real Vector Structure

In this section, we will implement a general vector structure consisting of only real double variables. Here, the column-major convention will be used to define the vector and matrix, which is different from the row-major convention used in WPF.

Start with a new C# Console application in Visual Studio and name it *RealVectorTest*. Right click the project in the Solution Explorer, select Add – Class... and name the new class *VectorR*. Change its namespace to *XuMath*.

Now, we need to add appropriate constructors to the *VectorR* class, which should be able to create an initialized vector for a given length (size), or a vector converted from a given double array. To deal with these aspects, we first write:

```
using System;

namespace XuMath
{
    public struct VectorR
    {
        private int size;
        public double[] vector;

        public VectorR(int size)
        {
            this.size = size;
            this.vector = new double[size];
            for (int i = 0; i < size; i++)
            {
                vector[i] = 0.0;
            }
        }

        public VectorR(double[] vector)
        {
            this.size = vector.Length;
            this.vector = vector;
        }
    }
}
```

As in the case of creating *Complex* structure, here we also change the class to the structure. The first constructor takes the integer size, the length of the vector, as an input parameter. It creates a vector, whose elements are initialized to zero.

The second constructor creates a vector to hold a given double precision float-point array. The size of the vector is the same as the length of the array.

Basic Definitions

Now, we need to define a number of operators and public methods in the *VectorR* struct. First, here is the indexing operation, which can be written as:

```
public double this[int n]
{
    get
    {
        if (n < 0 || n > size)
        {
            throw new ArgumentOutOfRangeException("n", n, "n is out of range!");
        }
        return vector[n];
    }
    set { vector[n] = value; }
}
```

This indexing operation allows you to access the nth element of the vector. For a vector v, we now can simply use $v[n]$ to access the nth element of v, which is similar to ordinary mathematical notations.

The next useful method is the *ToString* method, which returns a string representation of a real vector. Here we need to override the *ToString* method using the following code:

```
public override string ToString()
{
    string s = "(";
    for (int i = 0; i < size - 1; i++)
    {
        s += vector[i].ToString() + ", ";
    }
    s += vector[size - 1].ToString() + ")";
    return s;
}
```

In condition statements, we need to know if two vectors are equal or not. The *System.Object* type offers a virtual method, named *Equals*, whose purpose is to return *true* if two objects have the same "value".

In order to compare two vectors, we need to implement the *Equals* method in the *VectorR* structure using the following code snippet:

```
public override bool Equals(object obj)
{
    return (obj is VectorR) && this.Equals((VectorR)obj);
}

public bool Equals(VectorR v)
{
    return vector == v.vector;
}

public override int GetHashCode()
{
    return vector.GetHashCode();
}

public static bool operator ==(VectorR v1, VectorR v2)
{
    return v1.Equals(v2);
}

public static bool operator !=(VectorR v1, VectorR v2)
{
    return !v1.Equals(v2);
}
```

You can examine the above *Equals* methods by adding a new static method, *TestEquals*, to the *Program.cs* file:

```
static void TestEquals()
{
    double[] a1 = new double[3] { 1.0, 2.0, 3.0 };
    double[] a2 = new double[3] { 4.0, 5.0, 6.0 };
    VectorR v1 = new VectorR(a1);
    VectorR v2 = new VectorR(a2);
    bool b = v1 == v2;
    Console.WriteLine("\n b = {0}", b);
```

```
    }
```

The result of *b* will be *False*.

Then there are the unary plus and minus operators:

```
    public static VectorR operator +(VectorR v)
    {
        return v;
    }

    public static VectorR operator -(VectorR v)
    {
        double[] result = new double[v.GetSize()];
        for(int i = 0; i < v.GetSize(); i++)
        {
            result[i] = -v[i];
        }
        return new VectorR(result);
    }
```

Next, we define the norm and unit vectors:

```
    public double GetNorm()
    {
        double result = 0.0;
        for (int i = 0; i < size; i++)
        {
            result += vector[i] * vector[i];
        }
        return Math.Sqrt(result);
    }

    public double GetNormSquare()
    {
        double result = 0.0;
        for (int i = 0; i < size; i++)
        {
            result += vector[i] * vector[i];
        }
        return result;
    }

    public VectorR GetUnitVector()
    {
        VectorR result = new VectorR(vector);
        result.Normalize();
        return result;
    }

    public void Normalize()
    {
        double norm = GetNorm();
        if (norm == 0)
        {
            throw new DivideByZeroException("Normalize a vector
                                        with norm of zero!");
```

```
        }
        for (int i = 0; i < size; i++)
        {
            vector[i] /= norm;
        }
    }
```

For example, you can compute the norm for a vector by adding a new static method, *TestNorm*, to the *Program.cs* file:

```
    static void TestNorm()
    {
        VectorR v = new VectorR(10);
        for (int i = 0; i < 10; i++)
        {
            v[i] = 0.5 * i;
        }
        double result = v.GetNorm();
        Console.WriteLine("Norm of the Vector = {0}", result);
    }
```

This gives result = 8.44.

It is sometimes useful to be able to access the size or length of a vector directly using the code snippet:

```
    public int GetSize()
    {
        return size;
    }
```

You can exchange two elements of a vector using the following method:

```
    public VectorR GetSwap(int m, int n)
    {
        double temp = vector[m];
        vector[m] = vector[n];
        vector[n] = temp;
        return new VectorR(vector);
    }
```

Of course, the above method is only valid when both m and n are in the range of zero and the length of the vector.

Mathematical Operators

It is easy to implement various mathematical operations of vectors in C#. Add the following static methods to the *VectorR* struct:

```
    public static VectorR operator +(VectorR v1, VectorR v2)
    {
        VectorR result = new VectorR(v1.size);
        for (int i = 0; i < v1.size; i++)
        {
            result[i] = v1[i] + v2[i];
        }
        return result;
    }
```

```
public static VectorR operator +(VectorR v, double d)
{
    VectorR result = new VectorR(v.size);
    for (int i = 0; i < v.size; i++)
    {
        result[i] = v[i] + d;
    }
    return result;
}

public static VectorR operator +(double d, VectorR v)
{
    VectorR result = new VectorR(v.size);
    for (int i = 0; i < v.size; i++)
    {
        result[i] = v[i] + d;
    }
    return result;
}

public static VectorR operator -(VectorR v1, VectorR v2)
{
    VectorR result = new VectorR(v1.size);
    for (int i = 0; i < v1.size; i++)
    {
        result[i] = v1[i] - v2[i];
    }
    return result;
}

public static VectorR operator -(VectorR v, double d)
{
    VectorR result = new VectorR(v.size);
    for (int i = 0; i < v.size; i++)
    {
        result[i] = v[i] - d;
    }
    return result;
}

public static VectorR operator -(double d, VectorR v)
{
    VectorR result = new VectorR(v.size);
    for (int i = 0; i < v.size; i++)
    {
        result[i] = d - v[i];
    }
    return result;
}

public static VectorR operator *(VectorR v, double d)
{
    VectorR result = new VectorR(v.size);
    for (int i = 0; i < v.size; i++)
```

```
        {
            result[i] = v[i] * d;
        }
        return result;
    }

    public static VectorR operator *(double d, VectorR v)
    {
        VectorR result = new VectorR(v.size);
        for (int i = 0; i < v.size; i++)
        {
            result[i] = d * v[i];
        }
        return result;
    }

    public static VectorR operator /(VectorR v, double d)
    {
        VectorR result = new VectorR(v.size);
        for (int i = 0; i < v.size; i++)
        {
            result[i] = v[i] / d;
        }
        return result;
    }

    public static VectorR operator /(double d, VectorR v)
    {
        VectorR result = new VectorR(v.size);
        for (int i = 0; i < v.size; i++)
        {
            result[i] = d / v[i];
        }
        return result;
    }
```

Here, the standard vector addition, subtraction, multiplication, and division are performed using the above methods. Please note that the multiplication and division in the above methods are not for two-vector operations, but for a vector scaled by a scalar number.

You can test the math operations on the vectors by adding a new static method, *TestMathOperators*, to the *Program.cs* file:

```
static void TestMathOperators()
{
    VectorR v1 = new VectorR(new double[] { 1.0, 2.0, 3.0, 4.0, 5.0 });
    VectorR v2 = new VectorR(new double[] { 6.0, 7.0, 8.0, 9.0, 10.0 });
    double d = 20;
    Console.WriteLine("\n v1 = {0}", v1);
    Console.WriteLine(" v2 = {0}", v2);
    Console.WriteLine(" d = {0}", d);
    Console.WriteLine(" v2 + v1 = {0}", (v2 + v1));
    Console.WriteLine(" v2 - v1 = {0}", (v2 - v1));
    Console.WriteLine(" v1 * d = {0}", (v1 * d));
    Console.WriteLine(" v1 / d = {0}", (v1 / d));
}
```

This method produces the result shown in Figure 3-1.

```
v1 = (1, 2, 3, 4, 5)
v2 = (6, 7, 8, 9, 10)
d = 20
v2 + v1 = (7, 9, 11, 13, 15)
v2 - v1 = (5, 5, 5, 5, 5)
v1 * d = (20, 40, 60, 80, 100)
v1 / d = (0.05, 0.1, 0.15, 0.2, 0.25)
```

Figure 3-1 Math operations on vectors in C#.

Vector Multiplications

In this section, we will discuss various vector multiplications. Four different operations will be presented, including dot (or scalar or inner) and cross products for two vectors, as well as triple scalar and triple vector products for three vectors.

Scalar or Dot Product

The scalar or dot product of two vectors occurs frequently in physics. For instance:

$$work = force \cdot displacement \cdot \cos\theta$$

is usually interpreted as displacement times the projection of the force along the displacement. With such application in mind, we define the dot or scalar product of two vectors with n-dimensions in the form:

$$\mathbf{A} \cdot \mathbf{B} = \sum_{i=1}^{n} A_i B_i$$

The dot product of two vectors produces a scalar quantity. This can be performed using the following method:

```
public static double DotProduct(VectorR v1, VectorR v2)
{
    double result = 0.0;
    for (int i = 0; i < v1.size; i++)
    {
        result += v1[i] * v2[i];
    }
    return result;
}
```

For example, the following code snippet is used to calculate the dot product of two 10-dimensional vectors:

```
VectorR v1 = new VectorR(10);
VectorR v2 = new VectorR(10);
for (int i = 0; i < 10; i++)
{
    v1[i] = 0.5 * i;
    v2[i] = 1.5 * i;
}
```

```
double result = VectorR.DotProduct(v1, v2);
```

This gives output of 213.75.

Cross Product

A second form of vector multiplication employs the sine of the included angle instead of the cosine. The vector or cross product is defined as

$$\mathbf{C} = \mathbf{A} \times \mathbf{B} \quad \text{with} \quad C = AB\sin\theta$$

Unlike the previous case of the scalar product, the result is now a vector. We assign it a direction perpendicular to the plane of **A** and **B** such that **A**, **B**, and **C** form a right-handed system. Also, the cross product is defined only for vectors in three-dimension. An alternative definition of the vector product consists in specifying the components of the result:

$$\mathbf{A} \times \mathbf{B} = \mathbf{i}\left(A_y B_z - A_z B_y\right) + \mathbf{j}\left(A_z B_x - A_x B_z\right) + \mathbf{k}\left(A_x B_y - A_y B_x\right)$$

This can be achieved using the following method:

```
public static VectorR CrossProduct(VectorR v1, VectorR v2)
{
    if (v1.size !=3)
    {
        throw new ArgumentOutOfRangeException(
        "v1", v1, "Vector v1 must be 3 dimensional!");
    }
    VectorR result = new VectorR(3);
    result[0] = v1[1] * v2[2] - v1[2] * v2[1];
    result[1] = v1[2] * v2[0] - v1[0] * v2[2];
    result[2] = v1[0] * v2[1] - v1[1] * v2[0];
    return result;
}
```

Triple Scalar Product

The triple scalar product is defined as

$$\mathbf{A} \cdot (\mathbf{B} \times \mathbf{C}) = \mathbf{B} \cdot (\mathbf{C} \times \mathbf{A}) = \mathbf{C} \cdot (\mathbf{A} \times \mathbf{B})$$

$$= A_x\left(B_y C_z - B_z C_y\right) + A_y\left(B_z C_x - B_x C_z\right) + A_z\left(B_x C_y - B_y C_x\right)$$

A convenient representation of the component expansion is provided by a determinant

$$\mathbf{A} \cdot (\mathbf{B} \times \mathbf{C}) = \begin{vmatrix} A_x & A_y & A_z \\ B_x & B_y & B_z \\ C_x & C_y & C_z \end{vmatrix}$$

The rules for interchanging rows and columns of a determinant provide an immediate verification of the permutations and symmetry. This triple scalar product can be achieved by using the *TriScalarProduct* method:

```
public static double TriScalarProduct(VectorR v1, VectorR v2, VectorR v3)
{
    if (v1.size != 3)
    {
        throw new ArgumentOutOfRangeException(
        "v1", v1, "Vector v1 must be 3 dimensional!");
    }
    if (v1.size != 3)
    {
        throw new ArgumentOutOfRangeException(
        "v2", v2, "Vector v2 must be 3 dimensional!");
    }
    if (v1.size != 3)
    {
        throw new ArgumentOutOfRangeException(
        "v3", v3, "Vector v3 must be 3 dimensional!");
    }
    double result = v1[0] * (v2[1] * v3[2] - v2[2] * v3[1]) +
                    v1[1] * (v2[2] * v3[0] - v2[0] * v3[2]) +
                    v1[2] * (v2[0] * v3[1] - v2[1] * v3[0]);
    return result;
}
```

Again, this method is valid only for vectors in three dimensions.

Triple Vector Product

The second triple product of interest is defined in terms of the following formula:

$$\mathbf{A} \times (\mathbf{B} \times \mathbf{C}) = \mathbf{B}(\mathbf{A} \cdot \mathbf{C}) - \mathbf{C}(\mathbf{A} \cdot \mathbf{B})$$

Here the parentheses must be retained, as they may be seen by considering the unit vector operation:

$$\mathbf{i} \times (\mathbf{i} \times \mathbf{j}) = \mathbf{i} \times \mathbf{k} = -\mathbf{j} \quad \text{but} \quad (\mathbf{i} \times \mathbf{i}) \times \mathbf{j} = 0$$

The following method performs this triple vector product operation:

```
public static VectorR TriVectorProduct(VectorR v1,
            VectorR v2, VectorR v3)
{
    if (v1.size != 3)
    {
        throw new ArgumentOutOfRangeException(
        "v1", v1, "Vector v1 must be 3 dimensional!");
    }
    if (v1.size != 3)
    {
        throw new ArgumentOutOfRangeException(
        "v2", v2, "Vector v2 must be 3 dimensional!");
    }
    if (v1.size != 3)
    {
        throw new ArgumentOutOfRangeException(
        "v3", v3, "Vector v3 must be 3 dimensional!");
    }
```

```
        return v2 * VectorR.DotProduct(v1, v3) - v3 * VectorR.DotProduct(v1, v2);
    }
```

Again, this method is also defined for vectors in three-dimension. Be careful when using this method. The order of vectors **v1**, **v2**, and **v3** in this method always implies the following operation: $v1 \times (v2 \times v3)$.

You can examine vector multiplications by adding a new static method, *TestMultiplications*, to the *Program.cs* file:

```
static void TestMultiplications()
{
    VectorR v1 = new VectorR(new double[] { 1.0, 2.0, -1.0 });
    VectorR v2 = new VectorR(new double[] { 0.0, 1.0, 1.0 });
    VectorR v3 = new VectorR(new double[] { 1.0, -1.0, 0.0 });
    Console.WriteLine("\n v1 = {0}", v1);
    Console.WriteLine(" v2 = {0}", v2);
    Console.WriteLine(" v3 = {0}", v3);
    Console.WriteLine(" Dot product of v1 and v2 = {0}",
                        VectorR.DotProduct(v1, v2));
    Console.WriteLine(" Cross product of v1 and v2 = {0}",
                        VectorR.CrossProduct(v1, v2).ToString());
    Console.WriteLine(" Triple scalar product of v1, v2, and v3 = {0}",
                        VectorR.TriScalarProduct(v1, v2, v3));
    Console.WriteLine(" Triple vector product of v1, v2, and v3 = {0}",
                        VectorR.TriVectorProduct(v1, v2, v3));
}
```

This will produce the results shown in Figure 3-2.

```
v1 = <1, 2, -1>
v2 = <0, 1, 1>
v3 = <1, -1, 0>
Dot product of v1 and v2 = 1
Cross product of v1 and v2 = <3, -1, 1>
Triple scalar product of v1, v2, and v3 = 4
Triple vector product of v1, v2, and v3 = <-1, 0, -1>
```

Figure 3-2 Vector multiplications in C#.

In this section, we demonstrated how to implement the basic operations of vectors. Following the procedure presented here, you should be able to add more features to the *VectorR* structure according to the requirements of your applications.

Complex Vector Class

Following the same procedure as in the case of the real vectors in the previous section, we can easily construct a complex vector structure. Now create a new C# Console application and name it *ComplexVectorTest*. Add the *Complex.cs* class from the *ComplexNumberTest* project and a new class of *VectorC* to the current project. Here, I will omit the detailed implementation process, but only point out the difference between the complex and real vector structures.

Implementation

We first need to create constructors that are used to initialize a complex vector of a given length (size), or convert a complex vector from a given complex array, as shown in the following code snippet:

```
using System;

namespace XuMath
{
    public struct VectorC
    {
        public int size;
        public Complex[] vector;

        public VectorC(int size)
        {
            this.size = size;
            this.vector = new Complex[size];
            for (int i = 0; i < size; i++)
            {
                vector[i] = Complex.Zero;
            }
        }

        public VectorC(Complex[] vector)
        {
            this.size = vector.Length;
            this.vector = vector;
        }
    }
}
```

As in the case of creating a *VectorR* structure, here we also change the class to the structure. The first constructor takes the integer *n*, the length of the vector, as an input parameter. It creates a complex vector, whose elements are initialized to zero, for a given the length or size of the vector.

The second constructor creates a complex vector to hold a given complex array. The size of the vector is the same as the length of the array. Please note that the complex numbers are defined in the *Complex* structure. In order to create a complex vector, the *Complex* structure must have the namespace of *XuMath*.

The basic definitions and mathematical operations of complex vectors are similar to those of real vectors. I want to point out a minor difference in the dot product. For two complex vectors, their dot product is defined by taking the conjugate of the first vector (in some reference books, it takes the conjugate of the second vector) and then applying the dot product formula as in the case of real vectors. This is achieved using the following method:

```
public static Complex DotProduct(VectorC v1, VectorC v2)
{
    Complex result = Complex.Zero;
    for (int i = 0; i < v1.size; i++)
    {
        result += v1[i].Conjugate * v2[i];
    }
    return result;
```

```
        }
```

Here, I will not present the detail implementation for the *VectorC* class. Instead, I will only provide you the code listing of this class for your review:

```csharp
using System;

namespace XuMath
{
    public struct VectorC:ICloneable
    {
        public int size;
        public Complex[] vector;

         public VectorC(int size)
        {
            this.size = size;
            this.vector = new Complex[size];
            for (int i = 0; i < size; i++)
            {
                vector[i] = Complex.Zero;
            }
        }

        public VectorC(Complex[] vector)
        {
            this.size = vector.Length;
            this.vector = vector;
        }

        #region Make a deep copy
        public VectorC Clone()
        {
            // returns a deep copy of the vector
            VectorC v = new VectorC(vector);
            v.vector = (Complex[])vector.Clone();
            return v;
        }

        object ICloneable.Clone()
        {
            return Clone();
        }
        #endregion

        #region Equals and Hashing:
        public override bool Equals(object obj)
        {
            return (obj is VectorC) && this.Equals((VectorC)obj);
        }

        public bool Equals(VectorC v)
        {
            return vector == v.vector;
        }
```

```
public override int GetHashCode()
{
    return vector.GetHashCode();
}

public static bool operator ==(VectorC v1, VectorC v2)
{
    return v1.Equals(v2);
}

public static bool operator !=(VectorC v1, VectorC v2)
{
    return !v1.Equals(v2);
}
#endregion

#region Definition and basics:
public Complex this[int n]
{
    get
    {
        if (n < 0 || n > size)
        {
            throw new ArgumentOutOfRangeException(
              "n", n, "n is out of range!");
        }
        return vector[n];
    }
    set { vector[n] = value; }
}

public double GetNorm()
{
    Complex result = Complex.Zero;
    for (int i = 0; i < size; i++)
    {
        result += vector[i].Conjugate * vector[i];
    }
    return Math.Sqrt(result.Real);
}

public double GetNormSquare()
{
    Complex result = Complex.Zero;
    for (int i = 0; i < size; i++)
    {
        result += vector[i].Conjugate * vector[i];
    }
    return result.Real;
}

public void Normalize()
{
    double norm = GetNorm();
    if (norm == 0)
```

```
        {
            throw new DivideByZeroException("Normalize a vector
                    with norm of zero!");
        }
        for (int i = 0; i < size; i++)
        {
            vector[i] /= norm;
        }
    }

    public VectorC GetUnitVector()
    {
        VectorC result = new VectorC(vector);
        result.Normalize();
        return result;
    }

    public int GetSize()
    {
        return size;
    }

    public VectorC GetConjugate()
    {
        for (int i = 0; i < size; i++)
        {
            vector[i] = vector[i].Conjugate;
        }
        return new VectorC(vector);
    }

    public VectorC GetSwap(int m, int n)
    {
        Complex temp = vector[m];
        vector[m] = vector[n];
        vector[n] = temp;
        return new VectorC(vector);
    }
    #endregion

    #region Mathematical operations:

    public static VectorC operator +(VectorC v)
    {
        return v;
    }

    public static VectorC operator -(VectorC v)
    {
        Complex[] result = new Complex[v.GetSize()];
        for (int i = 0; i < v.GetSize(); i++)
        {
            result[i] = -v[i];
        }
        return new VectorC(result);
```

```
    }

    public static VectorC operator +(VectorC v1, VectorC v2)
    {
        VectorC result = new VectorC(v1.size);
        for (int i = 0; i < v1.size; i++)
        {
            result[i] = v1[i] + v2[i];
        }
        return result;
    }

    public static VectorC operator +(VectorC v, double d)
    {
        VectorC result = new VectorC(v.size);
        for (int i = 0; i < v.size; i++)
        {
            result[i] = v[i] + d;
        }
        return result;
    }

    public static VectorC operator +(double d, VectorC v)
    {
        VectorC result = new VectorC(v.size);
        for (int i = 0; i < v.size; i++)
        {
            result[i] = v[i] + d;
        }
        return result;
    }

    public static VectorC operator +(VectorC v, Complex c)
    {
        VectorC result = new VectorC(v.size);
        for (int i = 0; i < v.size; i++)
        {
            result[i] = v[i] + c;
        }
        return result;
    }

    public static VectorC operator +(Complex c, VectorC v)
    {
        VectorC result = new VectorC(v.size);
        for (int i = 0; i < v.size; i++)
        {
            result[i] = v[i] + c;
        }
        return result;
    }

    public static VectorC operator -(VectorC v1, VectorC v2)
    {
        VectorC result = new VectorC(v1.size);
```

```
        for (int i = 0; i < v1.size; i++)
        {
            result[i] = v1[i] - v2[i];
        }
        return result;
    }

    public static VectorC operator -(VectorC v, double d)
    {
        VectorC result = new VectorC(v.size);
        for (int i = 0; i < v.size; i++)
        {
            result[i] = v[i] - d;
        }
        return result;
    }

    public static VectorC operator -(double d, VectorC v)
    {
        VectorC result = new VectorC(v.size);
        for (int i = 0; i < v.size; i++)
        {
            result[i] = d - v[i];
        }
        return result;
    }

    public static VectorC operator -(VectorC v, Complex c)
    {
        VectorC result = new VectorC(v.size);
        for (int i = 0; i < v.size; i++)
        {
            result[i] = v[i] - c;
        }
        return result;
    }

    public static VectorC operator -(Complex c, VectorC v)
    {
        VectorC result = new VectorC(v.size);
        for (int i = 0; i < v.size; i++)
        {
            result[i] = c - v[i];
        }
        return result;
    }

    public static VectorC operator *(VectorC v, double d)
    {
        VectorC result = new VectorC(v.size);
        for (int i = 0; i < v.size; i++)
        {
            result[i] = v[i] * d;
        }
        return result;
```

```
    }

    public static VectorC operator *(double d, VectorC v)
    {
        VectorC result = new VectorC(v.size);
        for (int i = 0; i < v.size; i++)
        {
            result[i] = d * v[i];
        }
        return result;
    }

    public static VectorC operator *(VectorC v, Complex c)
    {
        VectorC result = new VectorC(v.size);
        for (int i = 0; i < v.size; i++)
        {
            result[i] = v[i] * c;
        }
        return result;
    }

    public static VectorC operator *(Complex c, VectorC v)
    {
        VectorC result = new VectorC(v.size);
        for (int i = 0; i < v.size; i++)
        {
            result[i] = c * v[i];
        }
        return result;
    }

    public static VectorC operator /(VectorC v, double d)
    {
        VectorC result = new VectorC(v.size);
        for (int i = 0; i < v.size; i++)
        {
            result[i] = v[i] / d;
        }
        return result;
    }

    public static VectorC operator /(double d, VectorC v)
    {
        VectorC result = new VectorC(v.size);
        for (int i = 0; i < v.size; i++)
        {
            result[i] = d / v[i];
        }
        return result;
    }

    public static VectorC operator /(VectorC v, Complex c)
    {
        VectorC result = new VectorC(v.size);
```

```
        for (int i = 0; i < v.size; i++)
        {
            result[i] = v[i] / c;
        }
        return result;
    }

    public static VectorC operator /(Complex c, VectorC v)
    {
        VectorC result = new VectorC(v.size);
        for (int i = 0; i < v.size; i++)
        {
            result[i] = c / v[i];
        }
        return result;
    }
    #endregion;

    #region Public methods:
    public static Complex DotProduct(VectorC v1, VectorC v2)
    {
        Complex result = Complex.Zero;
        for (int i = 0; i < v1.size; i++)
        {
            result += v1[i].Conjugate * v2[i];
        }
        return result;
    }

    public static VectorC Product(VectorC v1, VectorC v2)
    {
        VectorC result = new VectorC(v1.size);
        for (int i = 0; i < v1.size; i++)
        {
            result[i] = v1[i] * v2[i];
        }
        return result;
    }

    public static VectorC CrossProduct(VectorC v1, VectorC v2)
    {
        if (v1.size != 3)
        {
            throw new ArgumentOutOfRangeException(
            "v1", v1, "Vector v1 must be 3 dimensional!");
        }
        VectorC result = new VectorC(3);
        result[0] = v1[1] * v2[2] - v1[2] * v2[1];
        result[1] = v1[2] * v2[0] - v1[0] * v2[2];
        result[2] = v1[0] * v2[1] - v1[1] * v2[0];
        return result;
    }
    #endregion;

    public override string ToString()
```

```
        {
            string s = "(";
            for (int i = 0; i < size -1; i++)
            {
                s += vector[i].ToString() + ", ";
            }
            s += vector[size - 1].ToString() + ")";
            return s;
        }
    }
}
```

In this structure, we implement a variety of public properties and public methods, which allow you to perform various operations on complect vectors. Following the same procedure presented here, you can easily add more public properties and methods according to the requirements of your application.

Testing the VectorC Class

Now, you can test the *VectorC* class by using the following *Program.cs* file:

```
using System;
using XuMath;

namespace ComplexVectorTest
{
    class Program
    {
        static void Main(string[] args)
        {
            TestVectorC();
            Console.ReadLine();
        }

        static void TestVectorC()
        {
            VectorC v1 = new VectorC(new Complex[] {new Complex(1, 2),
                                                    new Complex(2, 3),
                                                    new Complex(3, 4),
                                                    new Complex(4, 5)});
            VectorC v2 = new VectorC(new Complex[] {new Complex(2, 1),
                                                    new Complex(3, 2),
                                                    new Complex(4, 3),
                                                    new Complex(5, 4)});
            Complex c = new Complex(5, 10);

            Console.WriteLine("\n v1 = {0}", v1);
            Console.WriteLine(" v2 = {0}", v2);
            Console.WriteLine(" c = {0}", c);
            Console.WriteLine(" v2 + v1 = {0}", (v2 + v1));
            Console.WriteLine(" v2 - v1 = {0}", (v2 - v1));
            Console.WriteLine(" v2 * c = {0}", (v2 * c));
            Console.WriteLine(" v2 / c = {0}", (v2 / c));
            Console.WriteLine(" Product of v1 and v2 = {0}",
                                VectorC.Product(v1, v2));
```

```
        Console.WriteLine(" Dot product of v1 and v2 = {0}",
                             VectorC.DotProduct(v1, v2));
        }
    }
}
```

Pressing F5 to run the application produces the results shown in Figure 3-3.

```
v1 = (1 + 2 i, 2 + 3 i, 3 + 4 i, 4 + 5 i)
v2 = (2 + 1 i, 3 + 2 i, 4 + 3 i, 5 + 4 i)
c = 5 + 10 i
v2 + v1 = (3 + 3 i, 5 + 5 i, 7 + 7 i, 9 + 9 i)
v2 - v1 = (1 - 1 i, 1 - 1 i, 1 - 1 i, 1 - 1 i)
v2 * c = (0 + 25 i, -5 + 40 i, -10 + 55 i, -15 + 70 i)
v2 / c = (0.16 - 0.12 i, 0.28 - 0.16 i, 0.4 - 0.2 i, 0.52 - 0.24 i)
Product of v1 and v2 = (0 + 5 i, 0 + 13 i, 0 + 25 i, 0 + 41 i)
Dot product of v1 and v2 = 80 - 24 i
```

Figure 3-3 Complex vector computation in C#.

Real Matrix Class

A matrix can be used to describe linear equations, to keep track of the coefficients of the linear transformations, and to record data that depend on two parameters. Matrices can be added, multiplied, and decomposed in various ways, making them a key concept in linear algebra and matrix theory. Although much of our work will involve square matrices, this is not always the case. We must therefore allow for the possibility that the number of rows is not the same as the number of columns. Thus, we need two integer numbers, *Rows* and *Cols*, to specify the dimensions of a matrix.

A real matrix structure can be constructed using a two-dimensional array. Now let's start with a new C# Console project and name it *RealMatrixTest*. Add the *VectorR* class from the previous project, *RealVectorTest*, and a new class, *MatrixR*, to the current project. The matrix constructors should be able to create an initialized matrix of a given dimension, a matrix converted from a given two-dimensional floating-point array, as described in the following code snippet:

```csharp
using System;

namespace XuMath
{
    public struct MatrixR
    {
        private int Rows;
        private int Cols;
        private double[,] matrix;

        public MatrixR(int Rows, int Cols)
        {
            this.Rows = Rows;
            this.Cols = Cols;
            this.matrix = new double[Rows, Cols];
            for (int i = 0; i < Rows; i++)
            {
                for (int j = 0; j < Cols; j++)
                {
```

```
                    matrix[i, j] = 0.0;
                }
            }
        }

        public MatrixR(double[,] matrix)
        {
            this.Rows = matrix.GetLength(0);
            this.Cols = matrix.GetLength(1);
            this.matrix = matrix;
        }
    }
}
```

The first constructor takes the integers, *Rows* and *Cols*, the dimension of the matrix, as input parameters. It creates a matrix, whose elements are initialized to zero, for a given dimension or size for the matrix.

The second constructor creates a matrix that holds a given two-dimensional double precision float-point array. The size of the matrix is the same as the dimension of the array.

Basic Definitions

We first define the indexing property:

```
public double this[int m, int n]
{
    get
    {
        if (m < 0 || m > Rows)
        {
            throw new ArgumentOutOfRangeException(
            "m", m, "m is out of range!");
        }
        if (n < 0 || n > Cols)
        {
            throw new ArgumentOutOfRangeException(
            "n", n, "n is out of range!");
        }
        return matrix[m, n];
    }
    set { matrix[m, n] = value; }
}
```

If *m* is a matrix, the $m[i, j]$ represents the element at the *i*th row and *j*th column, which is very similar to ordinary mathematical notations.

Next we need to override the *ToString* method:

```
public override string ToString()
{
    string ss = "(";
    for (int i = 0; i < Rows; i++)
    {
        string s = "";
        for (int j = 0; j < Cols - 1; j++)
```

```
        {
            s += matrix[i, j].ToString() + ", ";
        }
        s += matrix[i, Cols - 1].ToString();
        if (i != Rows - 1 && i == 0)
            ss += s + "\n";
        else if (i != Rows - 1 && i != 0)
            ss += " " + s + "\n";
        else
            ss += " " + s + ")";
    }
    return ss;
}
```

You can use this method to display a matrix on your screen. For example, the following code snippet will display a 2 by 3 matrix:

```
MatrixR m = new MatrixR(2, 3);
Console.WriteLine(m.ToString());
```

The matrix's output will be

```
        (0, 0, 0
         0, 0, 0)
```

Another important matrix is the identity matrix. We implement this identity matrix using the following code:

```
public MatrixR Identity()
{
    MatrixR m = new MatrixR(Rows, Cols);
    for (int i = 0; i < Rows; i++)
    {
        for (int j = 0; j < Cols; j++)
        {
            if (i == j)
            {
                m[i, j] = 1;
            }
        }
    }
    return m;
}
```

In condition statements, we need to know if two matrices are equal or not. The *System.Object* type offers a virtual method, named *Equals*, whose purpose is to return *true* if two objects have the same "value".

In order to compare two matrices, we need to implement the *Equals* method in the *MatrixR* structure using the following code snippet:

```
public override bool Equals(object obj)
{
    return (obj is MatrixR) && this.Equals((MatrixR)obj);
}

public bool Equals(MatrixR m)
{
```

```
        return matrix == m.matrix;
    }

    public override int GetHashCode()
    {
        return matrix.GetHashCode();
    }

    public static bool operator ==(MatrixR m1, MatrixR m2)
    {
        return m1.Equals(m2);
    }

    public static bool operator !=(MatrixR m1, MatrixR m2)
    {
        return !m1.Equals(m2);
    }
```

It is sometimes useful for us to access the dimensions of a matrix, and to know if a matrix is squared:

```
    public int GetRows()
    {
        return Rows;
    }

    public int GetCols()
    {
        return Cols;
    }

    public bool IsSquared()
    {
        if (Rows == Cols)
            return true;
        else
            return false;
    }
```

Sometimes we need to know whether or not two matrices have the same dimensions:

```
    public static bool CompareDimension(MatrixR m1, MatrixR m2)
    {
        if (m1.GetRows() == m2.GetRows() &&
            m1.GetCols() == m2.GetCols())
            return true;
        else
            return false;
    }
```

Matrix and Vector Manipulations

In this section, we will implement various methods for manipulating matrices and vectors. We will add some frequently used operations to the *MatrixR* structure. You can easily add more methods to the structure by following the procedure presented here.

Vector from Matrix

In some cases, we may want to extract a vector from a row or a column of a matrix. This can be achieved by using the methods:

```
public VectorR GetRowVector(int m)
{
    if (m < 0 || m > Rows)
    {
        throw new ArgumentOutOfRangeException(
        "m", m, "m is out of range!");
    }
    VectorR v = new VectorR(Cols);
    for (int i = 0; i < Cols; i++)
    {
        v[i] = matrix[m, i];
    }
    return v;
}

public VectorR GetColVector(int m)
{
    if (m < 0 || m > Cols)
    {
        throw new ArgumentOutOfRangeException(
        "m", m, "m is out of range!");
    }
    VectorR v = new VectorR(Rows);
    for (int i = 0; i < Rows; i++)
    {
        v[i] = matrix[i, m];
    }
    return v;
}
```

Swap, Transpose, and Trace

You may want to interchange two rows or two columns of a matrix:

```
public MatrixR GetRowSwap(int m, int n)
{
    double temp = 0.0;
    for (int i = 0; i < Cols; i++)
    {
        temp = matrix[m, i];
        matrix[m, i] = matrix[n, i];
        matrix[n, i] = temp;
    }
    return new MatrixR(matrix);
}

public MatrixR GetColSwap(int m, int n)
{
    double temp = 0.0;
    for (int i = 0; i < Rows; i++)
```

```
    {
        temp = matrix[i, m];
        matrix[i, m] = matrix[i, n];
        matrix[i, n] = temp;
    }
    return new MatrixR(matrix);
}
```

It is often necessary to transpose a matrix by interchanging rows and columns:

```
public MatrixR GetTranspose()
{
    MatrixR v = this;
    v.Transpose();
    return v;
}

public void Transpose()
{
    MatrixR m = new MatrixR(Cols, Rows);
    for (int i = 0; i < Rows; i++)
    {
        for (int j = 0; j < Cols; j++)
        {
            m[j, i] = matrix[i, j];
        }
    }
    this = m;
}
```

The trace of a matrix can be computed using the following method:

```
public double GetTrace()
{
    double d = 0.0;
    for (int i = 0; i < Rows; i++)
    {
        if (i < Cols)
            d += matrix[i, i];
    }
    return d;
}
```

Transformations

In solving linear equations and performing operation on graphic objects, we often need to carry out various matrix transformations. Here, we will implement transformation methods for *n*-dimensional vectors and *n* by *n* matrices in the *MatrixR* structure. We will present three methods for carrying out these tasks.

The first method returns a column vector obtained by multiplying a column vector by a matrix:

```
public static VectorR Transform(MatrixR m, VectorR v)
{
    if (!m.IsSquared())
    {
```

```
            throw new ArgumentOutOfRangeException(
            "Dimension", m.GetRows(), "The matrix must
            be squared!");
        }
        if (m.GetCols() != v.GetSize())
        {
            throw new ArgumentOutOfRangeException(
            "Size", v.GetSize(), "The size of the vector
            must be equal"
            + "to the number of rows of the matrix!");
        }
        VectorR result = new VectorR(v.GetSize());
        for (int i = 0; i < m.GetRows(); i++)
        {
            result[i] = 0.0;
            for (int j = 0; j < m.GetCols(); j++)
            {
                result[i] += m[i, j] * v[j];
            }
        }
        return result;
    }
```

Note that the transformation matrix must be squared.

The next method, which is the transpose of the first method, multiplies a row vector by a matrix to obtain a row vector:

```
public static VectorR Transform(VectorR v, MatrixR m)
{
    if (!m.IsSquared())
    {
        throw new ArgumentOutOfRangeException(
        "Dimension", m.GetRows(), "The matrix must be
        squared!");
    }
    if (m.GetRows() != v.GetSize())
    {
        throw new ArgumentOutOfRangeException(
        "Size", v.GetSize(), "The size of the vector
        must be equal"
        + "to the number of rows of the matrix!");
    }
    VectorR result = new VectorR(v.GetSize());
    for (int i = 0; i < m.GetRows(); i++)
    {
        result[i] = 0.0;
        for (int j = 0; j < m.GetCols(); j++)
        {
            result[i] += v[j] * m[j, i];
        }
    }
    return result;
}
```

The third method, which multiplies a column vector by a row vector to produce a matrix, can be written in the form:

```
public static MatrixR Transform(VectorR v1, VectorR v2)
{
    if (v1.GetSize() != v2.GetSize())
    {
        throw new ArgumentOutOfRangeException(
            "v1", v1.GetSize(), "The vectors must have the same size!");
    }
    MatrixR result = new MatrixR(v1.GetSize(), v1.GetSize());
    for (int i = 0; i < v1.GetSize(); i++)
    {
        for (int j = 0; j < v1.GetSize(); j++)
        {
            result[j, i] = v1[i] * v2[j];
        }
    }
    return result;
}
```

Matrix Multiplication

Matrix multiplication is defined between two matrices only if the number of columns of the first matrix is the same as the number of rows of the second matrix. If U is an m-by-n matrix and V is an n-by-p matrix, then their product is an m-by-p matrix denoted by UV (or sometimes $U \cdot V$). Their product is given by

$$\left(UV\right)_{ij} = \sum_{k=0}^{n-1} U_{ik} V_{kj}$$

We can compute this product directly by using the above equation, which involves multiple for-loops. Instead, here we will use the vector-lists method. The matrix product can be considered as a dot product of a column-list of vectors and a row-list of vectors. If U and V are matrices given by:

$$U = \begin{pmatrix} u_{11} & u_{12} & u_{13} & \cdots \\ u_{21} & u_{22} & u_{23} & \cdots \\ u_{31} & u_{32} & u_{33} & \cdots \\ \vdots & \vdots & \vdots & \ddots \end{pmatrix} = \begin{pmatrix} U_1 \\ U_2 \\ U_3 \\ \vdots \end{pmatrix}$$

$$V = \begin{pmatrix} v_{11} & v_{12} & v_{13} & \cdots \\ v_{21} & v_{22} & v_{23} & \cdots \\ v_{31} & v_{32} & v_{33} & \cdots \\ \vdots & \vdots & \vdots & \ddots \end{pmatrix} = \begin{pmatrix} V_1 & V_2 & V_3 & \cdots \end{pmatrix}$$

where U_i is the row vector from ith row of the U matrix and V_i is the column vector from the ith column of the V matrix, then, the matrix product becomes

$$UV = \begin{pmatrix} U_1 \\ U_2 \\ U_3 \\ \vdots \end{pmatrix} * \begin{pmatrix} V_1 & V_2 & V_3 & \cdots \end{pmatrix} = \begin{pmatrix} U_1 \cdot V_1 & U_1 \cdot V_2 & U_1 \cdot V_3 & \cdots \\ U_2 \cdot V_1 & U_2 \cdot V_2 & U_2 \cdot V_3 & \cdots \\ U_3 \cdot V_1 & U_3 \cdot V_2 & U_3 \cdot V_3 & \cdots \\ \vdots & \vdots & \vdots & \ddots \end{pmatrix}$$

Note that the matrix product is expressed in terms of dot products of vectors. This can be easily implemented in C#:

```
public static MatrixR operator *(MatrixR m1, MatrixR m2)
{
    if (m1.GetCols() != m2.GetRows())
    {
        throw new ArgumentOutOfRangeException("Columns", m1,
            "The numbers of columns of the first matrix must be " +
            "equal to the number of rows of the second matrix!");
    }
    MatrixR result = new MatrixR(m1.GetRows(), m2.GetCols());
    VectorR v1 = new VectorR(m1.GetCols());
    VectorR v2 = new VectorR(m2.GetRows());
    for (int i = 0; i < m1.GetRows(); i++)
    {
        v1 = m1.GetRowVector(i);
        for (int j = 0; j < m2.GetCols(); j++)
        {
            v2 = m2.GetColVector(j);
            result[i, j] = VectorR.DotProduct(v1, v2);
        }
    }
    return result;
}
```

Mathematical Operators

Here, we will implement various mathematical operations. Below is the code listing:

```
public static MatrixR operator +(MatrixR m)
{
    return m;
}

public static MatrixR operator -(MatrixR m)
{
    for (int i = 0; i < m.GetRows(); i++)
    {
        for (int j = 0; j < m.GetCols(); j++)
        {
            m[i, j] = -m[i, j];
        }
    }
    return m;
}

public static MatrixR operator +(MatrixR m1, MatrixR m2)
```

```
{
    if (!MatrixR.CompareDimension(m1, m2))
    {
        throw new ArgumentOutOfRangeException("Dimension", m1,
        "The dimensions of two matrices must be the same!");
    }
    MatrixR result = new MatrixR(m1.GetRows(), m1.GetCols());
    for (int i = 0; i < m1.GetRows(); i++)
    {
        for (int j = 0; j < m1.GetCols(); j++)
        {
            result[i, j] = m1[i, j] + m2[i, j];
        }
    }
    return result;
}

    public static MatrixR operator +(MatrixR m, double d)
    {
     MatrixR result = new MatrixR(m.GetRows(), m.GetCols());
     for (int i = 0; i < m.GetRows(); i++)
     {
         for (int j = 0; j < m.GetCols(); j++)
         {
             result[i, j] = m[i, j] + d;
         }
     }
     return result;
    }

public static MatrixR operator +(double d, MatrixR m)
{
    MatrixR result = new MatrixR(m.GetRows(), m.GetCols());
    for (int i = 0; i < m.GetRows(); i++)
    {
        for (int j = 0; j < m.GetCols(); j++)
        {
            result[i, j] = m[i, j] + d;
        }
    }
    return result;
}

public static MatrixR operator -(MatrixR m1, MatrixR m2)
{
    if (!MatrixR.CompareDimension(m1, m2))
    {
        throw new ArgumentOutOfRangeException("Dimension", m1,
        "The dimensions of two matrices must be the same!");
    }
    MatrixR result = new MatrixR(m1.GetRows(), m1.GetCols());
    for (int i = 0; i < m1.GetRows(); i++)
    {
        for (int j = 0; j < m1.GetCols(); j++)
        {
```

```
                result[i, j] = m1[i, j] - m2[i, j];
            }
        }
        return result;
}

public static MatrixR operator -(MatrixR m, double d)
{
    MatrixR result = new MatrixR(m.GetRows(), m.GetCols());
    for (int i = 0; i < m.GetRows(); i++)
    {
        for (int j = 0; j < m.GetCols(); j++)
        {
            result[i, j] = m[i, j] - d;
        }
    }
    return result;
}

public static MatrixR operator -(double d, MatrixR m)
{
    MatrixR result = new MatrixR(m.GetRows(), m.GetCols());
    for (int i = 0; i < m.GetRows(); i++)
    {
        for (int j = 0; j < m.GetCols(); j++)
        {
            result[i, j] = d - m[i, j];
        }
    }
    return result;
}

public static MatrixR operator *(MatrixR m, double d)
{
    MatrixR result = new MatrixR(m.GetRows(), m.GetCols());
    for (int i = 0; i < m.GetRows(); i++)
    {
        for (int j = 0; j < m.GetCols(); j++)
        {
            result[i, j] = m[i, j] * d;
        }
    }
    return result;
}

public static MatrixR operator *(double d, MatrixR m)
{
    MatrixR result = new MatrixR(m.GetRows(), m.GetCols());
    for (int i = 0; i < m.GetRows(); i++)
    {
        for (int j = 0; j < m.GetCols(); j++)
        {
            result[i, j] = m[i, j] * d;
        }
    }
}
```

```
            return result;
    }

    public static MatrixR operator /(MatrixR m, double d)
    {
        MatrixR result = new MatrixR(m.GetRows(), m.GetCols());
        for (int i = 0; i < m.GetRows(); i++)
        {
            for (int j = 0; j < m.GetCols(); j++)
            {
                result[i, j] = m[i, j] / d;
            }
        }
        return result;
    }

    public static MatrixR operator /(double d, MatrixR m)
    {
        MatrixR result = new MatrixR(m.GetRows(), m.GetCols());
        for (int i = 0; i < m.GetRows(); i++)
        {
            for (int j = 0; j < m.GetCols(); j++)
            {
                result[i, j] = d/ m[i, j];
            }
        }
        return result;
    }
```

Determinant

Another important quantity associated with a square matrix is its determinant, which can be computed using the recursion approach. Let $M^{(p,q)}$ be the matrix obtained by striking out the pth row and qth column of the matrix M. Then,

$$\det(M) = \sum_j (-1)^{i+j} M[i, j] \det\left(M^{(i,j)}\right)$$

for any row i. This is called Cramer's Rule. Usually, it should not be used to compute the determinant of large-sized matrices because it requires $n!$ multiplications to calculate the determinant of an $n \times n$ matrix. However, we still present this traditional method here as reference.

First, we need to implement a method for obtaining the minor matrix, $M^{(p,q)}$:

```
    public static MatrixR Minor(MatrixR m, int row, int col)
    {
        MatrixR mm = new MatrixR(m.GetRows() - 1, m.GetCols() - 1);
        int ii = 0, jj = 0;
        for (int i = 0; i < m.GetRows(); i++)
        {
            if (i == row)
                continue;
            jj = 0;
            for (int j = 0; j < m.GetCols(); j++)
            {
```

```
            if (j == col)
                continue;
            mm[ii, jj] = m[i, j];
            jj++;
        }
        ii++;
    }
    return mm;
}
```

Next, we can add the recursion method to the *MatrixR* structure:

```
    public static double Determinant(MatrixR m)
{
    double result = 0.0;
    if (!m.IsSquared())
    {
        throw new ArgumentOutOfRangeException(
        "Dimension", m.GetRows(), "The matrix must
        be squared!");
    }
    if (m.GetRows() == 1)
        result = m[0, 0];
    else
    {
        for (int i = 0; i < m.GetRows(); i++)
        {
            result += Math.Pow(-1, i)*m[0, i] *
                    Determinant(MatrixR.Minor(m, 0, i));
        }
    }
    return result;
}
```

Inverse

The inverse of a square matrix M with a non-zero determinant is the adjoint matrix divided by its determinant. This can be written

$$M^{-1} = adj(M)/\det(M)$$

The adjoint matrix is the transpose of the cofactor matrix. The cofactor matrix is the matrix of the determinants of the minors $M^{(p,q)}$ multiplied by $(-1)^{p+q}$. This can be easily implemented in C#:

```
public static MatrixR Inverse(MatrixR m)
{
    if (Determinant(m) == 0)
    {
        throw new DivideByZeroException(
        "Cannot inverse a matrix with a zero determinant!");
    }
    return (Adjoint(m) / Determinant(m));
}

public static MatrixR Adjoint(MatrixR m)
```

```
{
    if (!m.IsSquared())
    {
        throw new ArgumentOutOfRangeException(
        "Dimension", m.GetRows(), "The matrix must
        be squared!");
    }
    MatrixR ma = new MatrixR(m.GetRows(), m.GetCols());
    for (int i = 0; i < m.GetRows(); i++)
    {
        for (int j = 0; j < m.GetCols(); j++)
        {
            ma[i, j] = Math.Pow(-1, i + j) * (
                      Determinant(Minor(m, i, j)));
        }
    }
    return ma.GetTranspose();
}
```

Testing the Real Matrix

The *MatrixR* class presented in the above sections defines a variable with real numbers arranged in a 2D grid. You can use this class to perform various matrix operations. Add the following code to the *Program.cs* file of the *RealMatrixTest* project:

```
using System;
using XuMath;

namespace RealMatrixTest
{
    class Program
    {
        static void Main(string[] args)
        {
            TestMatrixR();
            Console.ReadLine();
        }

        static void TestMatrixR()
        {
            // Create a matrix using a 2D double array:
            MatrixR m1 = new MatrixR(new double[3, 3] { {1, 1, 1},
                                                        {1, 2, 3},
                                                        {1, 3, 6}});
            // Create a matrix by directly defining its elements:
            MatrixR m2 = new MatrixR(3, 3);
            m2[0, 0] = 8; m2[0, 1] = 1; m2[0, 2] = 6;
            m2[1, 0] = 3; m2[1, 1] = 5; m2[1, 2] = 7;
            m2[2, 0] = 4; m2[2, 1] = 9; m2[2, 2] = 2;
            VectorR v = new VectorR(new double[] { 2, 0, -1 });

            Console.WriteLine("\n Original matrix: m1 = \n{0}", m1);
            Console.WriteLine("\n Original matrix: m2 = \n{0}", m2);
            Console.WriteLine("\n v = {0}", (v));
```

```
Console.WriteLine("\n m1 + m2 = \n{0}", (m1 + m2));
Console.WriteLine("\n m1 - m2 = \n{0}", (m1 - m2));
Console.WriteLine("\n m1 * m2 = \n{0}", (m1 * m2));
Console.WriteLine("\n m2 * m1 = \n{0}", (m2 * m1));
Console.WriteLine("\n m1 * v = {0}", MatrixR.Transform(m1, v));
Console.WriteLine("\n v * m1 = {0}", MatrixR.Transform(v, m2));
Console.WriteLine("\n Inverse of m1 = \n{0}", MatrixR.Inverse(m1));
Console.WriteLine("\n Inverse of m2 = \n{0}", MatrixR.Inverse(m2));
Console.WriteLine("\n Determinant of m1 = {0}",
                  MatrixR.Determinant(m1));
Console.WriteLine("\n Determinant of m2 = {0}",
                  MatrixR.Determinant(m2));
        }
    }
}
```

This class shows you how to implement various matrix operations using the *MatrixR* class, including matrix addition, subtraction, multiplication, transformation, inverse, and determinant. It also demonstrates that a real matrix can be created using either a 2D double array or by directly defining its matrix elements.

Running this example creates the results shown in Figure 3-4.

Complex Matrix Class

As in the case of implementing the real matrix class, we can easily create the complex matrix structure. Start with a new C# Console application in Visual Studio and name the project *ComplexMatrixTest*. Add classes, *Complex* and *VectorC*, from previous projects to the current application. Add a new class named *MatrixC* to this project and change its namespace to *XuMath*. I will not present the detailed implementation procedure here; instead, I will only outline the difference between the *MatrixR* and *MatrixC* structures. First, we need to create constructors that are used to initialize a complex matrix of a given dimension, or convert a complex matrix from a given two-dimensional complex array:

```
using System;

namespace UniMathLib
{
    public struct MatrixC
    {
        private int Rows;
        private int Cols;
        private Complex[,] matrix;

        public MatrixC(int Rows, int Cols)
        {
            this.Rows = Rows;
            this.Cols = Cols;
            this.matrix = new Complex[Rows, Cols];
            for (int i = 0; i < Rows; i++)
            {
```

```
 Original matrix: m1 =
<1, 1, 1
 1, 2, 3
 1, 3, 6>

 Original matrix: m2 =
<8, 1, 6
 3, 5, 7
 4, 9, 2>

 v = <2, 0, -1>

 m1 + m2 =
<9, 2, 7
 4, 7, 10
 5, 12, 8>

 m1 - m2 =
<-7, 0, -5
 -2, -3, -4
 -3, -6, 4>

 m1 * m2 =
<15, 15, 15
 26, 38, 26
 41, 70, 39>

 m2 * m1 =
<15, 28, 47
 15, 34, 60
 15, 28, 43>

 m1 * v = <1, -1, -4>

 v * m1 = <12, -7, 10>

 Inverse of m1 =
<3, -3, 1
 -3, 5, -2
 1, -2, 1>

 Inverse of m2 =
<0.147222222222222, -0.144444444444444, 0.0638888888888889
 -0.0611111111111111, 0.0222222222222222, 0.105555555555556
 -0.0194444444444444, 0.188888888888889, -0.102777777777778>

Determinant of m1 = 1

Determinant of m2 = -360
```

Figure 3-4 Matrix operations in C#.

```
        for (int j = 0; j < Cols; j++)
            {
                matrix[i, j] = Complex.Zero;
            }
        }
    }
```

```
    public MatrixC(Complex[,] matrix)
    {
        this.Rows = matrix.GetLength(0);
        this.Cols = matrix.GetLength(1);
        this.matrix = matrix;
    }
}
}
```

The first constructor takes two integers *Rows* and *Cols*, the dimension of the matrix, as input parameters. It creates a complex matrix whose elements are initialized to zero for a given dimension or size of the matrix.

The second constructor creates a complex matrix to hold a given two-dimensional complex array. The size of the matrix is the same as the dimension of the array.

Implementation

The basic definitions and mathematical operations of complex matrices are similar to those of real matrices. However, there is a minor difference in the matrix multiplication. For two complex matrices, their product should be the direct multiplication without involving complex conjugate. So care must be taken when computing a complex matrix product using dot products of complex vectors that involve the complex conjugate. Here, I specifically list the corresponding code snippet for the matrix product:

```
public static MatrixC operator * (MatrixC m1, MatrixC m2)
{
    if (m1.GetCols() != m2.GetRows())
    {
        throw new ArgumentOutOfRangeException("Columns", m1,
            "The numbers of columns of the first matrix must be " +
            "equal to the number of rows of the second matrix!");
    }
    MatrixC result = new MatrixC(m1.GetRows(), m2.GetCols());
    VectorC v1 = new VectorC(m1.GetCols());
    VectorC v2 = new VectorC(m2.GetRows());
    for (int i = 0; i < m1.GetRows(); i++)
    {
        v1 = m1.GetRowVector(i);
        for (int j = 0; j < m2.GetCols(); j++)
        {
            v2 = m2.GetColVector(j);
            result[i, j] =
                VectorC.DotProduct(v1.GetConjugate(), v2);
        }
    }
    return result;
}
```

I won't present the detailed implementation procedure here. Instead, I will present the code of the *MatrixC* class here for you to review:

```csharp
using System;

namespace XuMath
{
    public struct MatrixC:ICloneable
    {
        private int Rows;
        private int Cols;
        private Complex[,] matrix;

        public MatrixC(int Rows, int Cols)
        {
            this.Rows = Rows;
            this.Cols = Cols;
            this.matrix = new Complex[Rows, Cols];
            for (int i = 0; i < Rows; i++)
            {
                for (int j = 0; j < Cols; j++)
                {
                    matrix[i, j] = Complex.Zero;
                }
            }
        }

        public MatrixC(Complex[,] matrix)
        {
            this.Rows = matrix.GetLength(0);
            this.Cols = matrix.GetLength(1);
            this.matrix = matrix;
        }

        #region Make a deep copy
        public MatrixC Clone()
        {
            // returns a deep copy of the matrix
            MatrixC m = new MatrixC(matrix);
            m.matrix = (Complex[,])matrix.Clone();
            return m;
        }

        object ICloneable.Clone()
        {
            return Clone();
        }
        #endregion

        #region Indexing:
        public Complex this[int m, int n]
        {
            get
            {
                if (m < 0 || m > Rows)
                {
                    throw new ArgumentOutOfRangeException(
                        "m", m, "m is out of range!");
```

```
            }
            if (n < 0 || n > Cols)
            {
                throw new ArgumentOutOfRangeException(
                    "n", n, "n is out of range!");
            }
            return matrix[m, n];
        }
        set { matrix[m, n] = value; }
    }
#endregion

#region Identity matrix:
public MatrixC Identity()
{
    MatrixC m = new MatrixC(Rows, Cols);
    for (int i = 0; i < Rows; i++)
    {
        for (int j = 0; j < Cols; j++)
        {
            if (i == j)
            {
                m[i, j] = new Complex(1, 0);
            }
        }
    }
    return m;
}
#endregion

#region Equals and Hashing:
public override bool Equals(object obj)
{
    return (obj is MatrixC) && this.Equals((MatrixC)obj);
}

public bool Equals(MatrixC m)
{
    return matrix == m.matrix;
}

public override int GetHashCode()
{
    return matrix.GetHashCode();
}

public static bool operator ==(MatrixC m1, MatrixC m2)
{
    return m1.Equals(m2);
}

public static bool operator !=(MatrixC m1, MatrixC m2)
{
    return !m1.Equals(m2);
}
```

```csharp
#endregion

#region Matrix dimension
public int GetRows()
{
    return Rows;
}

public int GetCols()
{
    return Cols;
}

public bool IsSquared()
{
    if (Rows == Cols)
        return true;
    else
        return false;
}

public static bool CompareDimension(MatrixC m1, MatrixC m2)
{
    if (m1.GetRows() == m2.GetRows() && m1.GetCols() == m2.GetCols())
        return true;
    else
        return false;
}
#endregion

#region Get row or column vector from a matrix:
public VectorC GetRowVector(int m)
{
    if (m < 0 || m > Rows)
    {
        throw new ArgumentOutOfRangeException(
        "m", m, "m is out of range!");
    }
    VectorC v = new VectorC(Cols);
    for (int i = 0; i < Cols; i++)
    {
        v[i] = matrix[m, i];
    }
    return v;
}

public VectorC GetColVector(int m)
{
    if (m < 0 || m > Cols)
    {
        throw new ArgumentOutOfRangeException(
        "m", m, "m is out of range!");
    }
    VectorC v = new VectorC(Rows);
    for (int i = 0; i < Rows; i++)
```

```
        {
            v[i] = matrix[i, m];
        }
        return v;
    }
#endregion

#region Matrix transpose and trace:
public MatrixC GetTranspose()
{
    MatrixC t = this;
    t.Transpose();
    return t;
}

public void Transpose()
{
    MatrixC m = new MatrixC(Cols, Rows);
    for (int i = 0; i < Rows; i++)
    {
        for (int j = 0; j < Cols; j++)
        {
            m[j, i] = matrix[i, j];
        }
    }
    this = m;
}

public Complex GetTrace()
{
    Complex c = Complex.Zero;
    for (int i = 0; i < Rows; i++)
    {
        for (int j = 0; j < Cols; j++)
        {
            if (i == j)
                c += matrix[i, j];
        }
    }
    return c;
}
#endregion

#region Matrix transformation:
public static VectorC Transform(MatrixC m, VectorC v)
{
    VectorC result = new VectorC(v.GetSize());
    if (!m.IsSquared())
    {
        throw new ArgumentOutOfRangeException(
          "Dimension", m.GetRows(), "The matrix must be squared!");
    }
    if (m.GetCols() != v.GetSize())
    {
        throw new ArgumentOutOfRangeException(
```

```
                    "Size", v.GetSize(), "The size of the vector must be equal"
                    + "to the number of rows of the matrix!");
        }
        for (int i = 0; i < m.GetRows(); i++)
        {
            result[i] = Complex.Zero;
            for (int j = 0; j < m.GetCols(); j++)
            {
                result[i] += m[i, j] * v[j];
            }
        }
        return result;
    }

    public static VectorC Transform(VectorC v, MatrixC m)
    {
        VectorC result = new VectorC(v.GetSize());
        if (!m.IsSquared())
        {
            throw new ArgumentOutOfRangeException(
                "Dimension", m.GetRows(), "The matrix must be squared!");
        }
        if (m.GetRows() != v.GetSize())
        {
            throw new ArgumentOutOfRangeException(
                "Size", v.GetSize(), "The size of the vector must be equal"
                + "to the number of rows of the matrix!");
        }
        for (int i = 0; i < m.GetRows(); i++)
        {
            result[i] = Complex.Zero;
            for (int j = 0; j < m.GetCols(); j++)
            {
                result[i] += v[j] * m[j, i];
            }
        }
        return result;
    }

    public static MatrixC Transform(VectorC v1, VectorC v2)
    {
        if (v1.GetSize() != v2.GetSize())
        {
            throw new ArgumentOutOfRangeException(
                "v1", v1.GetSize(), "The vectors must have the same size!");
        }
        MatrixC result = new MatrixC(v1.GetSize(), v1.GetSize());
        for (int i = 0; i < v1.GetSize(); i++)
        {
            for (int j = 0; j < v1.GetSize(); j++)
            {
                result[j, i] = v1[i] * v2[j];
            }
        }
        return result;
```

```csharp
}
#endregion

#region Exchange rows or columns
public MatrixC GetRowSwap(int m, int n)
{
    Complex temp = Complex.Zero;
    for (int i = 0; i < Cols; i++)
    {
        temp = matrix[m, i];
        matrix[m, i] = matrix[n, i];
        matrix[n, i] = temp;
    }
    return new MatrixC(matrix);
}

public MatrixC GetColSwap(int m, int n)
{
    Complex temp = Complex.Zero;
    for (int i = 0; i < Rows; i++)
    {
        temp = matrix[i, m];
        matrix[i, m] = matrix[i, n];
        matrix[i, n] = temp;
    }
    return new MatrixC(matrix);
}
#endregion

#region Mathematical operators
public static MatrixC operator +(MatrixC m)
{
    return m;
}

public static MatrixC operator -(MatrixC m)
{
    for (int i = 0; i < m.GetRows(); i++)
    {
        for (int j = 0; j < m.GetCols(); j++)
        {
            m[i, j] = -m[i, j];
        }
    }
    return m;
}

public static MatrixC operator +(MatrixC m1, MatrixC m2)
{
    if (!MatrixC.CompareDimension(m1, m2))
    {
        throw new ArgumentOutOfRangeException(
          "Dimension", m1, "The dimensions of two matrices
           must be the same!");
    }
```

```
            MatrixC result = new MatrixC(m1.GetRows(), m1.GetCols());
            for (int i = 0; i < m1.GetRows(); i++)
            {
                for (int j = 0; j < m1.GetCols(); j++)
                {
                    result[i, j] = m1[i, j] + m2[i, j];
                }
            }
            return result;
        }

        public static MatrixC operator +(MatrixC m, Complex c)
        {
            MatrixC result = new MatrixC(m.GetRows(), m.GetCols());
            for (int i = 0; i < m.GetRows(); i++)
            {
                for (int j = 0; j < m.GetCols(); j++)
                {
                    result[i, j] = m[i, j] + c;
                }
            }
            return result;
        }

        public static MatrixC operator +(Complex c, MatrixC m)
        {
            MatrixC result = new MatrixC(m.GetRows(), m.GetCols());
            for (int i = 0; i < m.GetRows(); i++)
            {
                for (int j = 0; j < m.GetCols(); j++)
                {
                    result[i, j] = m[i, j] + c;
                }
            }
            return result;
        }

        public static MatrixC operator -(MatrixC m1, MatrixC m2)
        {
            if (!MatrixC.CompareDimension(m1, m2))
            {
                throw new ArgumentOutOfRangeException(
                  "Dimension", m1, "The dimensions of two matrices
                   must be the same!");
            }
            MatrixC result = new MatrixC(m1.GetRows(), m1.GetCols());
            for (int i = 0; i < m1.GetRows(); i++)
            {
                for (int j = 0; j < m1.GetCols(); j++)
                {
                    result[i, j] = m1[i, j] - m2[i, j];
                }
            }
            return result;
        }
```

```
public static MatrixC operator -(MatrixC m, Complex c)
{
    MatrixC result = new MatrixC(m.GetRows(), m.GetCols());
    for (int i = 0; i < m.GetRows(); i++)
    {
        for (int j = 0; j < m.GetCols(); j++)
        {
            result[i, j] = m[i, j] - c;
        }
    }
    return result;
}

public static MatrixC operator -(Complex c, MatrixC m)
{
    MatrixC result = new MatrixC(m.GetRows(), m.GetCols());
    for (int i = 0; i < m.GetRows(); i++)
    {
        for (int j = 0; j < m.GetCols(); j++)
        {
            result[i, j] =  c - m[i, j];
        }
    }
    return result;
}

public static MatrixC operator *(MatrixC m1, MatrixC m2)
{
    if (m1.GetCols() != m2.GetRows())
    {
        throw new ArgumentOutOfRangeException(
         "Columns", m1, "The numbers of columns of the first
          matrix must be equal to" +
         " the number of rows of the second matrix!");
    }
    MatrixC result = new MatrixC(m1.GetRows(), m2.GetCols());
    VectorC v1 = new VectorC(m1.GetCols());
    VectorC v2 = new VectorC(m2.GetRows());
    for (int i = 0; i < m1.GetRows(); i++)
    {
        v1 = m1.GetRowVector(i);
        for (int j = 0; j < m2.GetCols(); j++)
        {
            v2 = m2.GetColVector(j);
            result[i, j] = VectorC.DotProduct(v1.GetConjugate(), v2);
        }
    }
    return result;
}

public static MatrixC operator *(MatrixC m, Complex c)
{
    MatrixC result = new MatrixC(m.GetRows(), m.GetCols());
    for (int i = 0; i < m.GetRows(); i++)
```

```csharp
        {
            for (int j = 0; j < m.GetCols(); j++)
            {
                result[i, j] = m[i, j] * c;
            }
        }
        return result;
    }

    public static MatrixC operator *(Complex c, MatrixC m)
    {
        MatrixC result = new MatrixC(m.GetRows(), m.GetCols());
        for (int i = 0; i < m.GetRows(); i++)
        {
            for (int j = 0; j < m.GetCols(); j++)
            {
                result[i, j] = m[i, j] * c;
            }
        }
        return result;
    }

    public static MatrixC operator /(MatrixC m, Complex c)
    {
        MatrixC result = new MatrixC(m.GetRows(), m.GetCols());
        for (int i = 0; i < m.GetRows(); i++)
        {
            for (int j = 0; j < m.GetCols(); j++)
            {
                result[i, j] = m[i, j] / c;
            }
        }
        return result;
    }

    public static MatrixC operator /(Complex c, MatrixC m)
    {
        MatrixC result = new MatrixC(m.GetRows(), m.GetCols());
        for (int i = 0; i < m.GetRows(); i++)
        {
            for (int j = 0; j < m.GetCols(); j++)
            {
                result[i, j] = c / m[i, j];
            }
        }
        return result;
    }
    #endregion

    #region Overrise ToString method
    public override string ToString()
    {
        string ss = "(";
        for (int i = 0; i < Rows; i++)
        {
```

```
            string s = "";
            for (int j = 0; j < Cols - 1; j++)
            {
                s += matrix[i, j].ToString() + ", ";
            }
            s += matrix[i, Cols - 1].ToString();
            if (i != Rows - 1 && i == 0)
                ss += s + "\n";
            else if (i != Rows - 1 && i != 0)
                ss += " " + s + "\n";
            else
                ss += " " + s + ")";
        }
        return ss;
    }
    #endregion
    }
}
```

Testing the Complex Matrix

Using the *MatrixC* class implemented in the previous section, you can perform various computations for complex matrices. To examine this class, you can add the following code to the *Program.cs* file of the *ComplexMatrixTest* project:

```
using System;
using XuMath;

namespace ComplexMatrixTest
{
    class Program
    {
        static void Main(string[] args)
        {
            TestMatrixC();
            Console.ReadLine();
        }

        static void TestMatrixC()
        {
            // Create a complex matrix using a 2D complex array:
            MatrixC m1 = new MatrixC(new Complex[,]{
                    {new Complex(1,1),new Complex(1,2), new Complex(1,3)},
                    {new Complex(2,1),new Complex(2,2), new Complex(2,3)},
                    {new Complex(3,1),new Complex(3,2), new Complex(3,3)}});

            //Create a complex matrix by directly defining its elements:
            MatrixC m2 = new MatrixC(3, 3);
            m2[0, 0] = new Complex(1, 2);
            m2[0, 1] = new Complex(7, -3);
            m2[0, 2] = new Complex(3, 4);
            m2[1, 0] = new Complex(6, -2);
            m2[1, 1] = new Complex(0, 9);
            m2[1, 2] = new Complex(4, 7);
```

```
m2[2, 0] = new Complex(2, 1);
m2[2, 1] = new Complex(3, -1);
m2[2, 2] = new Complex(3, 0);
VectorC v = new VectorC(new Complex[] { new Complex(1, 1),
                                        new Complex(1, 2),
                                        new Complex(1, 3) });
Console.WriteLine("\nOriginal matrix: m1 = \n" + m1.ToString());
Console.WriteLine("\nOriginal matrix: m2 = \n" + m2.ToString());
Console.WriteLine("\nv = " + (v).ToString());

Console.WriteLine("\nm1 + m2 = \n" + (m1 + m2).ToString());
Console.WriteLine("\nm1 - m2 = \n" + (m1 - m2).ToString());
Console.WriteLine("\nm1 * m2 = \n" + (m1 * m2).ToString());
Console.WriteLine("\nm2 * m1 = \n" + (m2 * m1).ToString());
Console.WriteLine("\nm1 * v = " +
                  MatrixC.Transform(m1, v).ToString());
Console.WriteLine("\nv * m1 = " +
                  MatrixC.Transform(v, m2).ToString());
Console.WriteLine("\nInverse of m2 = \n" +
                  MatrixC.Inverse(m2).ToString());
Console.WriteLine("\nDeterminant of m1 = " +
                  MatrixC.Determinant(m1).ToString());
Console.WriteLine("\nDeterminant of m2 = " +
                  MatrixC.Determinant(m2).ToString());
        }
    }
}
```

This file shows you how to implement various complex matrix operations using the *MatrixC* class, including matrix addition, subtraction, multiplication, transformation, inverse, and determinant. It also demonstrates that a complex matrix can be created using either a 2D complex array or by directly defining its matrix elements.

Running this example creates the results shown in Figure 3-5.

```
Original matrix: m1 =
<1 + 1 i, 1 + 2 i, 1 + 3 i
 2 + 1 i, 2 + 2 i, 2 + 3 i
 3 + 1 i, 3 + 2 i, 3 + 3 i>

Original matrix: m2 =
<1 + 2 i, 7 - 3 i, 3 + 4 i
 6 - 2 i, 0 + 9 i, 4 + 7 i
 2 + 1 i, 3 - 1 i, 3 + 0 i>

v = <1 + 1 i, 1 + 2 i, 1 + 3 i>

m1 + m2 =
<2 + 3 i, 8 - 1 i, 4 + 7 i
 8 - 1 i, 2 + 11 i, 6 + 10 i
 5 + 2 i, 6 + 1 i, 6 + 3 i>

m1 - m2 =
<0 - 1 i, -6 + 5 i, -2 - 1 i
 -4 + 3 i, 2 - 7 i, -2 - 4 i
 1 + 0 i, 0 + 3 i, 0 + 3 i>

m1 * m2 =
<8 + 20 i, 22 - 11 i, -8 + 31 i
 17 + 21 i, 32 - 6 i, 2 + 42 i
 26 + 22 i, 42 - 1 i, 12 + 53 i>

m2 * m1 =
<21 + 19 i, 30 + 14 i, 15 + 41 i
 4 + 47 i, 46 - 17 i, -24 + 67 i
 17 + 7 i, 17 + 17 i, 17 + 23 i>

m1 * v = <-11 + 12 i, -8 + 18 i, -5 + 24 i>

v * m1 = <8 + 20 i, -2 + 21 i, -8 + 31 i>

Inverse of m2 =
<-0.00116674955051452 + 0.0938946482537011 i, 0.052522856814965 + 0.068283539267
8168 i, 0.215657396427069 - 0.305937033778356 i
 0.0568264412226005 + 0.115393443250067 i, -0.0171378294632952 - 0.0142687731915
382 i, 0.0865881182816266 - 0.132148731877128 i
 -0.0632148731877128 - 0.158658811828163 i, 0.00964002907310356 - 0.054473815079
7598 i, 0.0450441834665851 + 0.293083661680884 i>

Determinant of m1 = 0 + 0 i

Determinant of m2 = 109 + 201 i
```

Figure 3-5 Complex matrix operations in C#.

Chapter 4
Linear Algebraic Equations

In this chapter, we will consider how to solve linear equations with an arbitrary number of unknowns. The solution of linear equations is one of the most commonly used operations in numerical analysis as well as scientific and engineering applications. A set of linear equations can be written in a matrix form as

$$A \cdot x = b \tag{4.1}$$

Here A is a square matrix of coefficients, and both x and b are column vectors representing unknowns and right-hand constants respectively. In this chapter, I will present several selected methods you can use to solve the above linear equations in C#.

Gauss-Jordan Elimination

Solving sets of linear equations using Gauss-Jordan elimination produces both the solution of the equations and the inverse of the coefficient matrix. The Gauss-Jordan elimination process requires two steps. The first step, called the forward elimination, reduces a given system to either triangular or echelon form, or results in a degenerate equation indicating that the system has no solution. This is accomplished through the use of elementary operations. The second step, called the backward elimination, uses back-substitution to find the solution of the linear equations.

In terms of matrix formalism, the first step reduces a matrix to row echelon form using elementary operations, while the second step reduces it to row canonical form.

Gauss-Jordan elimination computes the matrix decomposition using three elementary operations, including multiplying rows, switching rows, and adding multiples of rows to other rows. The first part of the algorithm computes the decomposition, while the second part writes the original matrix as the product of a uniquely determined invertible matrix and a uniquely determined reduced row-echelon matrix. The Gauss-Jordan method is generally as efficient as other techniques and is very stable.

Algorithms

The simplest method in Gauss-Jordan elimination is to eliminate $x[0]$ in equation (4.1) from all equations with $i > 0$; $x[1]$ from all the resulting equations with $i > 1$; and in general $x[j]$ from all equations with $i > j$. The process, called Gauss-Jordan elimination, produces a triangular matrix of coefficients with all zero elements below the principal diagonal.

Care must be taken when using this elimination method. It may fail because one of the diagonal elements becomes to zero or too small in the process of triangulation. In this case, we can avoid this problem by interchanging the two sets of equations. The order in which the equations are presented makes no difference to their solution, but it can make a great impact on how you solve the equations. This can be achieved by rearranging equations to make diagonal elements as large as possible. We'll use the *PivotGJ* method to perform this task.

Implementation

Now start with a C# Console application and name it *LinearSystemTest*. Add a new class, *LinearSystem*, to the current project and change its namespace to *XuMath*. Add the classes, *VectorR* and *MatrixR*, from the previous *RealMatrixTest* project, to the current project. Here we only consider the linear systems with real numbers. The following is the code listing of this class:

```
using System;

namespace XuMath
{
    public class LinearSystem
    {
        double epsilon = 1.0e-800;

        public VectorR GaussJordan(MatrixR A, VectorR b)
        {
            Triangulate(A, b);
            int n = b.GetSize();
            VectorR x = new VectorR(n);
            for (int i = n - 1; i >= 0; i--)
            {
                double d = A[i, i];
                if (Math.Abs(d) < epsilon)
                    throw new ArgumentException(
                        "Diagonal element is too small!");
                x[i] = (b[i] - VectorR.DotProduct(A.GetRowVector(i), x)) / d;
            }
            return x;
        }

        private void Triangulate(MatrixR A, VectorR b)
        {
            int n = A.GetRows();
            VectorR v = new VectorR(n);
            for (int i = 0; i < n - 1; i++)
            {
                double d = PivotGJ(A, b, i);
                if (Math.Abs(d) < epsilon)
```

```
                    throw new ArgumentException(
                            "Diagonal element is too small!");
                for (int j = i + 1; j < n; j++)
                {
                    double dd = A[j, i] / d;
                    for (int k = i + 1; k < n; k++)
                    {
                        A[j, k] -= dd * A[i, k];
                    }
                    b[j] -= dd * b[i];
                }
            }
        }

        private double PivotGJ(MatrixR A, VectorR b, int q)
        {
            int n = b.GetSize();
            int i = q;
            double d = 0.0;
            for (int j = q; j < n; j++)
            {
                double dd = Math.Abs(A[j, q]);
                if (dd > d)
                {
                    d = dd;
                    i = j;
                }
            }
            if (i > q)
            {
                A.GetRowSwap(q, i);
                b.GetSwap(q, i);
            }
            return A[q, q];
        }
    }
}
```

This class includes three methods for performing Gauss-Jordan elimination. The *PivotGJ* method is intended to return the largest available diagonal element obtained by rearranging the equations. If needed, we interchange the rows for both the matrix of coefficients A and the constant vector b by using the *Swap* method, defined in the *MatrixR* and *VectorR* classes. The definition of the vector and matrix structures have been discussed in the previous chapter.

The triangulation is achieved by the *Triangulate* method. In the first loop, instead of directly using the diagonal element $A[i, i]$, we use the largest possible element from the *PivotGJ* method.

Finally, the public *GaussJordan* method is implemented. In this method, we need to find the values of unknowns for triangulated equations, which can be done by using back-substitution. Starting with the last equation, which contains the single unknowns, we can immediately find its solution. This value may be substituted into the second last equation in order to calculate the next unknown. Repeat this process until we find all of unknowns.

Testing Gauss-Jordan Elimination

We can examine the above implementation for the Gauss-Jordan elimination. For example, if we want to solve the following equations

$$\begin{pmatrix} 2 & 1 & -1 \\ -3 & -1 & 2 \\ -2 & 1 & 2 \end{pmatrix} \begin{pmatrix} x_0 \\ x_1 \\ x_2 \end{pmatrix} = \begin{pmatrix} 8 \\ -11 \\ -3 \end{pmatrix}$$

we can add the following method to the *Program.cs* file:

```
static void TestGaussJordan()
{
    LinearSystem ls = new LinearSystem();
    MatrixR A = new MatrixR(new double[3, 3] { { 2, 1, -1 },
                                               { -3, -1, 2 },
                                               { -2, 1, 2 } });
    VectorR b = new VectorR(new double[3] { 8, -11, -3 });
    VectorR x = ls.GaussJordan(A, b);
    Console.WriteLine("Solution x = {0}", x);
}
```

Then, calling this method from the *Main* method gives the solution $x = (2, 3, -1)$.

LU Decomposition

The LU decomposition is a matrix decomposition which represents a matrix as the product of a lower and upper triangular matrix. This decomposition is used in numerical analysis to solve systems of linear equations and to find the inverse of a matrix.

Algorithms

Suppose we can write the matrix A as the product of two matrices

$$A = L \cdot U$$

where L is the lower triangular and U is the upper triangular. We can use this decomposition to solve the linear equations

$$A \cdot x = (L \cdot U) \cdot x = L \cdot (U \cdot x) = b$$

by first solving for the vector y such that

$$L \cdot y = b$$

and then solving

$$U \cdot x = y$$

The advantage of decomposing one linear system into two successive ones is that the solution of a triangular set of equations can be easily obtained just by a forward or backward substitution.

There are many ways you can perform a LU decomposition. For example, the *Doolittle* method sets the diagonal elements of *L* equal to unity, whereas the *Crout* method sets the diagonal elements of *U* equal to unity.

Notice that both *L* and *U* may be stored compactly within a single matrix, since there is no need to stored diagonal elements since their values have been set to unity. Also, once the constant vector *b* has been used to compute *y*, it is no longer needed and may be used to store the solution *x*. A single vector can be used to store first *b*, then *y*, then *x*.

In the following discussion, we will use the *Crout* method to perform the LU decomposition.

Implementation

We now implement the LU decomposition using the *Crout* approach. Add the following methods to the *LinearSystem* class:

```
public double LUCrout(MatrixR A, VectorR b)
{
    LUDecompose(A);
    return LUSubstitute(A, b);
}

private void LUDecompose(MatrixR m)
{
    int n = m.GetRows();
    for (int i = 0; i < n; i++)
    {
        for (int j = 0; j < n; j++)
        {
            double d = m[i, j];
            for (int k = 0; k < Math.Min(i, j); k++)
            {
                d -= m[i, k] * m[k, j];
            }
            if (j > i)
            {
                double dd = m[i, i];
                if (Math.Abs(d) < epsilon)
                    throw new ArgumentException(
                    "Diagonal element is too small!");
                d /= dd;
            }
            m[i, j] = d;
        }
    }
}

private double LUSubstitute(MatrixR m, VectorR v)
{
    int n = v.GetSize();
    double det = 1.0;
    for (int i = 0; i < n; i++)
    {
        double d = v[i];
```

```
            for (int j = 0; j < i; j++)
            {
                d -= m[i, j] * v[j];
            }
            double dd = m[i, i];
            if (Math.Abs(d) < epsilon)
                throw new ArgumentException(
                    "Diagonal element is too small!");
            d /= dd;
            v[i] = d;
            det *= m[i, i];
        }
        for (int i = n - 1; i >= 0; i--)
        {
            double d = v[i];
            for (int j = i + 1; j < n; j++)
            {
                d -= m[i, j] * v[j];
            }
            v[i] = d;
        }
        return det;
    }
```

The private *LUDecompose* method is used to perform the decomposition $A = LU$. Then, we use the *LUSubstitute* method to perform the forward substitution implied by $Ly = b$, followed by the backward substitution implied by $Ux = y$. Finally, we combine these methods into a single public method, *LUCrout*.

The determinant of an LU decomposed matrix in the *Crout* method is just the product of the diagonal of the L matrix, due to the fact that

$$\det(A) = \det(LU) = \det(L) \cdot \det(U) = \det(L) = \prod_i L[i,i]$$

So the *LUCrout* and *LUSubstitute* methods also return the determinant of the matrix A.

As in the case of the Gauss-Jordan elimination, you can also introduce pivoting to avoid possible zero divisors. This will be left as an exercise for the reader.

Matrix Inverse

Using the above LU decomposition and substitution methods, computing the inverse of a matrix column by column is straightforward. Add the following *LUInverse* method to the *LinearSystem* class:

```
public MatrixR LUInverse(MatrixR m)
{
    int n = m.GetRows();
    MatrixR u = m.Identity();
    LUDecompose(m);
    VectorR uv = new VectorR(n);
    for (int i = 0; i < n; i++)
    {
        uv = u.GetRowVector(i);
        LUSubstitute(m, uv);
        u.ReplaceRow(uv, i);
```

```
    }
    MatrixR inv = u.GetTranspose();
    return inv;
}
```

Testing LU Decomposition

We can use the above LU decomposition to find the solution of linear equations and obtain the inverse of a matrix. For example, we can use the LU method to solve the same example equations we used to test the Gauss-Jordan method.

```
static void TestLU()
{
    LinearSystem ls = new LinearSystem();
    MatrixR A = new MatrixR(new double[3, 3] { { 2, 1, -1 },
                                               { -3, -1, 2 },
                                               { -2, 1, 2 } });
    VectorR b = new VectorR(new double[3] { 8, -11, -3 });
    MatrixR AA = A.Clone();
    MatrixR BB = A.Clone();
    double d = ls.LUCrout(A, b);
    MatrixR inv = ls.LUInverse(AA);
    Console.WriteLine("\nInverse of A = \n{0}", inv);
    Console.WriteLine("\nSolution of the equations = {0}", b);
    Console.WriteLine("\nDeterminant of A = {0}", d);
    Console.WriteLine("\nTest Inverse: BB*Inverse = \n{0}", BB * inv);
}
```

Please note that you need to make a copy of the original matrix by using the *Clone* method when you perform the successive computations, because the original matrix will change during the LU decomposition process. The above method produces the solution of linear equations, the inverse of the coefficient matrix, and the determinant. The solution is (2, 3, −1), the determinant is −1, and the inverse matrix is given by

$$\begin{pmatrix} 4 & 3 & -1 \\ -2 & -2 & 1 \\ 5 & 4 & -1 \end{pmatrix}.$$

We also verify the result we got for the inverse matrix by multiplying the original matrix by its inverse. This should give us an identical matrix, as shown in Figure 4-1.

```
Inverse of A =
(4, 3, -1
-2, -2, 1
5, 4, -1)

Solution of the equations = (2, 3, -1)

Determinant of A = -1

Test Inverse: BB*Inverse =
(1, 0, 0
0, 1, 0
0, 0, 1)
```

Figure 4-1 Testing LU composition in C#.

Iteration Methods

The linear equations can also be solved by using iteration approaches. We can transform the linear system equations, $Ax = b$, into a different form:

$$x_0 = (b_0 - A_{01}x_1 - A_{02}x_2 - A_{03}x_3 - \cdots - A_{0n-1}x_{n-1})/A_{00}$$
$$x_1 = (b_1 - A_{10}x_0 - A_{12}x_2 - A_{13}x_3 - \cdots - A_{1n-1}x_{n-1})/A_{11}$$
$$x_2 = (b_2 - A_{20}x_0 - A_{21}x_1 - A_{23}x_3 - \cdots - A_{2n-1}x_{n-1})/A_{22}$$
$$\cdots\cdots\cdots\cdots$$
$$x_{n-1} = (b_{n-1} - A_{n-11}x_1 - A_{n-12}x_2 - A_{n-13}x_3 - \cdots - A_{n-1n-2}x_{n-2})/A_{n-1n-1}$$

This suggests a number of iteration schemes.

Gauss-Jacobi Iteration

The Gauss-Jacobi iteration is one of the simplest iteration schemes. It uses the following formula:

$$x_j^{(k+1)} = \frac{1}{A_{ii}}\left(b_i - \sum_{j \neq i} A_{ij}x_j^{(k)}\right)$$

Here we start with an initial solution, if known; otherwise, we can start with

$$x^{(0)} = \begin{pmatrix} 0 \\ 0 \\ \vdots \\ 0 \end{pmatrix}$$

We then iterate until the following condition is satisfied:

$$\sqrt{\left(x_1^{(k+1)} - x_1^{(k)}\right)^2 + \left(x_2^{(k+1)} - x_2^{(k)}\right)^2 + \cdots + \left(x_{n-1}^{(k+1)} - x_{n-1}^{(k)}\right)^2} < tol$$

where *tol* is the tolerance provided by the user. The last iteration gives the solution within that tolerance. It is, of course, assumed that none of the diagonal elements are zero. If some of the diagonal elements are zero, you need to rearrange the equations before you use the iteration approach.

We can implement the Gauss-Jacobi iteration by adding the following method to the *LinearSystem* class:

```
public VectorR GaussJacobi(MatrixR A, VectorR b,
                           int MaxIterations, double tolerance)
{
    int n = b.GetSize();
    VectorR x = new VectorR(n);

    for (int nIteration = 0; nIteration <  MaxIterations; nIteration++)
    {
        VectorR xOld = x.Clone();
        for (int i = 0; i < n; i++)
        {
            double db = b[i];
            double da = A[i, i];
            if (Math.Abs(da)<epsilon)
                throw new ArgumentException(
                "Diagonal element is too small!");
            for (int j = 0; j < n; j++)
            {
                if (j != i)
                {
                    db -= A[i, j] * xOld[j];
                }
            }
            x[i] = db / da;
        }
        VectorR dx = x - xOld;
        if (dx.GetNorm() < tolerance)
        {
            MessageBox.Show(nIteration.ToString());
            return x;
        }
    }
    return x;
}
```

Note that in the above implementation, two parameters are used to control the number of iterations: one is the tolerance and the other is the maximum number of iterations. If either one of these two conditions is satisfied, the computation is terminated to return the solution. We also need a new vector *xOld*, which is a shallow copy of the *x*, to retain the previous value of *x*.

Gauss-Seidel Iteration

The Gauss-Seidel iteration is simply an improved version of the Gauss-Jacobi method. The iteration equation has the form:

$$x_j^{(k+1)} = \frac{1}{A_{ii}}\left(b_i - \sum_{j \neq i} A_{ij} x_j^{(k)}\right)$$

Note that on the right-hand side, the components x_j with $j < i$ have already been updated, but their updated values are not used until all the components have been calculated. In fact, the correct iteration equation should be rewritten in the form

$$x_j^{(k+1)} = \frac{1}{A_{ii}} \left(b_i - \sum_{j<i} A_{ij} x_j^{(k+1)} - \sum_{j>i} A_{ij} x_j^{(k)} \right)$$

Using the above iteration scheme, we can easily implement the Gauss-Seidel method in C#. Add the following method to the *LinearSystem* class:

```csharp
public VectorR GaussSeidel(MatrixR A, VectorR b,
                           int MaxIterations, double tolerance)
{
    int n = b.GetSize();
    VectorR x = new VectorR(n);

    for (int nIteration = 0; nIteration < MaxIterations; nIteration++)
    {
        VectorR xOld = x.Clone();
        for (int i = 0; i < n; i++)
        {
            double db = b[i];
            double da = A[i, i];
            if (Math.Abs(da)<epsilon)
                throw new ArgumentException("Diagonal element is too small!");

            for (int j = 0; j < i; j++)
            {
                db -= A[i, j] * x[j];
            }
            for (int j = i + 1; j < n; j++)
            {
                db -= A[i, j] * xOld[j];
            }
            x[i] = db / da;
        }
        VectorR dx = x - xOld;
        if (dx.GetNorm() < tolerance)
        {
            MessageBox.Show(nIteration.ToString());
            return x;
        }
    }
    return x;
}
```

Testing Iteration Methods

We can test the Gauss-Jacobi and Gauss-Seidel iterations by solving the following set of linear equations

$$\begin{pmatrix} 5 & 1 & 2 \\ 1 & 4 & 1 \\ 2 & 1 & 3 \end{pmatrix} \begin{pmatrix} x_1 \\ x_2 \\ x_3 \end{pmatrix} = \begin{pmatrix} 8 \\ 6 \\ 6 \end{pmatrix}$$

This set of equations has an analytic solution of (1, 1, 1). We can find the solution by adding the following static method to the *Program.cs* file of the *LinearSystemTest* project:

```
static void TestIterations()
{
    LinearSystem ls = new LinearSystem();
    MatrixR A = new MatrixR(new double[3, 3] { { 5, 1, 2 },
                                               { 1, 4, 1 },
                                               { 2, 1, 3 } });
    VectorR b = new VectorR(new double[3] { 8, 6, 6 });
    MatrixR A1 = A.Clone();
    VectorR b1 = b.Clone();

    VectorR x = ls.GaussJacobi(A, b, 10, 1.0e-4);
    Console.WriteLine("\n Solusion from the Gauss-Jacobi iteration:");
    Console.WriteLine(" x[0] = {0}", x[0]);
    Console.WriteLine(" x[1] = {0}", x[1]);
    Console.WriteLine(" x[2] = {0}", x[2]);

    VectorR x1 = ls.GaussSeidel(A1, b1, 10, 1.0e-4);
    Console.WriteLine("\n Solusion from the Gauss-Seidel iteration:");
    Console.WriteLine(" x1[0] = {0}", x1[0]);
    Console.WriteLine(" x1[1] = {0}", x1[1]);
    Console.WriteLine(" x1[2] = {0}", x1[2]);
}
```

Here we set the tolerance equal to 1.0E-4 and the maximum iterations to 10. Running this application produces the results shown in Figure 4-2.

```
Solusion from the Gauss-Jacobi iteration:
x[0] = 0.971982222222222
x[1] = 0.976811851851852
x[2] = 0.962974814814815

Solusion from the Gauss-Seidel iteration:
x1[0] = 1.00000350062582
x1[1] = 1.00000320293873
x1[2] = 0.999996598603211
```

Figure 4-2 Solution of linear equations using iteration methods.

You can see from these results that the solution from the Gauss-Seidel method is much closer to the exact result than the solution from the Gauss-Jacobi iteration method, for a given number of iterations. In fact, for this example, the Gauss-Jacobi method needs about 30 iterations to achieve this tolerance. If you switch to the Gauss-Seidel iteration, it only takes about 8 iterations to reach the same level of accuracy. In general, the Gauss-Seidel method is about 1.5 times faster than the Gauss-Jacobi method.

I should point out here that both iteration methods have some limitations when it comes to solving linear equations. In some cases, if the initial guess is too far away from the real one, they may never find the solution,. In these situations, you may try to start with a different initial solution or use the other approaches, such as LU decomposition or Gauss elimination.

Chapter 5
Nonlinear Equations

Many scientific and engineering phenomena are characterized by nonlinear behavior. Therefore, determining the solutions of those nonlinear equations is a fundamental problem in scientific and engineering analysis. Numerical methods are often used to solve nonlinear equations when analytic solutions cannot be found. These numerical methods are all iterative in nature, and may be used for equations that contain one or several variables.

The simplest case is to find the single root of a single nonlinear function. The following equation represents the general form of such a function:

$$f(x) = 0$$

The above equation represents a single nonlinear function with one variable. The $f(x)$ is a continuous and differentiable function. In this chapter, I will present several methods you can use to solve nonlinear equations. We will concentrate on finding the real solutions of the nonlinear equations.

In general, a nonlinear equation may have any number of solutions, or no solutions at all. All approaches to soving nonlinear equations are iterative procedures that require a start point, i.e., an initial solution. This initial guess can be very critical; a bad initial value may fail to converge, or it may converge to a wrong solution.

For nonlinear equations, there is no general method for estimating the value of a solution. If the nonlinear equation is associated with a real-world scientific and engineering problem, then the physical insight of the problem might suggest the approximate location of the solution. Otherwise, you must carry out a systematic numerical search for the solutions. Sometimes, the solutions can also be found visually by plotting the function. In the following sections, several methods of solving nonlinear equations will be presented.

Incremental Method

The incremental method is the most basic numerical method for solving nonlinear equations. The basic idea behind this method is very simple: if a nonlinear function $f(x)$ has opposite signs at x_1 and x_2, then there is at least one solution in the interval (x_1, x_2). If the interval is small enough, it is likely to contain a single solution. Thus, the solutions of $f(x)$ can be detected by evaluating the function at intervals h and looking for a change in sign.

Implementation

Below is the procedure of the incremental method:

- Pick a starting point x_0 and a step size h. Using a positive h if you want to search to the right, and a negative h if you want to search to the left.

- Let $x_1 = x_0 + h$ and calculate $f(x_0)$ and $f(x_1)$.

- If the sign of $f(x)$ changes between x_0 and x_1, it indicates that a solution of $f(x)$ exists in the interval (x_0, x_1). The solution is given by $x = x_1 - h * f(x)/(f(x) - f(x-h))$.

- If the sign of $f(x)$ does not change between x_0 and x_1, let $x_2 = x_1 + h$ and repeat the process.

Now, start with a new C# Console application and name it *NonlinearSystemTest*. Add a new class, *NonlinearSystem*, to the current project, and change its namespace to *XuMath*. Here is the code listing of this class:

```
using System;

namespace XuMath
{
    public class NonlinearSystem
    {
        public delegate double Function(double x);

        public NonlinearSystem()
        {
        }

        public static double IncrementSearch(Function f, double x0,
                                    double h, int nMaxIncrements)
        {
            double f0 = f(x0);
            double fx = f0, p = 0;
            double x = x0;
            for (int i = 0; i < nMaxIncrements; i++)
            {
                x = x0 + i * h;
                fx = f(x);
                p = f0 * fx;
                if (p < 0)
                    break;
            }
            if (p > 0)
                throw new ArgumentException("Solution not found!");
            else
            {
                return x = x - h * fx / (fx - (f(x - h)));
            }
        }
    }
}
```

In this class, we first introduce a delegate function, *Function*(double x), which is a user-supplied nonlinear function. The *IncrementSearch* method takes the user-supplied function f, the starting search point $x0$, the incremental step size h, and the maximum number of increments as its input parameters.

This method searches for a solution for the nonlinear function. It returns the solution if the search is successful; otherwise, the method throws an arguement exception indicating that no solution was found.

After the first solution has been found, the *IncrementSearch* method can be called again with $x0$ replaced by $x1$, where $x1$ is the first solution, in order to find the next solution. This process can be repeated as long as the *IncrementSearch* method finds another solution.

Testing the Incremental Method

You can examine the incremental search method by solving the nonlinear equation $f(x) = x^3 - 3x + 1$. This can be done using the following *Program.cs* file:

```
using System;
using XuMath;

namespace NonlinearSystemTest
{
    class Program
    {
        static void Main(string[] args)
        {
            TestIncrementSearch();
            Console.ReadLine();
        }

        static void TestIncrementSearch()
        {
            double h = 0.1;
            int n = 50;
            double x = -3;
            for (int i = 1; i <= 3; i++)
            {
                x = NonlinearSystem.IncrementSearch(F, x, h, n);
                Console.WriteLine("\n Solution " +
                                  i.ToString() + " = " + x.ToString());
                Console.WriteLine(" Solution confirmation: f(x) = " +
                                  F(x).ToString());
            }
        }

        static double F(double x)
        {
            return x * x * x - 3.0 * x + 1.0;
        }
    }
}
```

Here we set the starting point $= -3$; the step size $h = 0.1$, and the maximum number of increments $= 50$. Here we try to find 3 solutions starting from -3. After the first solution is found, this solution is taken as the starting point for finding the second solution, which in turn is set as the starting point for finding the third solution. We also calculate the $f(x)$ to verify each solution's accuracy. If you find no solutions, you need to adjust the parameters of the starting point, the step size h, and the maximum number of increments. Note also how we define the delegate function and how we call this function from the *IncrementSearch* method.

Running this example produces the solutions shown in Figure 5-1.

```
Solution 1 = -1.87812929848693
Solution confirmation: f(x) = 0.00953158697450895

Solution 2 = 0.348097365582746
Solution confirmation: f(x) = -0.00211252076553792

Solution 3 = 1.53045890155392
Solution confirmation: f(x) = -0.00657601001137476
```

Figure 5-1 Solutions of a nonlinear equation using the incremental search method.

You can improve the accuracy of the solutions by reducing the step size and increasing the maximum number of increments. For example, if you set $h = 0.01$ and $n = 500$, you will obtain the results shown in Figure 5-2. You can see that the accuracy of solutions has been improved.

```
Solution 1 = -1.87938093786853
Solution confirmation: f(x) = 3.26919733666386E-05

Solution 2 = 0.347305077883933
Solution confirmation: f(x) = -2.30113661365827E-05

Solution 3 = 1.53206052188977
Solution confirmation: f(x) = -0.000114641850942565
```

Figure 5-2 More accurate solutions of the nonlinear equation using the incremental search method.

I should point out here that the incremental search methods has several limitations:

- You may miss two closely spaced solutions if the search increment h is too large (e.g., larger than the spacing between solutions).

- The degenerated solution (two or more solutions that coincide) cannot be detected.

- This method can only find solutions where $f(x)$ acrosses the x axis. It cannot find solutions where $f(x)$ is tangent to the x axis.

- This method may give you wrong solutions if singularities exist in $f(x)$. For example, the $\tan(x)$ function changes sign at $x = \pm 0.5n\pi$, with $n = 1, 3, 5,...$ However, these locations are not true solutions, since the function does not cross the x-axis.

Fixed Point Method

The idea behind the fixed point method is very simple: to find the solutions of a given nonlinear function $f(x) = 0$, we can algebraically rearrange $f(x) = 0$ into the form $x = g(x)$, where $g(x)$ denotes a function of x. Then, we simply start with an initial guess, $x0$, and apply the iteration process:

$$x_n = g(x_{n-1}) \quad \text{for } n = 1, 2, 3,...$$

For example, if we want to find solutions for the following nonlinear equation

$$f(x) = \sin x + e^x + x^2 - 2x = 0$$

There are several ways to rewrite the above equation into the form $x = g(x)$. For example, you can rearrange the above equation into the following form:

$$x = \frac{1}{2}\left(\sin x + e^x + x^2\right)$$

Or you can also rearrange the original problem as

$$x = \sqrt{2x - \sin x - e^x}$$

Implementation

For a given function $g(x)$, you can compute it successively with a starting value x_0. Then a sequence of values $\{x_n\}$ is obtained using the iteration rule $x_{n+1} = g(x_n)$. The sequence has the pattern

$$x_0$$
$$x_1 = g(x_0)$$
$$x_2 = g(x_1)$$
$$\vdots$$
$$x_n = g(x_{n-1})$$
$$\vdots$$

You can control the iteration process either by means of a tolerance parameter or by setting a maximum number of iterations.

Now, add a new static method, *FixedPoint*, to the *NonlinearSystem* class. Here is the code listing of this new method:

```
public static double FixedPoint(Function f, double x0,
                                double tolerance, int nMaxIterations)
{
    double x1 = x0, x2 = x0;
    double tol = 0.0;
    for (int i = 0; i < nMaxIterations; i++)
    {
        x2 = f(x1);
        tol = Math.Abs(x1 - x2);
        x1 = x2;
        if (tol < tolerance)
            break;
    }
    if (tol > tolerance)
    {
        throw new ArgumentException("Solution not found!");
    }
    return x1;
}
```

Here, we use a for-loop to control iterations. Inside this for-loop, we check the solution to see if it meets the tolerance requirement. You can further improve the fixed point method by limiting the solution in a centain range of x.

Testing the Fixed Point Method

Here, we will use an example to demonstrate how to solve a nonlinear equation by using the fixed point method. In this example, we want to find the solution of the following equation:

$$x^2 - 2x - 3 = 0$$

This equation has an analytic solution $x = 3$. In order to use the fixed point method, we must rewrite the above equation into the form

$$x = \sqrt{2x+3} \quad \Rightarrow \quad g(x) = \sqrt{2x+3}$$

Now, add the following code to the *Program.cs* file:

```
static void TestFixedPoint()
{
    double tol = 0.0001;
    int n = 10000;
    double x0 = 1.6;
    double x = NonlinearSystem.FixedPoint(G, x0, tol, n);
    Console.WriteLine("solution from the fixed point method: " + x.ToString());
}

static double G(double x)
{
    return Math.Sqrt(2 * x + 3);
}
```

Here, we use a static method to define $g(x)$, which will be called from the *FixedPoint* method. Inside the *TestFixedPoint* method, we specify the tolerance = 0.0001 and a larger maximum number of iterations, so that the solution is controlled by the tolerance parameter.

Running this application gives us the solution = 2.99997271109998, which is very close to the exact solution of 3.

Bisection Method

The bisection method is one of the bracking methods, which is used to solve for the root of a nonlinear function $f(x)$ by selecting an initial interval which contains the root.

For a give inverval $[xa, xb]$, you can use the sign of $f(xa)f(xb)$ to determine whether there are roots in the interval. If $f(xa)f(xb) < 0$, as shown in Figure 5-3, $f(x)$ has at least one root between xa and xb.

If $f(xa)f(xb) > 0$, there may or may not be any root between xa and xb, as shown in Figure 5-4.

As shown in the above, there is at least one root in the range of xa and xb if $f(xa)f(xb) < 0$. In this case we can always find the mid-point, xm, between xa and xb. This gives us two new intervals, (xa, xm) amd (xm, xb). To find out which interval, (xa, xm) or (xm, xb), contains the root, we can find the sign of $f(xa)f(xm)$. If $f(xa)f(xm) < 0$, then the root is in the interval (xa, xm). Otherwise, it is in between xm

and *xb*. As we repeat the process, the width of the interval that contains the root becomes smaller and smaller, and eventually we reach the root of the equation $f(x) = 0$.

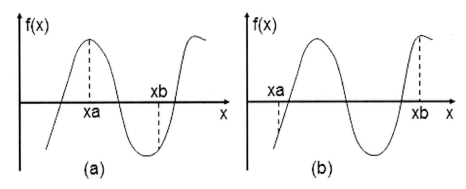

Figure 5-3 f(xa)f(xb) < 0, at least one root exists between xa and xb: (a) one root and (b) three roots.

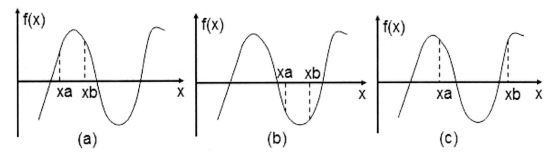

Figure 5-4 f(xa)f(xb) > 0, there are may or may not be any root between xa and xb: (a) and (b) no root; (c) two roots.

Implementation

The steps to apply the bisection method to find the root of the equation $f(x) = 0$ are listed here:

- Choose *xa* and *xb* as two guesses for the root such that $f(xa)f(xb) < 0$.
- Calculate the root, *xm*, of the equation $f(x) = 0$ as the mid-point between *xa* and *xb*, as $xm = (xa + xb)/2$.
- Check the sign of $f(xa)f(xm)$. If this quantity is less than zero, set $xb = xm$; otherwise set $xa = xm$.
- Repeat the above steps until the specified accuracy is reached.

The above algorithm for the bisection method is very simple. Note that this method is a slow-converging approach, but it is not affected by the slope of the function $f(x)$. So it is a very reliable method for solving nonlinear equations.

Add a new static method, *Bisection*, to the *NonlinearSystem* class:

```
public static double Bisection(Function f, double xa, double xb,
                               double tolerance)
{
    double x1 = xa;
    double x2 = xb;
    double fb = f(xb);
    while (Math.Abs(x2 - x1) > tolerance)
    {
        double xm = 0.5 * (x1 + x2);
        if (fb * f(xm) > 0)
            x2 = xm;
        else
            x1 = xm;
    }
    return x2 - (x2 - x1) * f(x2) / (f(x2) - f(x1));
}
```

Testing the Bisection Method

Now, we can use the bisection method to solve nonlinear equations. Here we will solve the same equation that was used in testing the incremental method:

$$f(x) = x^3 - 3x + 1$$

We know that this equation has a solution in the range (1, 2). Add a new static method, *TestBisection*, to the *Program.cs* file:

```
static void TestBisection()
{
    double x = NonlinearSystem.Bisection(F, 1.0, 2.0, 0.0001);
    Console.WriteLine("Solution from the bisection method: " + x.ToString());
}

static double F(double x)
{
    return x * x * x - 3.0 * x + 1.0;
}
```

Here, the function definition is the same as that used in the incremental method. We also set the parameters, $xa = 1.0$, $xb = 2.0$, and *tolerance* = 0.001 in calling the *Bisection* method.

Running this application generates the result $x = 1.53208888543173$.

False Position Method

The false position method, like the bisection method, starts with an interval (xa, xb) such that $f(xa)f(xb) < 0$, meaning that the function $f(x)$ has at least one root in this range. This method uses the relative values of function at end points to make better guesses, as shown in Figure 5-5.

You can see that given the heights of $f(xa)$ and $f(xb)$ as specified in the figure, it is apparent that the root is closer to xa than xb, and not in the middle. The false position method uses this fact.

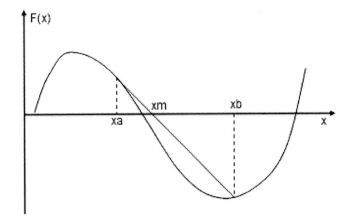

Figure 5-5 Illustration of the false position method.

Implamentation

The algorithm for the false position method is listed below:

1. Define the initial interval (*xa*, *xb*), so that $f(xa)$ and $f(xb)$ have different signs.
2. Find the first guess *xm*, which is the intersection of a line from $f(xa)$ to $f(xb)$ with the *x*-axis (see Figure 5-5):

$$xm = \frac{xa \cdot f(xb) - xb \cdot f(xa)}{f(xb) - f(xa)}$$

3. Evaluate the error $= |f(xm)|$ and compare it with the tolerance. If the error is too big, decide the location of the root. If $f(xa)f(xb) < 0$, set $xb = xm$; otherwise, set $xa = xm$. Re-compute the new guess for *xm* using the formula given in step 2.
4. Repeat the iteration process until the error is less than the tolerance.

Now we can implement the false position method in C#. Add a new static method, *FalsePosition*, to the *NonlinearSystem* class:

```
public static double FalsePosition(Function f, double xa, double xb,
                                   double tolerance)
{
    double x1 = xa;
    double x2 = xb;
    double fb = f(xb);
    while (Math.Abs(x2 - x1) > tolerance)
    {
        double xm = x2 - (x2 - x1) * f(x2) / (f(x2) - f(x1));
        if (fb * f(xm) > 0)
            x2 = xm;
        else
            x1 = xm;
        if (Math.Abs(f(xm)) < tolerance)
            break;
    }
```

```
            return x2 - (x2 - x1) * f(x2) / (f(x2) - f(x1));
    }
```

In this method, we use the tolerance parameter to control iterations.

Testing the False Position Method

Now, we can use the false position method to solve nonlinear equations. Here we will solve the same equation that was used in testing the incremental and bisection methods:

$$f(x) = x^3 - 3x + 1$$

We know that this equation has a solution in the range (1, 2). Add a new static method, *TestFalsePosition*, method to the *Program.cs* file:

```
static void TestBisection()
{
    double x = NonlinearSystem.FalsePosition(F, 1.0, 2.0, 0.0001);
    Console.WriteLine("Solution from the false position method: " +
                x.ToString());
}
```

The function definition is the same as that used in the incremental method. We also set the parameters *xa* = 1.0, *xb* = 2.0, and *tolerance* = 0.001 when calling the *FalsePosition* method.

Running this application generates the result *x* = 1.53208217650463.

Newton-Raphson Method

The Newton-Raphson method is the most popular method for solving nonlinear equations. This method is, in general, very suitable for its convenience and computation efficiency. In this method, the root is not bracketed and only one initial guess value of the solution is needed to start the iterative process. Thus, unlike the bracketing-based methods such as the bisection method and the false position method, which always give the convergent solutions, the Newton-Raphson method may fail to converge if the initial value is too far away from the true solution.

The Newton-Raphson method uses the following iteration relation:

$$x_{n+1} = x_n - \frac{f(x_n)}{f'(x_n)}$$

where $f'(x)$ is the first derivative of function $f(x)$.

Starting with an initial value, x_n, we can find the next guess, x_{n+1}, by using the above iterative relation. This process can be repeated until the solution within a given tolerance is found.

Implementation

The following lists the steps needed to apply the Newton-Raphson method to find the root of a nonlinear equation $f(x) = 0$:

1. Specify the initial guess x for the root of function $f(x)$, and the tolerance which controls the iterative process.
2. Evaluate $f'(x)$ symbolically.
3. Use the initial guess of the root, x_n, to calculate the new value of the root using the formula
    $$x_{n+1} = x_n - f(x_n)/f'(x_n).$$
4. Repeat the above steps until the solution is within the specified tolerance.

Add a new static method, *NewtonRaphson*, to the *NonlinearSystem* class. Here is the code listing of this method:

```
public static double NewtonRaphson(Function f, Function f1,
                                   double x0, double tolerance)
{
    double f0 = f(x0);
    double x = x0;
    while (Math.Abs(f(x)) > tolerance)
    {
        x -= f0 / f1(x);
        f0 = f(x);
    }
    return x;
}
```

This method requires an analytic expression for the first derivative of the function $f(x)$ as an input delegate function $f1$. Instead of using the analytic first derivative of the function as input, you can also implement a routine that computes the first derivative of the function numerically.

As mentioned previously, the Newton-Raphson method may fail to converge if the initial guess is too far from the true solution. Because of this, you can implement a modified version of this method in practical applications. For example, you can specify an interval (xa, xb) that is known to contain the root. Then you can choose the midpoint of this interval as the initial guess and stop the iteration if the new guess solution generated by the Newton-Raphson method lies outside the interval. This way, you can avoid the method's divergent issue.

Testing the Newton-Raphson Method

Now, we can use the Newton-Raphson method to solve nonlinear equations. Here we will solve the nonlinear equation:

$$f(x) = \cos x - x^3$$

The first derivative of this function is given by

$$f'(x) = -\sin x - 3x^2$$

Add a new static method, *TestNewtonRaphson*, method to the *Program.cs* file:

```
static void TestNewtonRaphson()
{
    double x = NonlinearSystem.NewtonRaphson(FF, FF1, 1.0, 0.0001);
    Console.WriteLine("Solution from the Newton-Raphson method: " +
                      x.ToString());
```

```
    }

    static double FF(double x)
    {
        return Math.Cos(x) - x * x * x;
    }

    static double FF1(double x)
    {
        return -Math.Sin(x) - 3 * x * x;
    }
```

Running this application gives the solution x = 0.865474075952977.

Secant Method

The Newton-Raphson method presented in the previous section requires you to supply the first derivative of the function. This requirement can be a tedious process for some of the more complicated nonlinear functions. In order to overcome this drawback, we can use the following formula to estimate the derivative, $f'(x)$:

$$f'(x_n) = \frac{f(x_n) - f(x_{n-1})}{x_n - x_{n-1}}$$

Substituting the above equation into the Newton-Raphson inerative relation, we have

$$x_{n+1} = x_n - \frac{f(x_n)}{f'(x_n)} = x_n - \frac{(x_n - x_{n-1})f(x_n)}{f(x_n) - f(x_{n-1})}$$

The above relation is called the secant method. You need to specify an initial interval for this method, but unlike with the bisection method, this initial interval is not required to contain the root. The secant method may or may not converge, but when it converges, it converges faster than the bisection method.

Comparing the secant method with the Newton-Raphson method, we see that since the derivative is approximated, the secant method usually converges slower. However, the Newton-Raphson method requires the evaluation of both $f(x)$ and its derivative at every step, while the secant method only requires the evaluation of $f(x)$. Therefore, the secant method may be faster in practice.

Note that the iteration formula in the secant method is the same as that used in the false position method. However, the secant method always retains the last two computed points, while the false position method retains two points that must contain the root.

Implamentation

The implementation of the secant method is very similar to that of the false position method. Add a new static method, *Secant*, to the *NonlinearSystem* class:

```
public static double Secant(Function f, double xa, double xb, double tolerance)
{
    double x1 = xa;
    double x2 = xb;
    double fb = f(xb);
```

```
        while(Math.Abs(f(x2)) > tolerance)
        {
            double xm = x2 - (x2 - x1) * fb / (fb - f(x1));
            x1 = x2;
            x2 = xm;
            fb = f(x2);
        }
        return x2;
    }
}
```

Inside this method, you can see how we keep the last two computed points after each iteration, which is different from the false position method.

Testing the Secant Method

Now, we can use the secant method to solve nonlinear equations. We will solve the same equation that we used previously:

$$f(x) = x^3 - 3x + 1$$

Here, we choose an initial interval of (1, 1.5). We know from the previous computation that this range does not contain any root. Thus, if you use the false position method with this intial interval, you will fail to find the root of the above equation. However, this interval works for the secant method because the secant method does not require the root to be bracketed in the initial interval. Add a new static method, *TestSecant*, method to the *Program.cs* file:

```
static void TestSecant()
{
    double x = NonlinearSystem.Secant(F, 1.0, 1.5, 0.0001);
    Console.WriteLine("Solution from the secant method: " + x.ToString());
}
```

The function definition is the same as that used in the incremental method. Here the tolerance is set to 0.001 when the Secant method is called.

Running this application gives the result $x = 1.53208898626515$.

Newton Multiroot Method

The Newton-Raphson method is usually used for solving for one root. It is possible to extend this method to find all of the real roots of a nonlinear equation. We know that the Newton-Raphson method uses the following iterative relation:

$$x = x - f(x)/f'(x)$$

Now, we can calculate the first derivative f′ (x) using the approximation:

$$f'(x) \approx \frac{\Delta f(x)}{\Delta x} = \frac{[f(x) - f(x-h) + (f(x+h) - f(x)]/2}{h} = \frac{f(x+h) - f(x-h)}{2h}$$

Here, $h = \Delta x$ is a very small interval in the x-axis. In the Newton multiroot method, the above formula will be used to update the new solution for each root.

Implementation

If there is more than one root, we need to develop an algorithm that avoids recomputing roots that have been already obtained. This can be achieved by dividing the function $f(x)$ by the product of x minus a previously determined root. Consequently, the modified function steers away from previous roots, because it becomes infinity at the locations of these roots.

Add a new method, *NewtonMultiroot*, to the *NonlinearSystem* class. Here is the code listing of this new method:

```
public static double[] NewtonMultiRoots(Function f, double x0, int nRoots,
                    int nIterations, double tolerance)
{
    double h, delta = 10*tolerance, f1, f2, f3, x = x0;
    double[] roots = new double[nRoots];
    int nroot = 0, i =0;
    while (i < nIterations && nroot < nRoots)
    {
        i = 0;
        while (i < nIterations && Math.Abs(f(x)) > tolerance)
        {
            if (Math.Abs(x) > 1)
                h = 0.01 * x;
            else
                h = 0.01;
            f1 = f(x - h);
            f2 = f(x);
            f3 = f(x + h);
            if (nroot > 0)
            {
                for (int j = 0; j < nroot; j++)
                {
                    f1 /= (x - h - roots[j]);
                    f2 /= (x - roots[j]);
                    f3 /= (x + h - roots[j]);
                }
            }
            delta = 2 * h * f2 / (f3 - f1);
            x -= delta;
            i++;
        }
        if (Math.Abs(f(x)) <= tolerance)
        {
            roots[nroot] = x;
            nroot++;
            if (x < 0)
                x *= 0.95;
            else if (x > 0)
                x *= 1.05;
            else
                x = 0.05;
        }
    }
    return roots;
}
```

In this method, input parameters include the maximum number of roots, maximum number of iterations, and tolerance.

Testing the Newton Multiroot Method

Now, we can use the Newton multiroot method to solve nonlinear equations. Here we will solve the same equation that was used previously:

$$f(x) = x^3 - 3x + 1$$

From the results calculated using the incremental method, the above equation has three roots. We can find all of the roots using the Newton multiroot method. Add a new static method, *TestNewtonMultiroot*, to the *Program.cs* file:

```
static void TestNewtonMultiRoots()
{
    double[] x = NonlinearSystem.NewtonMultiRoots(F, 0.0, 3, 1000, 0.0001);
    Console.WriteLine(x[0].ToString());
    Console.WriteLine(x[1].ToString());
    Console.WriteLine(x[2].ToString());
}
```

The function definition is the same as that used in the incremental method. We also set the initial value $x0 = 0.0$, number of roots = 3, tolerance = 0.0001, and the maximum number of iterations = 1000 when calling the Newton multiroot method.

Running this application produces results shown in Figure 5-6.

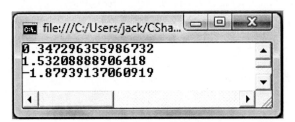

Figure 5-6 Roots from the Newton multiroot method.

Birge-Vieta Method

The Birge-Vieta method is also an iterative method for finding real roots of the nth degree polynomial equation $f(x) = P_n(x) = 0$ of the form

$$a_n x^n + a_{n-1} x^{n-1} + a_{n-2} x^{n-2} + ... + a_1 x + a_0 = 0$$

If x_1 is a real root of $P_n(x) = 0$, then

$$Q_{n-1}(x) = P_n(x)/(x - x_1)$$

is a reduced $(n-1)$th degree polynimial of the form

$$Q_{n-1}(x) = b_{n-1} x^{n-1} + b_{n-2} x^{n-2} + ... + b_1 x + b_0$$

If x_1 is any approximation to the true root, then we should have

$$P_n(x) = (x - x_1)Q_{n-1}(x) + R$$

where R is the residue which is a function of x_1. We can get the following relations by comparing the coefficients of $P_n(x)$ and $(x - x_1)Q_{n-1}(x) + R$:

$$b_{n-1} = a_n$$
$$b_{n-2} = a_{n-1} + x_1 b_{n-1}$$
$$b_{n-3} = a_{n-2} + x_1 b_{n-2}$$
$$\vdots$$
$$b_0 = a_1 + x_1 b_1$$
$$R = a_0 + x_1 b_0$$

Also note that $P_n'(x_1) = Q_{n-1}(x_1)$. Now, we can using Newton-Raphson technique to find the root:

$$x_{i+1} = x_i - \frac{P_n(x_i)}{P_n'(x_i)} = x_i - \frac{P_n(x_i)}{Q_{n-1}(x_i)}$$

We can use the above coeffients and the Newton-Raphson iteration to find the roots of polynomials. This algorithm is known as the Birge-Vieta method.

Implementation

Add a new static method, *BirgeVieta*, to the *NonlinearSystem* class:

```
public static double[] BirgeVieta(double[] a, double x0, int nOrder, int nRoots,
                                  int nIterations, double tolerance)
{
    double x = x0;
    double[] roots = new double[nRoots];
    double[] a1 = new double[nOrder + 1];
    double[] a2 = new double[nOrder + 1];
    double[] a3 = new double[nOrder + 1];
    for (int j = 0; j <= nOrder; j++)
        a1[j] = a[j];
    double delta = 10 * tolerance;
    int i = 1, n = nOrder + 1, nroot = 0;
    while (i++ < nIterations && n > 1)
    {
        double x1 = x;
        a2[n-1] = a1[n-1];
        a3[n-1] = a1[n-1];
        for (int j = n - 2; j > 0; j--)
        {
            a2[j] = a1[j] + x1 * a2[j + 1];
            a3[j] = a2[j] + x1 * a3[j + 1];
        }
        a2[0] = a1[0] + x1 * a2[1];
        delta = a2[0] / a3[1];
        x -= delta;
```

```
            if (Math.Abs(delta) < tolerance)
            {
                i = 1;
                n--;
                roots[nroot] = x;
                nroot++;
                for (int j = 0; j < n; j++)
                {
                    a1[j] = a2[j + 1];
                    if (n == 2)
                    {
                        n--;
                        roots[nroot] = -a1[0];
                        nroot++;
                    }
                }
            }
        }
        return roots;
}
```

This method takes the coefficients of the polynomial as input:

```
double [] a = double[n + 1]{a₀, a₁, a₂, ... aₙ}
```

To use this method, you also need to specify an initial guess $x0$, order of the polynomial, number of roots, maximum number of iterations, and tolerance.

Testing the BirgeVieta Method

Now, we can examine the *BirgeVieta* method by using it to solve nonlinear equations. Here, we will solve the same equation that we used previously:

$$f(x) = x^3 - 3x + 1$$

This is a third degree polynomial and it should have three real roots.. We can find all three roots by using the *BirgeVieta* method. Add a new static method, *TestBirgeVieta*, to the *Program.cs* file:

```
static void TestbirgeVieta()
{
    double[] x = NonlinearSystem.BirgeVieta(new double[4] { 1, -3, 0, 1.0 },
                                      0.0, 3, 3, 1000, 0.0001);
    Console.WriteLine(x[0].ToString());
    Console.WriteLine(x[1].ToString());
    Console.WriteLine(x[2].ToString());
}
```

The function definition is the same as that used in the incremental method. We also set the initial value $x0 = 0.0$, order of the polynomial = 3, number of roots = 3, tolerance = 0.0001, and the maximum number of iterations = 1000.

Running this application produces results shown in Figure 5-7.

Figure 5-7 Results from the BirgeVieta method.

System of Equations

In previous sections, we focused our attention on solving single nonlinear equations. In this section, we consider the n-dimensional version of the same problem, namely

$$f_1(x_1, x_2, \cdots, x_n) = 0$$
$$f_2(x_1, x_2, \cdots, x_n) = 0$$
$$\vdots$$
$$f_n(x_1, x_2, \cdots, x_n) = 0$$

The solution of n-coupled nonlinear equations is a much more difficult task than finding the root of a single equation. The problem is the lack of a reliable method for bracketing the solution $x_1, x_2, \cdots, x_n$. Therefore, we cannot provide the solution algorithm with guaranteed good starting values, unless such values are suggested by the physics of the problem.

The most effective method for finding solutions for a system of nonlinear equations is the Newton-Raphson method. It works well with simultaneous equations, provided that it is supplied with good starting values. In this section, we will only discusss the Newton-Raphson method.

Algorithm

In order to derive the Newton-Raphson method for a system of nonlinear equations, we start with the Taylor series expension of $f_i(x)$ about the point $x = (x_1, x_2, \cdots, x_n)$:

$$f_i(x + \Delta x) = f_i(x) + \sum_{j=0}^{n-1} \frac{\partial f_i}{\partial x_j} \Delta x_j + O(\Delta x^2)$$

Dropping terms of order Δx^2, we can rewrite the above equation as

$$f(x + \Delta x) = f(x) + J(x)\Delta x$$

where $J(x)$ is the Jacobian matrix of size $n \times n$ made up of the partial derivatives

$$J_{ij} = \frac{\partial f_i}{\partial x_j}$$

Suppose $\mathbf{x}$ is the current approximation of the solution of $\mathbf{f}(\mathbf{x}) = 0$, and $\mathbf{x} + \Delta\mathbf{x}$ is the improved solution. In order to find the correct $\Delta\mathbf{x}$, we set $\mathbf{f}(\mathbf{x} + \Delta\mathbf{x}) = 0$ in the above equation. The result is a set of linear equations for $\Delta\mathbf{x}$:

$$\mathbf{J}(\mathbf{x})\Delta\mathbf{x} = -\mathbf{f}(\mathbf{x})$$

The Newton-Raphson method for a system of coupled nonlinear equations is based on the above equation. Here is the procedure of the method:

- Estimate the solution vector $\mathbf{x}$.

- Evaluate $\mathbf{f}(\mathbf{x})$.

- Compute the Jacobian matrix $\mathbf{J}(\mathbf{x})$.

- Set up the simultaneous equations and solve for $\Delta\mathbf{x}$.

- Let $\mathbf{x} = \Delta\mathbf{x}$ and repeat the above steps.

The above process is continued until $|\Delta\mathbf{x}| <$ tolerance. As in the one-dimensional case, success of the Newton-Raphson method depends on the initial estimate of $\mathbf{x}$. If a good starting point is used, convergence to the solution is very rapid. Otherwise, the results are unpredictable.

Unlike with a single equation, the analytical derivatives $\partial f_i / \partial x_j$ for a system of nonlinear equations are usually difficult to calculate. Here we will compute the partial derivatives using the finite difference approximation:

$$\frac{\partial f_i}{\partial x_j} \approx \frac{f_i(\mathbf{x} + \mathbf{e}_j h) - f_i(\mathbf{x})}{h}$$

where h is a small increment and $\mathbf{e}_j$ represents a unit vector in the direction of $\mathbf{x}_j$.

Implementation

In this section, we will implement the Newton-Raphson method for numerically solving a system of nonlinear equations. Add a new static method, *NewtonMultiEquation*, to the *NonlinearSystem* class:

```
public delegate VectorR MFunction(VectorR x);

public static VectorR NewtonMultiEquations(MFunction f, VectorR x0,
                                           double tolerance)
{
    LinearSystem ls = new LinearSystem();
    VectorR dx = new VectorR(x0.GetSize());
    do
    {
        MatrixR A = Jacobian(f, x0);
        if (Math.Sqrt(VectorR.DotProduct(f(x0), f(x0)) / x0.GetSize()) <
                tolerance)
            return x0;
        dx = ls.GaussJordan(A, -f(x0));
        x0 = x0 + dx;
    }
    while (Math.Sqrt(VectorR.DotProduct(dx, dx)) > tolerance);
    return x0;
```

```
    }

    private static MatrixR Jacobian(MFunction f, VectorR x)
    {
        double h = 0.0001;
        int n = x.GetSize();
        MatrixR jacobian = new MatrixR(n, n);
        VectorR x1 = x.Clone();
        for (int j = 0; j < n; j++)
        {
            x1[j] = x[j] + h;
            for (int i = 0; i < n; i++)
            {
                jacobian[i, j] = (f(x1)[i] - f(x)[i]) / h;
            }
        }
        return jacobian;
    }
```

Here, we first define a vector delegate function that allows the user to supply a system of equations. The private static method, *Jacobian*, is used to compute the Jacobian matrix $J(x)$. Note that the Jacobian matrix needs to be recomputed during each iteration run.

Inside the *NewtonMultiEquations* method, the linear equations for Δx are solved using the *GaussJordan* method, so you need to add the *LinearSystem.cs* file from the *LinearSystemTest* project of Chapter 4 to the current project.

Testing the NewtonMultiEquations Method

Now, we can examine the *NewtonMultiEquations* method by using it to solve a system of nonlinear equations. Here we will consider the following example:

$$2x - y - e^{-2x} = 0$$
$$-x + 2y - e^{-y} = 0$$

Add a new method, *TestNewtonMultiEquations*, to the *Program.cs* file:

```
static void TestNewtonMultiEquations()
{
    VectorR x0 = new VectorR(new double[] { 0.0, 0.0 });
    VectorR x = NonlinearSystem.NewtonMultiEquations(FV, x0, 1e-5);
    Console.WriteLine("\n x[0] =  {0,8:n6}, x[1] = {1,8:n6} \n    f1 = {2,8:n6},
                      f2 = {3,8:n6}", x[0], x[1],FV(x)[0], FV(x)[1]);
}

static VectorR FV(VectorR x)
{
    VectorR result = new VectorR(2);
    result[0] = 2 * x[0] - x[1] - Math.Exp(-2 * x[0]);
    result[1] = -x[0] + 2 * x[1] - Math.Exp(-x[1]);
    return result;
}
```

Pay attention to how we define the vector delegate function. Inside the *TestNewtonMultiEquations* method, we set the tolerance = 1e-5 and print out both the solution of **x** and f(**x**) to check out if the results are within the specified tolerance.

Running this example generates results shown in Figure 5-8. You can see the results indeed have 5-digit accuracy.

```
x[0] =   0.461707, x[1] = 0.526260
 f1 = -0.000006,   f2 = 0.000002
```

Figure 5-8 Results from the NewtonMultiEquations method.

Chapter 6
Special Functions

Special functions are simply specialized functions beyond the familiar trigonometric and exponential functions. Many special functions arise naturally in areas of analysis, statistics, number theory, physics, and other fields. In this chapter, we will implement a special function class, which contains popular special functions, such as the gamma function, beta function, error function, elliptic intergral, Bessel function, and etc.

Gamma and Beta Functions

The gamma and beta functions appear occasionally in physical problems, such as probability distribution in statistical mechanics. In general, however, they have less direct physical applications and interpretations than some other special functions, such as the Legendre and Bessel functions. Rather, their importance originates from their usefelness in developing other functions that have direct physical applications.

Gamma Function

The gamma function can be defined by either the integral or infinite series. The series of the gamma function has the form:

$$\Gamma(x) = \lim_{n \to \infty} \frac{1 \cdot 2 \cdot 3 \cdots n}{x(x+1)(x+2)\cdots(x+n)} n^x \quad x \neq 0, -1, -2, -3, \ldots$$

From the above equation, we can obtain the recurrence relation by replacing x with $x + 1$:

$$\Gamma(x+1) = x\Gamma(x)$$

Also, from the series definition of the gamma function:

$$\Gamma(1) = \lim_{n \to \infty} \frac{1 \cdot 2 \cdot 3 \cdots n}{1 \cdot 2 \cdot 3 \cdots n(n+1)} n = 1$$

Now, application of the recurrence relation gives

$$\Gamma(2) = 1$$
$$\Gamma(3) = 2\Gamma(2) = 2$$
$$\Gamma(n) = 1 \cdot 2 \cdot 3 \cdots (n-1) = (n-1)!$$

This means that if the argument x is an integer, the gamma function is simply a factorial function.

The gamma function can be also defined in the integral form:

$$\Gamma(x) = \int_0^\infty t^{x-1} e^{-t} dt$$

When $x = \frac{1}{2}$, the above integral is just the Gausss error function, and then we have the interesting result:

$$\Gamma(1/2) = \sqrt{\pi}$$

If the gamma function is known for arguments $x > 1$, it can be obtained for $x < 1$ by the reflection formula:

$$\Gamma(-x) = -\frac{\pi}{x\Gamma(x)\sin \pi x}$$

Anticipating the recurrence relation, we have $-x\Gamma(-x) = \Gamma(1 - x)$. The above reflection relation may be rewritten as

$$\Gamma(1-x) = \frac{\pi}{\Gamma(x)\sin \pi x}$$

Several numerical methods exist for computing the gamma function. The best known method is based on the Lanczos technique. We will not derive the approximation, but only list the resulting formula. For certain integer choices of g and N, and for certain coefficients $c0$, c_1, c_2,..., c_N, the gamma function can be expressed in terms of the following approximation:

$$\Gamma(x+1) = \left(x + g + \frac{1}{2}\right)^{x+\frac{1}{2}} e^{-\left(x+g+\frac{1}{2}\right)} \sqrt{2\pi} \left(c_0 + \frac{c_1}{x+1} + \frac{c_2}{x+2} + \cdots + \frac{c_N}{z+N} + \varepsilon \right) \quad (x > 0)$$

You can see that the above formula is similar to the Stirling's approximation for factorials, but with a series of corrections. The above equation is also applicable when x is a complex arguement. The error term is described by ε. For $g = 7$, $N = 9$, and a certain set of c's, the error is smaller than $|\varepsilon| < 10^{-15}$.

Implementation

Here, we will only consider a gamma function with a real argument. Start with a new C# Console application and name it *SpecialFunctionsTest*. Add a new class, *SpecialFunctions*, to the project and change its namespace to *XuMath*.

Here is the implementation of the *Gamma* method based on the Lanczos approximation:

```
using System;

namespace XuMath
```

```
{
    public class SpecialFunctions
    {
        public SpecialFunctions()
        {
        }

        public static double Gamma(double x)
        {
            const int g = 7;
            double[] coef = new double[9]{0.99999999999980993,
                                          676.5203681218851,
                                          -1259.1392167224028,
                                          771.32342877765313,
                                          -176.61502916214059,
                                          12.507343278686905,
                                          -0.13857109526572012,
                                          9.9843695780195716e-6,
                                          1.5056327351493116e-7};
            if (x < 0.5)
            {
                return Math.PI / (Math.Sin(Math.PI * x) * Gamma(1.0 - x));
            }
            x -= 1.0;
            double y = coef[0];
            for (int i = 1; i < g + 2; i++)
            {
                y += coef[i] / (x + 1.0 * i);
            }
            double z = x + (g + 0.5);
            return Math.Sqrt(2 * Math.PI) * Math.Pow(z, x + 0.5) *
                    Math.Exp(-z) * y;
        }
    }
}
```

Testing the Gamma Method

You can examine the *Gamma* method by calculating the gamma functions with known results. Here, we will calculate $\Gamma(1/2) = \sqrt{\pi}$ and $\Gamma(5) = 4! = 24$ using the *Gamma* method. The following is the code list:

```
using System;
using XuMath;

namespace SpecialFunctionsTest
{
    class Program
    {
        static void Main(string[] args)
        {
            TestGamma();
            Console.ReadLine();
        }
```

```
static void TestGamma()
{
    Console.WriteLine("Gamma(5) = {0}", SpecialFunctions.Gamma(5));
    Console.WriteLine("Gamma(1/2) = {0}", SpecialFunctions.Gamma(0.5));
}
}
}
```

Running this application generates the results shown in Figure 6-1. You can see that the results from the Gamma method are very accurate.

Figure 6-1 Gamma function results calculated using the Gamma method.

The *Gamma* method can also be used to calculate the factorial by specifying an integer argument.

Beta Function

The beta function, also called the Euler integral of the first kind, is a special function defined by

$$B(x, y) = \int_0^1 t^{x-1}(1-t)^{y-1} dt$$

The beta function is symmetric, meaning that $B(x, y) = B(y, x)$. It is related to the gamma function:

$$B(x, y) = \frac{\Gamma(x)\Gamma(y)}{\Gamma(x+y)}$$

Implementation

Add a new method, *Beta*, to the *SpecialFunctions* class:

```
public static double Beta(double x, double y)
{
    return Gamma(x) * Gamma(y) / Gamma(x + y);
}
```

Testing the Beta Method

Add a new method, *TestBeta*, to the *Program.cs* file:

```
static void TestBeta()
{
```

```
    Console.WriteLine("Beta(2, 3) = {0}", SpecialFunctions.Beta(2, 3));
}
```

This gives us the result of 0.0833333333333334.

Error Function

The error function and complementary error function are defined as

$$\operatorname{erf}(x) = \frac{2}{\sqrt{\pi}} \int_0^x e^{-t^2} \, dt$$

and

$$\operatorname{erfc}(x) = 1 - \operatorname{erf}(x) = \frac{2}{\sqrt{\pi}} \int_x^\infty e^{-t^2} \, dt$$

The above functions have the following properties:

$$\operatorname{erf}(0) = 0, \quad \operatorname{erf}(\infty) = 1, \quad \operatorname{erf}(-x) = -\operatorname{erf}(x)$$

$$\operatorname{erfc}(0) = 1, \quad \operatorname{erfc}(\infty) = 0, \quad \operatorname{erfc}(-x) = 2 - \operatorname{erfc}(x)$$

The error function has no singularities (except at infinity) and its Taylor series always converges. The integral cannot be evaluated in closed form in terms of elementary functions, but by expanding the integrand in a Taylor series, we can obtain the Taylor series for the error function as follows:

$$\operatorname{erf}(x) = \frac{2}{\sqrt{\pi}} \left(x - \frac{x^3}{3} + \frac{x^5}{10} - \frac{x^7}{42} + \cdots \right) = \frac{2}{\sqrt{\pi}} \sum_{n=0}^\infty \frac{(-1)^n x^{2n+1}}{n!\,(2n+1)}$$

This series holds not only for the real argument but also for the complex argument.

We can express erfc(x) in terms of continued fraction:

$$\operatorname{erfc}(x) = \frac{e^{-x^2}}{x\sqrt{\pi}} \cfrac{1}{x + \cfrac{1/2}{x + \cfrac{3/2}{x + \cfrac{4/2}{x + \cfrac{5/2}{x + \cfrac{7/2}{x + \cdots}}}}}} \qquad (\text{for } x \geq 0)$$

Implementation

From the discussion in the previous section, we see that the Taylor series converges fast for a small x, while the continued fraction expression works better for a large argument x. Therefore, we will use the series expansion to implement error functions erf(x) and erfc(x) = 1 − erf(x) in C# for a smaller x (< 2.2). For a larger x (≥ 2.2) argument, we will use the continued fraction expression for erfc(x) instead.

Add two new methods, *Erf* and *Erfc*, to the *SpecialFunctions* class. Here are their code listing:

```
public static double Erf(double x)
{
    if (Math.Abs(x) > 2.2)
        return 1.0 - Erfc(x);
    double sum = 0.0;
    double sum0 = 0.0;
    int i = 0;
    do
    {
        sum0 = sum;
        sum += Math.Pow(-1, i) * Math.Pow(x, 2 * i + 1) /
            Gamma(i + 1) / (2 * i + 1);
        i++;
    }
    while (Math.Abs((sum - sum0) / sum0) > 1e-12);
    return 2.0 * sum / Math.Sqrt(Math.PI);
}

public static double Erfc(double x)
{
    if (Math.Abs(x) <= 2.2)
        return 1.0 - Erf(x);
    if (x < 0)
        return 2.0 - Erfc(-x);
    double x1 = Math.Exp(-x * x) / Math.Sqrt(Math.PI);
    double c1 = 1.0;
    double c2 = x;
    double c3 = x;
    double c4 = x * x + 0.5;
    double c5 = 0.0;
    double c6 = c2 / c4;
    double c7 = 1.0;
    double c8 = 0.0;

    do
    {
        c8 = c1 * c7 + c2 * x;
        c1 = c2;
        c2 = c8;
        c8 = c3 * c7 + c4 * x;
        c3 = c4;
        c4 = c8;
        c7 += 0.5;
        c5 = c6;
        c6 = c2 / c4;
    }
    while (Math.Abs(c5 - c6) / c6 > 1.0e-12);
    return c6 * x1;
}
```

Testing the Erf and Erfc Methods

We can calculate the error functions using the *Erf* and *Erfc* methods. Add a method, *TestError*, to the *Program.cs* file:

```
static void TestEror()
{
    for (int i = 0; i < 21; i++)
    {
        double x = (i - 10.0) / 2.0;
        Console.WriteLine("x = {0,5:n2}, erf(x) = {1,20:e12},
                erfc(x) = {2,20:e12}",
                x, SpecialFunctions.Erf(x), SpecialFunctions.Erfc(x));
    }
}
```

Running this example generates results shown in Figure 6-2. The error functions calculated using the *Erf* and *Erfc* methods provide results with at least 12-digit accuracy.

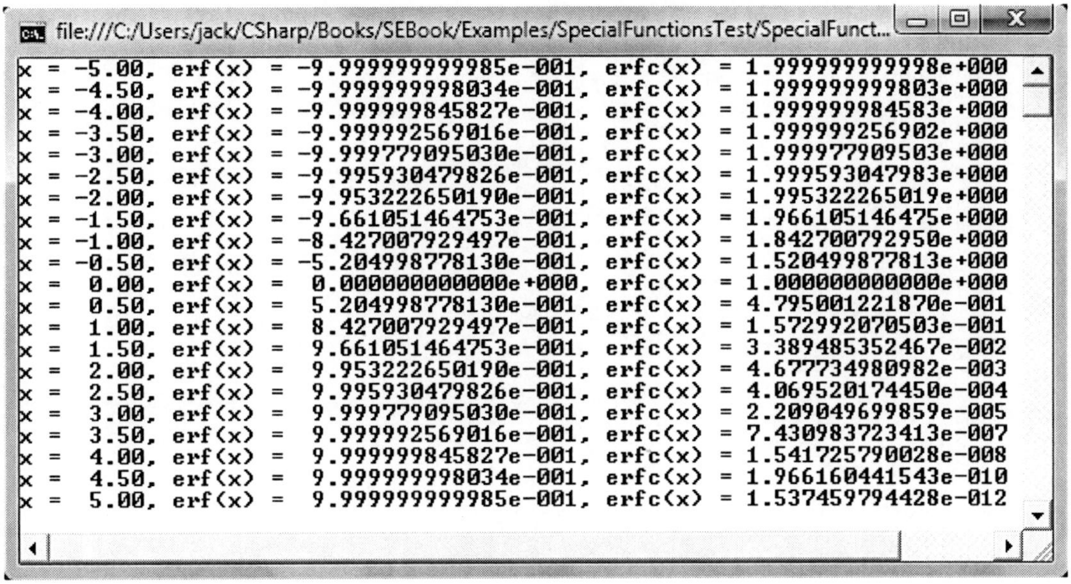

Figure 6-2 Results of error functions.

Sine and Cosine Integral Functions

The sine integral function is defined as

$$\mathrm{Si}(x) = \int_0^x \frac{\sin t}{t}\, dt$$

and the cosine integral function is defined as

$$Ci(x) = -\int_{x}^{\infty} \frac{\cos t}{t}\, dt = \gamma + \ln x + \int_{0}^{x} \frac{\cos t - 1}{t}\, dt$$

where $\gamma = 0.5772156649$ is the Euler constant.

We will calculate the values for the sine and cosine integral functions using the following series expansions:

$$Si(x) = \sum_{n=0}^{\infty} \frac{(-1)^n x^{2n+1}}{(2n+1)(2n+1)!}$$

$$Ci(x) = \gamma + \ln x + \sum_{n=1}^{\infty} \frac{(-1)^n x^{2n}}{(2n)(2n)!}$$

Implementation

In this section, we will implement a C# method for computing the sine and cosine integrals based on their Taylor series expansions. Add two new methods, *Si* and *Ci*, to the *SpecialFunctions* class. Here is their code listing:

```
public static double Si(double x)
{
    double sum = 0.0;
    double term = 0.0;
    int i = 0;
    do
    {
        term = Math.Pow(-1, i) * Math.Pow(x, 2 * i + 1) /
               (2 * i + 1) / Gamma(2 * i + 2);
        sum += term;
        i++;
    }
    while (Math.Abs(term) > 1.0e-12);
    return sum;
}

public static double Ci(double x)
{
    double sum = 0.0;
    double term = 0.0;
    int i = 1;
    do
    {
        term = Math.Pow(-1, i) * Math.Pow(x, 2 * i) /
               (2 * i) / Gamma(2 * i + 1);
        sum += term;
        i++;
    }
    while (Math.Abs(term) > 1.0e-12);
    return 0.5772156649 + Math.Log(x) + sum;
}
```

Testing the Si and Ci Methods

We can calculate the sine and cosine functions using the *Si* and *Ci* methods. Add a method, *TestSiCi*, to the *Program.cs* file:

```
static void TestSiCi()
{
    for (int i = 1; i < 21; i++)
    {
        double x = i / 2.0;
        Console.WriteLine("x = {0,5:n2}, Si(x) = {1,20:e12},
                Ci(x) = {2,20:e12}",
                x, SpecialFunctions.Si(x), SpecialFunctions.Ci(x));
    }
}
```

Running this example generates the results shown in Figure 6-3. The sine and cosine integral functions calculated using the *Si* and *Ci* methods provide results with at least 12-digit accuracy.

```
file:///C:/Users/jack/CSharp/Books/SEBook/Examples/SpecialFunctionsTest/SpecialFun...
x =    0.50, Si(x) = 4.931074180431e-001, Ci(x) = -1.777840788081e-001
x =    1.00, Si(x) = 9.460830703672e-001, Ci(x) =  3.374039228994e-001
x =    1.50, Si(x) = 1.324683531172e+000, Ci(x) =  4.703563171939e-001
x =    2.00, Si(x) = 1.605412976803e+000, Ci(x) =  4.229808287733e-001
x =    2.50, Si(x) = 1.778520173444e+000, Ci(x) =  2.858711963639e-001
x =    3.00, Si(x) = 1.848652527999e+000, Ci(x) =  1.196297860065e-001
x =    3.50, Si(x) = 1.833125398666e+000, Ci(x) = -3.212854851400e-002
x =    4.00, Si(x) = 1.758203138949e+000, Ci(x) = -1.409816978885e-001
x =    4.50, Si(x) = 1.654140414379e+000, Ci(x) = -1.934911221033e-001
x =    5.00, Si(x) = 1.549931244945e+000, Ci(x) = -1.900297496582e-001
x =    5.50, Si(x) = 1.468724072665e+000, Ci(x) = -1.420529475530e-001
x =    6.00, Si(x) = 1.424687551281e+000, Ci(x) = -6.805724389473e-002
x =    6.50, Si(x) = 1.421794274436e+000, Ci(x) =  1.110151951341e-002
x =    7.00, Si(x) = 1.454596614248e+000, Ci(x) =  7.669527848069e-002
x =    7.50, Si(x) = 1.510681530943e+000, Ci(x) =  1.156332032364e-001
x =    8.00, Si(x) = 1.574186821707e+000, Ci(x) =  1.224338825305e-001
x =    8.50, Si(x) = 1.629597099590e+000, Ci(x) =  9.943135857196e-002
x =    9.00, Si(x) = 1.665040075829e+000, Ci(x) =  5.534753133162e-002
x =    9.50, Si(x) = 1.674463342281e+000, Ci(x) =  2.678058834155e-003
x =   10.00, Si(x) = 1.658347594218e+000, Ci(x) = -4.545643300602e-002
```

Figure 6-3 Results of sine and cosine integrals.

Laguerre Polynomials

The Laguerre polynomial is the canonical solution of the Laguerre equation:

$$xy'' + (1 - x)y' + ny = 0$$

which is a second-order linear differential equation. This equation has nonsingular solutions only if n is a positive integer.

The polynomials from the canonical solutions of the above equation, usually denoted L_0, L_1, L_2,..., form a polynomial sequence which can be defined by the Rodrigues formula:

$$L_n(x) = \frac{e^x}{n!} \frac{d^n}{dx^n}\left(e^{-x}x^n\right)$$

We can easily derive the first few polynomials from the above equation:

$$L_0(x) = 1$$
$$L_1(x) = -x + 1$$
$$L_2(x) = \frac{1}{2}\left(x^2 - 4x + 2\right)$$

For any $n \geq 1$, the Laguerre polynomials satisfy the following recurrence relation:

$$L_{n+1}(x) = \frac{1}{n+1}\left[(2n+1-x)L_n(x) - nL_{n-1}(x)\right]$$

Implementation

In this section, we will implement a C# method for computing the Laguerre polynomials based on their recurrence relation. Add a new method, *Laguerre*, to the *SpecialFunctions* class. Here is the code listing of this method:

```
public static double Laguerre(double x, int n)
{
    double L0 = 1;
    double L1 = -x + 1;
    double L2 = (x * x - 4 * x + 2) / 2;
    int i = 1;
    if (n < 0)
        return -1;
    if (n == 0)
        return L0;
    else if (n == 1)
        return L1;
    else
    {
        while (i < n)
        {
            L2 = ((2.0 * i + 1.0 - x) * L1 - i * L0) / (i + 1);
            L0 = L1;
            L1 = L2;
            i++;
        }
        return L2;
    }
}
```

Testing the Laguerre Method

We can calculate Laguerre polynomials using the *Laguerre* method. Add a method, *TestLaguerre*, to the *Program.cs* file:

```
static void TestLaguerre()
{
    for (int i = 0; i < 7; i++)
    {
        double x = 1.0 * i - 1.0;
        Console.WriteLine("x = {0,5:n2}, L3(x) = {1,20:e12}",
                x, SpecialFunctions.Laguerre(x, 20));
    }
}
```

Running this example generates the results shown in Figure 6-4.

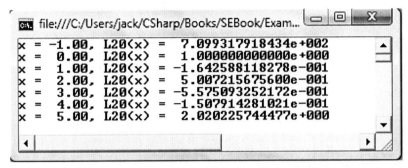

Figure 6-4 Results of the 20th order Laguere polynomial.

Hermite Polynomials

The Hermite polynomials are defined by

$$H_n(x) = (-1)^n e^{x^2} \frac{d^n}{dx^n} e^{-x^2}$$

The first three Hermite polynomials are:

$$H_0(x) = 1$$
$$H_1(x) = 2x$$
$$H_3(x) = 4x^2 - 2$$

The Hermite polynomials satisfy the following recurrence relation:

$$H_{n+1}(x) = 2xH_n(x) - 2nH_{n-1}(x)$$

Implementation

In this section, we will implement a C# method for computing the Hermite polynomials based on their recurrence relation. Add a new method, *Hermite*, to the *SpecialFunctions* class. Here is the code listing for this method:

```
public static double Hermite(double x, int n)
{
    double H0 = 1.0;
```

```
        double H1 = 2 * x;
        double H2 = 4 * x * x - 2;
        int i = 1;
        if (n < 0)
            return -1;
        if (n == 0)
            return H0;
        else if (n == 1)
            return H1;
        else
        {
            while (i < n)
            {
                H2 = 2.0 * x * H1 - 2.0 * i * H0;
                H0 = H1;
                H1 = H2;
                i++;
            }
            return H2;
        }
    }
}
```

Testing the Hermite Method

We can calculate Hermite polynomials using the *Hermite* method. Add a method, *TestHermite*, to the *Program.cs* file:

```
static void TestHermite()
{
    for (int i = 0; i < 7; i++)
    {
        double x = 1.0 * i - 1.0;
        Console.WriteLine("x = {0,5:n2}, H10(x) = {1,20:e12}",
                x, SpecialFunctions.Hermite(x, 10));
    }
}
```

Running this example generates the results shown in Figure 6-5.

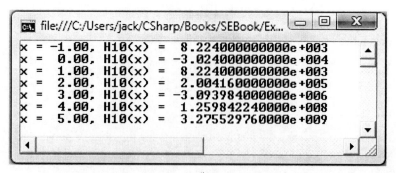

Figure 6-5 Results of the 10th order Hermite polynomial.

Chebyshev Polynomials

Chebyshev polynomials are important in numerical analysis. There are two types of Chebyshev polynomials, both of which are solutions to the Chebyshev different equations

$$(1-x^2)y''-xy'+n^2 y = 0 \quad \text{(first kind)}$$

$$(1-x^2)y''-3xy'+n(n+2)y = 0 \quad \text{(second kind)}$$

The Chebyshev polynomials of the first kind are defined by the recurrence relation

$$T_0(x) = 1$$
$$T_1(x) = x$$
$$T_2(x) = 2x^2 - 1$$
$$T_{n+1}(x) = 2xT_n(x) - T_{n-1}(x)$$

and the Chebyshev polynomials of the second kind are defined by

$$U_0(x) = 1$$
$$U_1(x) = 2x$$
$$U_2(x) = 4x^2 - 1$$
$$U_{n+1}(x) = 2xU_n(x) - U_{n-1}(x)$$

Implementation

In this section, we will implement C# methods for computing the Chebyshev polynomials based on their recurrence relations. Add two new method, *ChebyshevT* (first kind) and *ChebyshevU* (second kind), to the *SpecialFunctions* class. Here is the code listing of these two methods:

```
public static double ChebyshevT(double x, int n)
{
    double T0 = 1.0;
    double T1 = x;
    double T2 = 2 * x * x - 1;
    int i = 1;
    if (n < 0)
        return -1;
    if (n == 0)
        return T0;
    else if (n == 1)
        return T1;
    else
    {
        while (i < n)
        {
            T2 = 2.0 * x * T1 - T0;
            T0 = T1;
            T1 = T2;
            i++;
        }
    }
```

```
            return T2;
        }
    }

    public static double ChebyshevU(double x, int n)
    {
        double U0 = 1.0;
        double U1 = 2 * x;
        double U2 = 4 * x * x - 1;
        int i = 1;
        if (n < 0)
            return -1;
        if (n == 0)
            return U0;
        else if (n == 1)
            return U1;
        else
        {
            while (i < n)
            {
                U2 = 2.0 * x * U1 - U0;
                U0 = U1;
                U1 = U2;
                i++;
            }
            return U2;
        }
    }
```

Testing Chebyshev Methods

We can calculate Chebyshev polynomials using the *ChebyshevT* and *ChebyshevU* methods. Add a method, *TestChebyshev*, to the *Program.cs* file:

```
static void TestChebyshev()
{
    for (int i = 0; i < 11; i++)
    {
        double x = 0.25 * (i - 5.0);
        Console.WriteLine("x = {0:n2}, T15(x) = {1:e12}, U15(x) = {2:e12}",
                x, SpecialFunctions.ChebyshevT(x, 15),
                    SpecialFunctions.ChebyshevU(x, 15));
    }
}
```

Running this example generates the results shown in Figure 6-6.

```
file:///C:/Users/jack/CSharp/Books/SEBook/Examples/SpecialFunctionsTest/SpecialFunct...
x = -1.25, T15(x) = -1.638400001526e+004, U15(x) = -4.369066665649e+004
x = -1.00, T15(x) = -1.000000000000e+000, U15(x) = -1.600000000000e+001
x = -0.75, T15(x) =  1.539459228516e-001, U15(x) =  1.274322509766e+000
x = -0.50, T15(x) =  1.000000000000e+000, U15(x) =  1.000000000000e+000
x = -0.25, T15(x) = -6.040802001953e-001, U15(x) = -8.098449707031e-001
x =  0.00, T15(x) =  0.000000000000e+000, U15(x) =  0.000000000000e+000
x =  0.25, T15(x) =  6.040802001953e-001, U15(x) =  8.098449707031e-001
x =  0.50, T15(x) = -1.000000000000e+000, U15(x) = -1.000000000000e+000
x =  0.75, T15(x) = -1.539459228516e-001, U15(x) = -1.274322509766e+000
x =  1.00, T15(x) =  1.000000000000e+000, U15(x) =  1.600000000000e+001
x =  1.25, T15(x) =  1.638400001526e+004, U15(x) =  4.369066665649e+004
```

Figure 6-6 Results of Chebyshev polynomials.

Legendre Polynomials

Legendre polynomials appear in many different mathematical and physical problems. They can be defined by a generating function:

$$\frac{1}{\sqrt{1-2xt+t^2}} = \sum_{n=0}^{\infty} P_n(x)t^n$$

Where $P_n(x)$ are the Legendre polynomials. These polynomials have the following recurrence relation:

$$P_0(x) = 1$$
$$P_1(x) = x$$
$$P_2(x) = (3x^2 - 1)/2$$
$$P_{n+1}(x) = 2xP_n(x) - P_{n-1}(x) - [xP_n(x) - P_{n-1}(x)]/(n+1)$$

Implementation

In this section, we will implement a C# method for computing the Legendre polynomials based on their recurrence relation. Add a new method, *Legendre*, to the *SpecialFunctions* class. Here is the code listing of this method:

```csharp
public static double Legendre(double x, int n)
{
    double P0 = 1.0;
    double P1 = x;
    double P2 = (3.0 * x * x - 1) / 2.0;
    int i = 1;
    if (n < 0)
        return -1;
    if (n == 0)
        return P0;
    else if (n == 1)
        return P1;
    else
    {
```

```
        while (i < n)
        {
            P2 = 2.0 * x * P1 - P0 - (x * P1 - P0) / (i + 1);
            P0 = P1;
            P1 = P2;
            i++;
        }
        return P2;
    }
}
```

Testing the Legendre Method

We can calculate Legendre polynomials using the *Legendre* method. Add a method, *TestLegendre*, to the *Program.cs* file:

```
static void TestLegendre()
{
    for (int i = 0; i < 9; i++)
    {
        double x = 0.25 * (i - 4.0);
        Console.WriteLine("x = {0,5:n2}, P10(x) = {1,20:e12}",
                x, SpecialFunctions.Legendre(x, 10));
    }
}
```

Running this example generates the results shown in Figure 6-7.

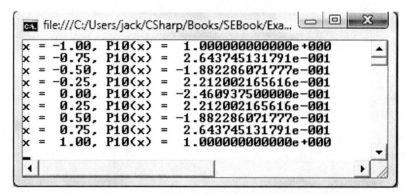

Figure 6-7 Results of the 10th order Legendre polynomial.

Bessel Functions

Bessel functions are solutions of the following Bessel's different equation

$$x^2 \frac{d^2 y}{dx^2} + x \frac{dy}{dx} + (x^2 - v^2)y = 0$$

Here v is the order of the Bessel function and can be any real or complex number. Since the above equation is a second-order differential equation, there must be two linearly independent solutions.

The Bessel function of the first kind, denoted as $J_v(x)$, is a solution of Bessel's differential equation. It is finite at $x = 0$ for a non-negative v, and diverges as x approaches zero for a negative non-integer v. This function can be expanded using the Taylor series:

$$J_v(x) = \sum_{n=0}^{\infty} \frac{(-1)^n}{n!\,\Gamma(n+v+1)} \left(\frac{x}{2}\right)^{2n+v}$$

The Bessel functions of the second kind, denoted by $Y_v(x)$, are solutions of the Bessel differential equation. They are singular at $x = 0$. For a non-integer v, these functions are related to the Bessel functions of the first kind by:

$$Y_v(x) = \frac{J_v(x)\cos v\pi - J_{-v}(x)}{\sin v\pi}$$

Implementation

In this section, we will implement two C# methods for computing the Bessel functions of the first and second kinds. Add two new methods, *BesselJ* and *BesselY*, to the *SpecialFunctions* class. Here is the code listing for these two methods:

```
public static double BesselJ(double x, double v)
{
    double sum = 0.0;
    double term = 0.0;
    int i = 0;
    do
    {
        term = Math.Pow(-1, i) * Math.Pow(0.5 * x, 2 * i + v) /
            Gamma(i + 1) / Gamma(i + v + 1);
        sum += term;
        i++;
    }
    while (Math.Abs(term) > 1.0e-12);
    return sum;
}

public static double BesselY(double x, double v)
{
    return (BesselJ(x, v) * Math.Cos(v * Math.PI) - BesselJ(x, -v)) /
        Math.Sin(v * Math.PI);
}
```

Note that the *BesselY* method can only be used to compute the Bessel functions of second kind for a non-integer v. If you want to use this method for the integer v case, you can set a v value which is very close to that integer. For example, you can use $v = 1 + 1.0e-5$ to approximately calculate $Y_1(x)$.

Testing the Bessel Method

I will show you how to calculate the zero-order Bessel functions using the *BesselJ* and *BesselY* methods. Add a method, *TestBessel*, to the *Program.cs* file:

```
static void TestBessel()
{
    for (int i = 1; i < 21; i++)
    {
        double x = 1.0 * i;
        Console.WriteLine("x = {0,5:n0}, J0(x) = {1,20:e10},
                          Y0(x) = {2,20:e10}",
                          x, SpecialFunctions.BesselJ(x, 0),
                          SpecialFunctions.BesselY(x, 0.0+1.0e-5));
    }
}
```

Running this example generates the results shown in Figure 6-8.

```
file:///C:/Users/jack/CSharp/Books/SEBook/Examples/SpecialFunctionsTest/Spec...

x =  1, J0(x) =    7.6519768656e-001, Y0(x) =    8.8244944499e-002
x =  2, J0(x) =    2.2389077914e-001, Y0(x) =    5.1037215569e-001
x =  3, J0(x) =   -2.6005195490e-001, Y0(x) =    3.7685409478e-001
x =  4, J0(x) =   -3.9714980986e-001, Y0(x) =   -1.6934501098e-002
x =  5, J0(x) =   -1.7759677131e-001, Y0(x) =   -3.0851483579e-001
x =  6, J0(x) =    1.5064525725e-001, Y0(x) =   -2.8819705089e-001
x =  7, J0(x) =    3.0007927052e-001, Y0(x) =   -2.5954458406e-002
x =  8, J0(x) =    1.7165080714e-001, Y0(x) =    2.2351879471e-001
x =  9, J0(x) =   -9.0333611182e-002, Y0(x) =    2.4993812454e-001
x = 10, J0(x) =   -2.4593576445e-001, Y0(x) =    5.5675072044e-002
x = 11, J0(x) =   -1.7119030040e-001, Y0(x) =   -1.6884451791e-001
x = 12, J0(x) =    4.7689310810e-002, Y0(x) =   -2.2523773168e-001
x = 13, J0(x) =    2.0692610241e-001, Y0(x) =   -7.8210301452e-002
x = 14, J0(x) =    1.7107347622e-001, Y0(x) =    1.2719164862e-001
x = 15, J0(x) =   -1.4224472521e-002, Y0(x) =    2.0546850566e-001
x = 16, J0(x) =   -1.7489907303e-001, Y0(x) =    9.5821306408e-002
x = 17, J0(x) =   -1.6985424935e-001, Y0(x) =   -9.2614119609e-002
x = 18, J0(x) =   -1.3355798484e-002, Y0(x) =   -1.8748851565e-001
x = 19, J0(x) =    1.4662945940e-001, Y0(x) =   -1.0936110749e-001
x = 20, J0(x) =    1.6702471259e-001, Y0(x) =    6.3193771290e-002
```

Figure 6-8 Results of the zero-order Bessel functions.

Chapter 7
Random Numbers and Distribution Functions

Random numbers are used to simulate different chaotic circumstances that can be found in the real world. C# provides a system for generating random numbers. In this chapter we will discuss a variety of random number generators and some useful probability distribution functions.

Built-in Random Number Generators

Mathematical definitions of randomness use notions of information content, non-computability, and stochasticity. The various definitions, however, do not always agree on which sequences are random and which are not. Practical methods for generating random numbers from specific distributions usually start with uniform random numbers. Once you have a uniform random number generator, you can create random numbers from different distributions.

C# has implemented a uniform random number generator in the *System.Random* class. In order to create a random number using this built-in generator, you need to create an instance, like this:

```
System.Random rand = new System.Random(0);
```

The integer passed into the constructor is called a seed value. The seed value determines where to start generating random numbers. If you give a seed to a random number generator and pull off numbers from it, and then give the same starting seed to a totally different generator, then the random numbers generated from these two generators will be identical. For example:

```
System.Random rand1 = new System.Random(0);
int x = rand1.Next();      // Should hold 1559595546
System.Random.rand2 = new System.Random(0);
int y = rand2.Next();      // Should hold 1559595546
```

Whenever you want a generator to create the same sequence of numbers, you should give it a common seed.

However, you are not required to provide a seed. In fact, most of the time you do not want to set a hard seed because that will make you numbers seem less random, as the same sequence of random numbers will be generated every time you use the same seed.

Instead, you can create a random generator like this:

```
System.Random rand = new System.Random();
```

This way, the system automatically uses a seed based on the current time, so you do not have to worry about it. C# provides several other ways to generate random numbers with a bit more meaning to them. For example, you can generate numbers from 0 to a specified amount:

```
int x = rand.Next(100);
```

This generates random numbers between 0 and 100. There is another version of the method as well, which allows you to specify both the lower and upper bounds:

```
x = rand.Next (20, 100);
```

This time, you are generating random numbers from 20 to 100.

There are two other ways you can generate random numbers using the *Random* class: one way is to generate double-precision floating-point numbers, and the other way is to generate arrays of random bytes. Both are fairly straightforward:

```
double d = rand.NextDouble();
```

This generates floating-point numbers between the values of 0.0 and 1.0.

And finally, you can fill an array with random bytes:

```
byte[] b = new byte[5];
rand.NextBytes(b);
```

Here, you learned how to use the C# built-in generator to create random numbers with a uniform probability distribution. In next few sections, I will show you how to generate random numbers with a prescribed probability distribution function.

Normal Distribution

The normal distribution, also called Gaussian distribution, is a probability distribution of great importance in many fields.

Probability Density Function

There are various ways to characterize a probability distribution. The commonly used method is the probability density function. The normal distribution is a two-parameter family of curves. The first parameter, μ, is the mean, and the second parameter, σ, is the standard deviation. The probability density function for the normal distribution is defined by the following formula:

$$f(x;\mu,\sigma) = \frac{1}{\sigma\sqrt{2\pi}} e^{-\frac{(x-\mu)^2}{2\sigma^2}}$$

The standard normal distribution is the special case with $\mu = 0$ and $\sigma = 1$.

Start with a new C# Console application and name it *DistributionFunctionsTest*. Add a new class named *DistributionFunctions* to the current project. Then add the following code to the class:

```
using System;

namespace XuMath
{
    public static class DistributionFunctions
    {
        public static double Normal(double x, double mu, double sigma)
        {
            double x1, x2;
            x1 = 1 / sigma / Math.Sqrt(2 * Math.PI);
            x2 = (x - mu) * (x - mu) / (2 * sigma * sigma);
            return x1 * Math.Exp(-x2);
        }

        public static double[] Normal(double[] x, double mu, double sigma)
        {
            double[] y = new double[x.Length];
            for (int i = 0; i < x.Length; i++)
            {
                y[i] = NormalFunction(x[i], mu, sigma);
            }
            return y;
        }
    }
}
```

In this class, we implement an overloaded method, *Normal*. For a given set of parameters, μ and σ, this function returns a single double number or a double array, depending on the input variable *x*. You can then use this method to calculate the probability density function for a normal distribution.

Normal Random Number Generator

A normal random distribution can be generated using the C# built-in uniform random generator. Here, we will use a polar algorithm for the normal random generator. This algorithm generates two random values at a time. It involves finding a random point in the unit circle by generating uniformly distributed points in a $[-1, 1] \times [-1, 1]$ square and rejecting any points outside the circle.

The algorithm of the polar method is listed below:

- Generate two random numbers *v1* and *v2*.

- Let $v1 = 2 * v1 - 1$, $v2 = 2 * v2 - 1$, and $v12 = v1 * v1 + v2 * v2$.

- If $v12 > 1$, regenerate *v1* and *v2*.

- Generate two independent standard normal random numbers:

$$x1 = \sqrt{\frac{-2\log v12}{v12}} v1 , \quad x2 = \sqrt{\frac{-2\log v12}{v12}} v2$$

- Return $x1 * \sigma + \mu$.

Now add a new public class, *RandomGenerators*, to the current project. Then, implement a public method, *NextNormal*, to this newly created class:

```
using System;

namespace XuMath
{
    public static class RandomGenerators
    {
        private static Random rand = new Random();

        public static double NextNormal(double mu, double sigma)
        {
            double v1 = 0.0, v2 = 0.0, v12 = 0.0, y = 0.0;
            while (v12 >= 1.0 || v12 == 0.0)
            {
                v1 = 2.0 * rand.NextDouble() - 1.0;
                v2 = 2.0 * rand.NextDouble() - 1.0;
                v12 = v1 * v1 + v2 * v2;
            }
            y = Math.Sqrt(-2.0 * Math.Log(v12) / v12);

            return v1 * y * sigma + mu;
        }
    }
}
```

In this class, we first declare a private field member, *rand*, which is used to generate uniformly distributed random numbers. You can use the *NextNormal* method to create a normal random distribution by specifying input parameters, σ and μ. For example, the following code snippet creates a standard normal distribution with $\sigma = 1$ and $\mu = 0$:

```
double y = RandomGenerators.NextNormal(0, 1);
```

You can also add the following overloaded *NextNormal* method to the class, which is used to generate a normal random distribution array:

```
public static double[] NextNormal(double mu, double sigma, int nLength)
{
    double[] array = new double[nLength];
    for (int i = 0; i < nLength; i++)
    {
        array[i] = NextNormal(mu, sigma);
    }
    return array;
}
```

Here you need to specify three input parameters μ, σ, and the length of the array.

Testing the Normal Distribution

In this section, we will examine the normal distribution and the normal random number generator created in the previous section. This testing program also shows you how to use the random generators in your C# applications. In order to use the normal random generator, we need to add a few utility methods to the *RandomGenerators* class:

```
public static ArrayList HistogramData(double[] data,
                        double min, double max, int nBins)
{
    ArrayList aList = new ArrayList();
    double dataSpacing = (max - min) / nBins;
    for (int i = 1; i < nBins + 1; i++)
    {
        int nCounts = 0;
        for (int j = 0; j < data.Length; j++)
        {
            if (data[j] >= min + (i - 1) * dataSpacing &&
                        data[j] < min + i * dataSpacing)
            {
                nCounts++;
            }
        }
        aList.Add((double)nCounts);
    }
    return aList;
}

public static ArrayList HistogramData(double[] data, int nBins)
{
    ArrayList aList = new ArrayList();
    for (int i = 0; i < nBins + 1; i++)
    {
        int nCounts = 0;
        for (int j = 0; j < data.Length; j++)
        {
            if (data[j] == i)
                nCounts++;
        }
        aList.Add((double)nCounts);
    }
    return aList;
}

public static double ArrayMax(double[] array)
{
    double max = array[0];
    for (int i = 1; i < array.Length; i++)
    {
        max = Math.Max(max, array[i]);
    }
    return max;
}
public static double ArrayMin(double[] array)
{
    double min = array[0];
    for (int i = 1; i < array.Length; i++)
    {
        min = Math.Min(min, array[i]);
    }
    return min;
}
```

The overloaded method, *HistogramData*, is used to create histogram data from a randomly generated data array. The histogram data is constructed by segmenting the range of the data into equal-sized bins. The vertical axis of the histogram is the number of counts for each bin, and the horizontal axis of the histogram is labeled with the range of the response variable. There are two overloaded *HistogramData* methods: one for continuous data and the other for discrete data.

Two other methods, *ArrayMax* and *ArrayMin*, are used to compute the maximum and minimum values of a double array.

The best approach to examine the normal distribution is to display the results graphically on a chart. The creation of C# chart applications is beyond the scope of this book. If you are interested in C# chart and graphics applications, you can read my other book – "*Practical C# Charts and Graphics*". Here, I will show the result using a C# Console application and plot the results through other charting program, such as Microsoft Excel and Matlab.

Add a *TestNormal* method to the *Program.cs* file:

```
using System;
using System.Collections;
using XuMath;

namespace DistributionFunctionsTest
{
    class Program
    {
        static void Main(string[] args)
        {
            TestNormal();
            Console.ReadLine();
        }

        static void TestNormal()
        {
            int nBins = 20;
            int nPoints = 1000;
            double xmin = -1;
            double xmax = 5;

            double[] rand = RandomGenerators.NextNormal(2.0, 1.0, nPoints);
            ArrayList aList =
                    RandomGenerators.HistogramData(rand, xmin, xmax, nBins);
            double[] xdata = new double[nBins];
            double[] ydata = new double[nBins];

            double[] ydistribution = new double[nBins];
            for (int i = 0; i < nBins; i++)
            {
                xdata[i] = xmin + (i + 0.5) * (xmax - xmin) / nBins;
                ydata[i] = (double)aList[i];
                ydistribution[i] =
                    DistributionFunctions.Normal(xdata[i], 2.0, 1.0);
            }
            double normalizeFactor = RandomGenerators.ArrayMax(ydata) /
                            RandomGenerators.ArrayMax(ydistribution);
            Console.WriteLine("");
```

```
        for (int i = 0; i < nBins; i++)
        {
            Console.WriteLine("x = {0,4:n1}, Normal random data = {1,3:n0},
                    Normal distribution = {2,3:n0}",
                    xdata[i], ydata[i],
                    Math.Round(ydistribution[i] * normalizeFactor, 0));
        }
    }
}
}
```

In the *TestNormal* method, we first set two parameters using the code snippet:

```
    int nBins = 20;
    int nPoints = 1000;
```

where the parameter *nBins* is the number of bins in the histogram and *nPoints* is the number of random data points. We then create a random array with a normal distribution. Finally we compare the histogram of the random data and the theoretical probability density function of the normal distribution. Note that the probability density function is normalized so that its maximum overlaps with the maximum of the normal random distribution, which allows us to compare them directly. The parameter *nPoints*, the number of random data points, cannot be set too small. Otherwise, the random distribution will deviate dramatically from the theoretical probability function.

Running this example generates the results shown in Figure 7-1. These results are also plotted in Figure 7-2. You can see that the results from the normal random generator are very close to the theoretical normal distribution function.

Note that the random distribution data may be different for different runs due to their random nature.

Exponential Distribution

Exponential distribution is used to model events that occur randomly over time. In particular, this distribution is very powerful for studying lifetimes.

Probability Density Function

The probability density function of the exponential distribution can be presented in the following formula:

$$f(x;\alpha) = \alpha\, e^{-\alpha x}$$

where $\alpha > 0$ is a parameter of the distribution, often called the rate parameter. This distribution is defined in the range of $0 \leq x < \infty$.

```
file:///C:/Users/jack/CSharp/Books/SEBook/Examples/DistributionFunctionsTe...

x = -0.9,  Normal  random  data  =    1,  Normal  distribution  =    2
x = -0.6,  Normal  random  data  =    4,  Normal  distribution  =    5
x = -0.3,  Normal  random  data  =    8,  Normal  distribution  =   10
x =  0.1,  Normal  random  data  =   15,  Normal  distribution  =   18
x =  0.4,  Normal  random  data  =   31,  Normal  distribution  =   31
x =  0.7,  Normal  random  data  =   47,  Normal  distribution  =   49
x =  1.0,  Normal  random  data  =   62,  Normal  distribution  =   71
x =  1.3,  Normal  random  data  =   82,  Normal  distribution  =   92
x =  1.6,  Normal  random  data  =  121,  Normal  distribution  =  111
x =  1.9,  Normal  random  data  =  119,  Normal  distribution  =  121
x =  2.2,  Normal  random  data  =  121,  Normal  distribution  =  121
x =  2.5,  Normal  random  data  =  103,  Normal  distribution  =  111
x =  2.8,  Normal  random  data  =   84,  Normal  distribution  =   92
x =  3.1,  Normal  random  data  =   66,  Normal  distribution  =   71
x =  3.4,  Normal  random  data  =   58,  Normal  distribution  =   49
x =  3.7,  Normal  random  data  =   34,  Normal  distribution  =   31
x =  4.0,  Normal  random  data  =   18,  Normal  distribution  =   18
x =  4.3,  Normal  random  data  =   10,  Normal  distribution  =   10
x =  4.6,  Normal  random  data  =   11,  Normal  distribution  =    5
x =  4.9,  Normal  random  data  =    1,  Normal  distribution  =    2
```

Figure 7-1 Results of the normal distributions.

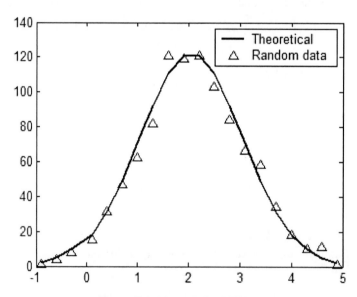

Figure 7-2 Normal distributions.

Add an overloaded method, *Exponential*, to the *DistributionFunctions* class:

```
public static double ExponentialFunction(double x, double alpha)
{
    return alpha * Math.Exp(-alpha * x);
}

public static double[] ExponentialFunction(double[] x, double alpha)
{
    double[] y = new double[x.Length];
    for (int i = 0; i < x.Length; i++)
    {
        y[i] = ExponentialFunction(x[i], alpha);
    }
    return y;
}
```

These two overloaded methods are used to create a single value and a value array, respectively.

Exponential Random Number Generator

The random numbers from the exponential variate Exp(α) can be created using the C# built-in uniformly distributed random numbers of the unit rectangular variate $u(0, 1)$ with the relationship:

$$\exp(\alpha) \sim -\frac{1}{\alpha}\ln u(0,1)$$

Now add a new public method, *NextExponential*, to the *RandomGenerators* class:

```
public static double NextExponential(double alpha)
{
    if (alpha <= 0.0)
    {
        throw new ArgumentOutOfRangeException(
        "alpha", alpha, "alpha must be > zero!");
    }
    return -Math.Log(rand.NextDouble()) / alpha;
}

public static double[] NextExponential(double alpha, int nLength)
{
    double[] array = new double[nLength];
    for (int i = 0; i < nLength; i++)
    {
        array[i] = NextExponential(alpha);
    }
    return array;
}
```

You can use these two overloaded methods to generate exponential random distributions. For example, you can create an exponentially distributed random array with 2000 data points simply by using the following code snippet:

```
double[] y = RandomGenerators.NextExponential(0.5, 2000);
```

Testing the Exponential Distribution

Here, we will examine the exponential distribution. Add a *TestExponential* method to the *Program.cs* file:

```
static void TestExponential()
{
    int nBins = 20;
    int nPoints = 2000;
    double xmin = 0;
    double xmax = 5;

    double[] rand = RandomGenerators.NextExponential(1.5, nPoints);
    ArrayList aList = RandomGenerators.HistogramData(rand, xmin, xmax, nBins);
    double[] xdata = new double[nBins];
    double[] ydata = new double[nBins];

    double[] ydistribution = new double[nBins];
    for (int i = 0; i < nBins; i++)
    {
        xdata[i] = xmin + (i + 0.5) * (xmax - xmin) / nBins;
        ydata[i] = (double)aList[i];
        ydistribution[i] = DistributionFunctions.Exponential(xdata[i], 1.5);
    }
    double normalizeFactor = RandomGenerators.ArrayMax(ydata) /
            RandomGenerators.ArrayMax(ydistribution);
    Console.WriteLine("");
    for (int i = 0; i < nBins; i++)
    {
        Console.WriteLine(" x = {0,4:n1}, Random data = {1,3:n0},
                            density distribution = {2,3:n0}",
                            xdata[i], ydata[i],
                            Math.Round(ydistribution[i] * normalizeFactor, 0));
    }
}
```

Here, we create a random array using the exponential distribution, and then compare the histogram of the random data with the theoretical probability density function of the exponential distribution. Note that the probability density function is normalized so that its maximum overlaps with the maximum of the random distribution, which allows us to compare them directly. Again, the number of random data points cannot be set too small. Otherwise, the random distribution will deviate dramatically from the theoretical probability function.

Running this example generates the results shown in Figure 7-3. The results are also plotted in Figure 7-4. You can see that the results from the exponential random generator are very close to the theoretical exponential distribution function.

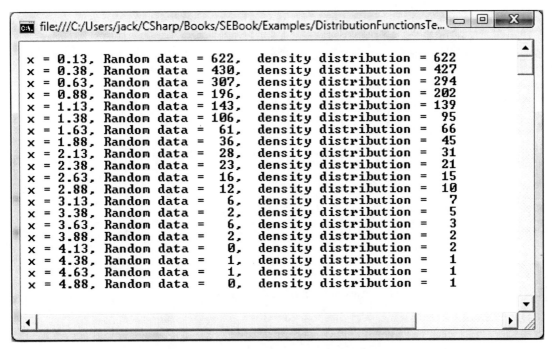

```
file:///C:/Users/jack/CSharp/Books/SEBook/Examples/DistributionFunctionsTe...

x = 0.13,  Random data =  622,   density distribution =  622
x = 0.38,  Random data =  430,   density distribution =  427
x = 0.63,  Random data =  307,   density distribution =  294
x = 0.88,  Random data =  196,   density distribution =  202
x = 1.13,  Random data =  143,   density distribution =  139
x = 1.38,  Random data =  106,   density distribution =   95
x = 1.63,  Random data =   61,   density distribution =   66
x = 1.88,  Random data =   36,   density distribution =   45
x = 2.13,  Random data =   28,   density distribution =   31
x = 2.38,  Random data =   23,   density distribution =   21
x = 2.63,  Random data =   16,   density distribution =   15
x = 2.88,  Random data =   12,   density distribution =   10
x = 3.13,  Random data =    6,   density distribution =    7
x = 3.38,  Random data =    2,   density distribution =    5
x = 3.63,  Random data =    6,   density distribution =    3
x = 3.88,  Random data =    2,   density distribution =    2
x = 4.13,  Random data =    0,   density distribution =    2
x = 4.38,  Random data =    1,   density distribution =    1
x = 4.63,  Random data =    1,   density distribution =    1
x = 4.88,  Random data =    0,   density distribution =    1
```

Figure 7-3 Results of the exponential distribution.

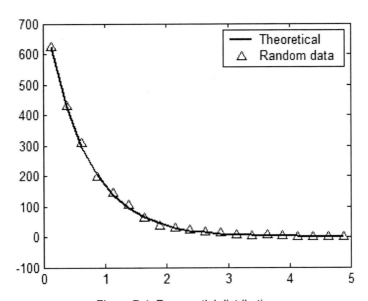

Figure 7-4 Exponential distributions.

Chi and Chi-Square Distributions

Chi and Chi-square distributions are continuous probability distributions. The distribution usually arises when a n-dimensional vector's orthogonal components are independent and each follows a standard normal distribution. They can be created using the normal distribution.

Probability Density Function

The Chi (or χ) probability density function is expressed in terms of the formula:

$$f(x; n, \sigma) = \frac{2\left(\dfrac{n}{2}\right)^{n/2} x^{n-1} e^{-\frac{n}{2\sigma^2}x^2}}{\Gamma\left(\dfrac{n}{2}\right)\sigma^n}$$

where Γ is the gamma function.

The Chi square (or χ^2) probability density function is defined by

$$f(x; n, \sigma) = \frac{x^{n/2-1} e^{-\frac{n}{2\sigma^2}x^2}}{2^{n/2}\Gamma\left(\dfrac{n}{2}\right)\sigma^n}$$

Since the Chi and Chi-square distributions contain the gamma function, we need to add the *SpecialFunction.cs* from the previous *SpecialFunctionsTest* project in Chapter 6 to the current project. We can then implement the Chi and Chi-square functions in C# by adding the following methods to the *DistributionFunctions* class:

```
public static double Chi(double x, int n)
{
    double gamma = SpecialFunctions.Gamma(n / 2.0);
    double exp = Math.Exp(-n * x * x / 2.0);
    return 2.0 * Math.Pow(n / 2.0, n / 2) * Math.Pow(x, n - 1) *
            exp / Math.Pow(2, n / 2) / gamma;
}

public static double[] Chi(double[] x, int n)
{
    double[] y = new double[x.Length];
    for (int i = 0; i < x.Length; i++)
    {
        y[i] = Chi(x[i], n);
    }
    return y;
}

public static double Chi(double x, int n, double sigma)
{
    double gamma = SpecialFunctions.Gamma(n / 2.0);
    double exp = Math.Exp(-n*x*x / 2.0 / sigma / sigma);
```

```
        return 2.0*Math.Pow(n/2.0, n / 2)*Math.Pow(x,n-1) *
                   exp / Math.Pow(2, n / 2) / gamma / Math.Pow(sigma, n);
}

public static double[] Chi(double[] x, int n, double sigma)
{
    double[] y = new double[x.Length];
    for (int i = 0; i < x.Length; i++)
    {
        y[i] = Chi (x[i], n, sigma);
    }
    return y;
}

public static double ChiSquare(double x, int n)
{
    double gamma = SpecialFunctions.Gamma(n / 2.0);
    double exp = Math.Exp(-x / 2.0);
    return Math.Pow(x, n / 2 - 1) * exp / Math.Pow(2, n / 2) / gamma;
}

public static double[] ChiSquare(double[] x, int n)
{
    double[] y = new double[x.Length];
    for (int i = 0; i < x.Length; i++)
    {
        y[i] = ChiSquare(x[i], n);
    }
    return y;
}

public static double ChiSquare(double x, int n, double sigma)
{
    double gamma = SpecialFunctions.Gamma(n / 2.0);
    double exp = Math.Exp(-x / 2.0 / sigma / sigma);
    return Math.Pow(x, n / 2 - 1) * exp / Math.Pow(2, n / 2) /
               gamma / Math.Pow(sigma, n);
}

public static double[] ChiSquare(double[] x, int n, double sigma)
{
    double[] y = new double[x.Length];
    for (int i = 0; i < x.Length; i++)
    {
        y[i] = ChiSquare(x[i], n, sigma);
    }
    return y;
}
```

Here, we implement four overloaded methods for each function. These methods are used to compute the Chi or Chi square function with single (corresponding to $\sigma = 1$) or double parameters. These methods can return a double number or a double array, depending on the input variable x.

Chi and Chi-Square Random Number Generators

The Chi square random distribution with parameters n and σ is equal to the sum of squares of n independent normal random distribution:

$$\chi^2(n,\sigma) = \sum_{i=1}^{n} \left[N_i(0,\sigma^2) \right]^2$$

The Chi random distribution can be calculated from the above Chi square random distribution:

$$\chi(n,\sigma) = \sqrt{\chi^2(n,\sigma)/n}$$

Using the above equations, we can easily create Chi and Chi square random distributions in C# by adding the following methods to the *RandomGenerators* class:

```
public static double NextChi(int n, double sigma)
{
    return Math.Sqrt(NextChiSquare(n, sigma) / n);
}

public static double[] NextChi(int n, double sigma, int nLength)
{
    double[] array = new double[nLength];
    for (int i = 0; i < nLength; i++)
    {
        array[i] = NextChi(n, sigma);
    }
    return array;
}

public static double NextChi(int n)
{
    return Math.Sqrt(NextChiSquare(n) / n);
}

public static double[] NextChi(int n, int nLength)
{
    double[] array = new double[nLength];
    for (int i = 0; i < nLength; i++)
    {
        array[i] = NextChi(n);
    }
    return array;
}

public static double NextChiSquare(int n, double sigma)
{
    if (n <= 0)
    {
        throw new ArgumentOutOfRangeException("n", n, "n must be positive!");
    }
    if (sigma <= 0.0)
    {
        throw new ArgumentOutOfRangeException(
```

```
                    "sigma", sigma, "sigma must be positive!");
    }

    double result = 0.0;
    for (int i = 0; i < n; i++)
    {
        result += Math.Pow(NextNormal(0, sigma * sigma), 2);
    }
    return result;
}

public static double[] NextChiSquare(int n, double sigma, int nLength)
{
    double[] array = new double[nLength];
    for (int i = 0; i < nLength; i++)
    {
        array[i] = NextChiSquare(n, sigma);
    }
    return array;
}

public static double NextChiSquare(int n)
{
    if (n <= 0)
    {
        throw new ArgumentOutOfRangeException("n", n, "n must be positive!");
    }

    double result = 0.0;
    for (int i = 0; i < n; i++)
    {
        result += Math.Pow(NextNormal(0, 1), 2);
    }
    return result;
}

public static double[] NextChiSquare(int n, int nLength)
{
    double[] array = new double[nLength];
    for (int i = 0; i < nLength; i++)
    {
        array[i] = NextChiSquare(n);
    }
        return array;
}
```

Here we implement the one-parameter and two-parameter Chi and Chi-square distributions. These overloaded methods can return a single double number or a double array, depending on whether the parameter *nLength* is specified or not.

Testing the Chi and Chi-Square Distributions

Here, we will examine the Chi and Chi-square distributions. Add a *TestChi* method to the *Program.cs* file:

```
static void TestChi()
{
    int nBins = 20;
    int nPoints = 2000;
    double xmin = 0;
    double xmax = 4;

    double[] rand1 = RandomGenerators.NextChi(2, 1.0, nPoints);
    double[] rand2 = RandomGenerators.NextChiSquare(2, 1.0, nPoints);
    ArrayList aList1 = RandomGenerators.HistogramData(rand1, xmin, xmax, nBins);
    ArrayList aList2 = RandomGenerators.HistogramData(rand2, xmin, xmax, nBins);
    double[] xdata = new double[nBins];
    double[] ydata1 = new double[nBins];
    double[] ydata2 = new double[nBins];

    double[] ychi = new double[nBins];
    double[] ychisquare = new double[nBins];
    for (int i = 0; i < nBins; i++)
    {
        xdata[i] = xmin + (i + 0.5) * (xmax - xmin) / nBins;
        ydata1[i] = (double)aList1[i];
        ydata2[i] = (double)aList2[i];
        ychi[i] = DistributionFunctions.Chi(xdata[i], 2, 1.0);
        ychisquare[i] = DistributionFunctions.ChiSquare(xdata[i], 2, 1.0);
    }
    double normalizeFactor1 = RandomGenerators.ArrayMax(ydata1) /
                              RandomGenerators.ArrayMax(ychi);
    double normalizeFactor2 = RandomGenerators.ArrayMax(ydata2) /
                              RandomGenerators.ArrayMax(ychisquare);
    Console.WriteLine("\n Chi distribution");
    for (int i = 0; i < nBins; i++)
    {
        Console.WriteLine(" x = {0:n1}, Random data = {1:n0},
                          Density distribution = {2:n0}",
                          xdata[i], ydata1[i],
                          Math.Round(ychi[i] * normalizeFactor1, 0));

    }

    Console.WriteLine("\n Chi-square distribution");
    for (int i = 0; i < nBins; i++)
    {
        Console.WriteLine(" x = {0:n1}, Random data = {1:n0},
                          Density distribution = {2:n0}",
                          xdata[i], ydata2[i],
                          Math.Round(ychisquare[i] * normalizeFactor2, 0));
    }
}
```

Here, we create two random arrays using the Chi and Chi-square distributions with parameters $n = 2$ and $\sigma = 1$, and then compare the histogram of the random data with the theoretical probability density functions of the Chi and Chi-square distributions. Note that the probability density function is normalized so that its maximum overlaps with the maximum of the corresponding random distribution, which allows us to compare them directly. Again, the number of random data points cannot be set too small. Otherwise, the random distribution will deviate dramatically from the theoretical probability function.

Running this example generates the results shown in Figure 7-5. These results are also plotted in Figure 7-6. You can see that the results from the random generators are very close to the theoretical distribution functions.

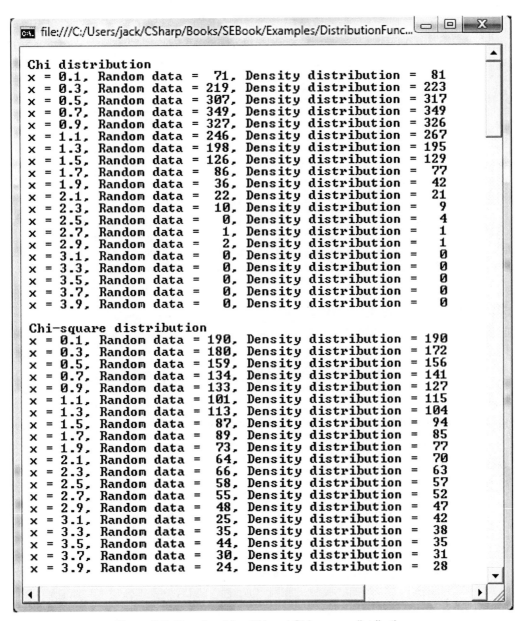

```
file:///C:/Users/jack/CSharp/Books/SEBook/Examples/DistributionFunc...

Chi distribution
x = 0.1, Random data =  71, Density distribution =  81
x = 0.3, Random data = 219, Density distribution = 223
x = 0.5, Random data = 307, Density distribution = 317
x = 0.7, Random data = 349, Density distribution = 349
x = 0.9, Random data = 327, Density distribution = 326
x = 1.1, Random data = 246, Density distribution = 267
x = 1.3, Random data = 198, Density distribution = 195
x = 1.5, Random data = 126, Density distribution = 129
x = 1.7, Random data =  86, Density distribution =  77
x = 1.9, Random data =  36, Density distribution =  42
x = 2.1, Random data =  22, Density distribution =  21
x = 2.3, Random data =  10, Density distribution =   9
x = 2.5, Random data =   0, Density distribution =   4
x = 2.7, Random data =   1, Density distribution =   1
x = 2.9, Random data =   2, Density distribution =   1
x = 3.1, Random data =   0, Density distribution =   0
x = 3.3, Random data =   0, Density distribution =   0
x = 3.5, Random data =   0, Density distribution =   0
x = 3.7, Random data =   0, Density distribution =   0
x = 3.9, Random data =   0, Density distribution =   0

Chi-square distribution
x = 0.1, Random data = 190, Density distribution = 190
x = 0.3, Random data = 180, Density distribution = 172
x = 0.5, Random data = 159, Density distribution = 156
x = 0.7, Random data = 134, Density distribution = 141
x = 0.9, Random data = 133, Density distribution = 127
x = 1.1, Random data = 101, Density distribution = 115
x = 1.3, Random data = 113, Density distribution = 104
x = 1.5, Random data =  87, Density distribution =  94
x = 1.7, Random data =  89, Density distribution =  85
x = 1.9, Random data =  73, Density distribution =  77
x = 2.1, Random data =  64, Density distribution =  70
x = 2.3, Random data =  66, Density distribution =  63
x = 2.5, Random data =  58, Density distribution =  57
x = 2.7, Random data =  55, Density distribution =  52
x = 2.9, Random data =  48, Density distribution =  47
x = 3.1, Random data =  25, Density distribution =  42
x = 3.3, Random data =  35, Density distribution =  38
x = 3.5, Random data =  44, Density distribution =  35
x = 3.7, Random data =  30, Density distribution =  31
x = 3.9, Random data =  24, Density distribution =  28
```

Figure 7-5 Results of the Chi and Chi-square distributions.

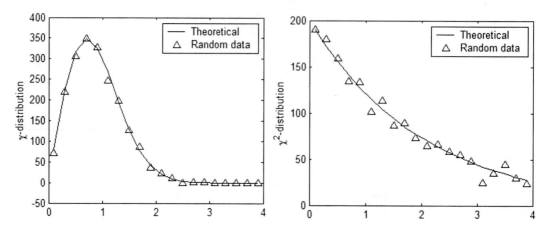

Figure 7-6 Chi and Chi-square distributions.

Cauchy Distribution

The Cauchy distribution, also called the Lorentzian distribution, is also a continuous distribution describing resonance behavior. This distribution pays an important role in physics because it is the solution to the differential equation describing forced resonance. In spectroscopy, it describes the line shape of spectral lines which are broadened by many mechanisms, including resonance broadening.

Probability Density Function

The probability density function for the Cauchy distribution can be written in the form:

$$f(x; a, b) = \frac{b}{\pi[b^2 + (x-a)]^2}$$

Here, a is the location parameter, specifying the location of the resonance peak of the distribution, and b is the scale parameter which specifies the half width at the half maximum.

Add the following methods to the DistributionFunctions class:

```
public static double CauchyFunction(double x, double a, double b)
{
    return b / (Math.PI * (b * b + (x - a) * (x - a)));
}

public static double[] CauchyFunction(double[] x, double a, double b)
{
    double[] y = new double[x.Length];
    for (int i = 0; i < x.Length; i++)
    {
        y[i] = CauchyFunction(x[i], a, b);
    }
    return y;
}
```

These overloaded methods can be used to create probability density function for Cauchy distributions.

Cauchy Random Number Generator

We can use the following formula, based on an inverse transformation, to create Cauchy random distribution:

$$Cauchy(a,b) = a + b[\tan(\pi u(0,1) - 0.5]$$

where $u(0,1)$ is the uniform random distribution defined in the range of [0 ,1]. Using this relation, we can easily create the Cauchy distribution by adding the following methods to the *RandomGenerators* class:

```
public static double NextCauchy(double a, double b)
{
    return a + b * (Math.Tan(Math.PI * rand.NextDouble() - 0.5));
}

public static double[] NextCauchy(double a, double b, int nLength)
{
    double[] array = new double[nLength];
    for (int i = 0; i < nLength; i++)
    {
        array[i] = NextCauchy(a, b);
    }
    return array;
}
```

You can use these two overloaded methods to generate Cauchy random distributions. For example, the Cauchy distributed random array with 2000 data points can simply be created using the following code snippet:

```
double[] y = RandomGenerators.NextCauchy(0.0, 1.0, 2000);
```

Test the Cauchy Distribution

Here, we will examine the Cauchy distribution. Add a *TestCauchy* method to the *Program.cs* file:

```
static void TestCauchy()
{
    int nBins = 20;
    int nPoints = 2000;
    double xmin = -4;
    double xmax = 4;

    double[] rand = RandomGenerators.NextCauchy(0.0, 0.5, nPoints);
    ArrayList aList = RandomGenerators.HistogramData(rand, xmin, xmax, nBins);
    double[] xdata = new double[nBins];
    double[] ydata = new double[nBins];

    double[] ydistribution = new double[nBins];
    for (int i = 0; i < nBins; i++)
    {
        xdata[i] = xmin + (i + 0.5) * (xmax - xmin) / nBins;
```

```
            ydata[i] = (double)aList[i];
            ydistribution[i] = DistributionFunctions.Cauchy(xdata[i], 0.0, 0.5);
    }
    double normalizeFactor = RandomGenerators.ArrayMax(ydata) /
                             RandomGenerators.ArrayMax(ydistribution);
    Console.WriteLine("");
    for (int i = 0; i < nBins; i++)
    {
        Console.WriteLine(" x = {0:n1}, Random data = {1:n0},
                           density distribution = {2:n0}",
                           xdata[i], ydata[i],
                           Math.Round(ydistribution[i] * normalizeFactor, 0));
    }
}
```

Here, we create the random array with the Cauchy distribution, and then compare the histogram of the random data and the theoretical probability density function of the Cauchy distribution. Note that the probability density function is normalized so that its maximum overlaps with the maximum of the random distribution, which allows us to compare them directly. Again, the number of random data points cannot be set too small. Otherwise, the random distribution will deviate dramatically from the theoretical probability function.

Running this example generates the results shown in Figure 7-7. These results are also plotted in Figure 7-8. You can see that the results from the Cauchy random generator are very close to the theoretical Cauchy distribution function.

Student T Distribution

In probability and statistics, the student-t distribution is a probability distribution arising from the problem of estimating the mean of a normally distributed population when the sample size is small. This distribution often arises when the population of the standard deviation is unknown and has to be estimated from the data.

Probability Density Function

The probability density function for the student-t distributions is given by

$$f(x;n) = \frac{\Gamma(n+1)/2)}{\Gamma(1/2)\Gamma(n/2)} n^{-1/2} \left(1 + \frac{x^2}{n}\right)^{-(n+1)/2}$$

where the parameter n is the degrees of freedom and $\Gamma(.)$ is the gamma function.

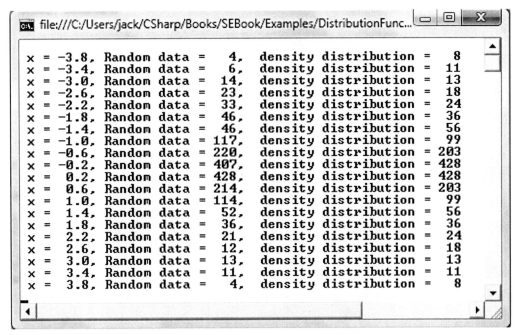

```
ᴄɪ  file:///C:/Users/jack/CSharp/Books/SEBook/Examples/DistributionFunc...

x = -3.8, Random data =    4,  density distribution =    8
x = -3.4, Random data =    6,  density distribution =   11
x = -3.0, Random data =   14,  density distribution =   13
x = -2.6, Random data =   23,  density distribution =   18
x = -2.2, Random data =   33,  density distribution =   24
x = -1.8, Random data =   46,  density distribution =   36
x = -1.4, Random data =   46,  density distribution =   56
x = -1.0, Random data =  117,  density distribution =   99
x = -0.6, Random data =  220,  density distribution =  203
x = -0.2, Random data =  407,  density distribution =  428
x =  0.2, Random data =  428,  density distribution =  428
x =  0.6, Random data =  214,  density distribution =  203
x =  1.0, Random data =  114,  density distribution =   99
x =  1.4, Random data =   52,  density distribution =   56
x =  1.8, Random data =   36,  density distribution =   36
x =  2.2, Random data =   21,  density distribution =   24
x =  2.6, Random data =   12,  density distribution =   18
x =  3.0, Random data =   13,  density distribution =   13
x =  3.4, Random data =   11,  density distribution =   11
x =  3.8, Random data =    4,  density distribution =    8
```

Figure 7-7 Results of the Cauchy distribution.

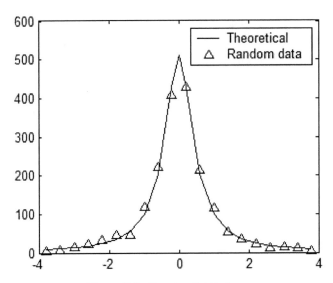

Figure 7-8 Cauchy distributions.

Using the above equation, we can implement this distribution in C#. Add the following methods to the *DistributionFunction* class:

```
public static double StudentT(double x, int n)
{
    double gamma1 = SpecialFunctions.Gamma((n + 1.0) / 2.0);
```

```
    double gamma2 = SpecialFunctions.Gamma(1.0 / 2.0);
    double gamma3 = SpecialFunctions.Gamma(n / 2.0);
    return Math.Pow(n, -0.5) * Math.Pow(1 + x * x / n,
                    -(n + 1) / 2) * gamma1 / gamma2 / gamma3;
}

public static double[] StudentT(double[] x, int n)
{
    double[] y = new double[x.Length];
    for (int i = 0; i < x.Length; i++)
    {
        y[i] = StudentT(x[i], n);
    }
    return y;
}
```

These methods can be used to create the probability density function for student-*t* distributions.

Student T Random Number Generator

The relationship of the student-*t* random distribution, denoted by *Stt(n)*, to the standard normal random distribution $N(0,1)$ and the Chi distribution results in

$$Stt(n) = \frac{N(0,1)}{\chi(n,1)}$$

Using this equation, we can easily implement the student-*t* random distribution in C# by adding the following methods to the *RandomGenerators* class:

```
public static double NextStudentT(int n)
{
    if (n <= 0)
    {
        throw new ArgumentOutOfRangeException("n", n, "n must be positive!");
    }
    return NextNormal(0, 1) / NextChi(n);
}

public static double[] NextStudentT(int n, int nLength)
{
    double[] array = new double[nLength];
    for (int i = 0; i < nLength; i++)
    {
        array[i] = NextStudentT(n);
    }
    return array;
}
```

You can use these methods to generate a random student-*t* distribution.

Testing the Student T Distribution

Here, we will examine the student-*t* distribution. Add a *TestStudentT* method to the *Program.cs* file:

```
static void TestStudentT()
{
    int nBins = 20;
    int nPoints = 2000;
    double xmin = -5;
    double xmax = 5;

    double[] rand = RandomGenerators.NextStudentT(5, nPoints);
    ArrayList aList = RandomGenerators.HistogramData(rand, xmin, xmax, nBins);
    double[] xdata = new double[nBins];
    double[] ydata = new double[nBins];

    double[] ydistribution = new double[nBins];
    for (int i = 0; i < nBins; i++)
    {
        xdata[i] = xmin + (i + 0.5) * (xmax - xmin) / nBins;
        ydata[i] = (double)aList[i];
        ydistribution[i] = DistributionFunctions.StudentT(xdata[i], 5);
    }
    double normalizeFactor = RandomGenerators.ArrayMax(ydata) /
                        RandomGenerators.ArrayMax(ydistribution);
    Console.WriteLine("");
    for (int i = 0; i < nBins; i++)
    {
        Console.WriteLine(" x = {0:n1}, Random data = {1:n0},
                        density distribution = {2:n0}",
                        xdata[i], ydata[i],
                        Math.Round(ydistribution[i] * normalizeFactor, 0));

    }
}
```

Here, we create the random array with the student-*t* distribution, and compare the histogram of the random data with the theoretical probability density function of the student-*t* distribution. Note that the probability density function is normalized so that its maximum overlaps with the maximum of the random distribution, which allows us to compare them directly. Again, the number of random data points cannot be set too small. Otherwise, the random distribution will deviate dramatically from the theoretical probability function.

Running this example generates the results shown in Figure 7-7. These results are also plotted in Figure 7-10. You can see that the results from the student-*t* random generator are very close to the theoretical student-*t* distribution function.

```
file:///C:/Users/jack/CSharp/Books/SEBook/Examples/DistributionFunction...

x = -4.8, Random data =    3,  density distribution =    2
x = -4.3, Random data =    6,  density distribution =    4
x = -3.8, Random data =   14,  density distribution =    7
x = -3.3, Random data =    9,  density distribution =   13
x = -2.8, Random data =   28,  density distribution =   24
x = -2.3, Random data =   48,  density distribution =   47
x = -1.8, Random data =   87,  density distribution =   92
x = -1.3, Random data =  170,  density distribution =  170
x = -0.8, Random data =  262,  density distribution =  280
x = -0.3, Random data =  337,  density distribution =  371
x =  0.3, Random data =  371,  density distribution =  371
x =  0.8, Random data =  282,  density distribution =  280
x =  1.3, Random data =  191,  density distribution =  170
x =  1.8, Random data =   81,  density distribution =   92
x =  2.3, Random data =   50,  density distribution =   47
x =  2.8, Random data =   24,  density distribution =   24
x =  3.3, Random data =   13,  density distribution =   13
x =  3.8, Random data =   11,  density distribution =    7
x =  4.3, Random data =    4,  density distribution =    4
x =  4.8, Random data =    2,  density distribution =    2
```

Figure 7-9 Results of the student t distribution.

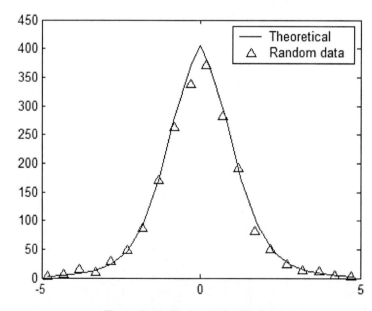

Figure 7-10 Student-t distributions.

Gamma Distribution

The gamma distribution is a two-parameter family of continuous probability distributions, representing the sum of exponentially-distributed random variables,

Probability Density Function

The probability density function of the gamma distribution can be expressed in terms of the gamma function:

$$f(x;r,\alpha) = \frac{\alpha^r}{\Gamma(r)} x^{r-1} e^{-\alpha x}$$

This function is defined in the parameter range $r > 0$, $\alpha > 0$, and $x > 0$.

This distribution function can be easily implemented in C# by adding the following methods to the *DistributionFunctions* class:

```
public static double GammaFunction(double x, int r, double alpha)
{
    return Math.Pow(alpha, r) * Math.Pow(x, r - 1) *
           Math.Exp(-alpha * x) / SpecialFunctions.Gamma(r);
}

public static double[] GammaFunction(double[] x, int r, double alpha)
{
    double[] y = new double[x.Length];
    for (int i = 0; i < x.Length; i++)
    {
        y[i] = GammaFunction(x[i], r, alpha);
    }
    return y;
}
```

Gamma Random Number Generator

If the parameter r is an integer, the random numbers from the gamma distribution can be calculated by the random numbers from uniform distribution, using the following formula:

$$Gamma(r,\alpha) = -\frac{1}{\alpha} \sum_{i=1}^{r} \ln U_i(0,1)$$

where $U_i(0,1)$ is the uniform random distribution defined in the range of $[0,1]$. Using this relation, we can easily implement the gamma random distribution in C# by adding the following methods to the *RandomGenerators* class:

```
public static double NextGamma(int r, double alpha)
{
    double result = 0.0;
    for (int i = 0; i < r; i++)
    {
        result += -Math.Log(rand.NextDouble()) / alpha;
```

```
        }
        return result;
    }

    public static double[] NextGamma(int r, double alpha, int nLength)
    {
        double[] array = new double[nLength];
        for (int i = 0; i < nLength; i++)
        {
            array[i] = NextGamma(r,alpha);
        }
        return array;
    }
```

These methods can be used to generate a single random number or a random number array with the gamma distribution.

Testing the Gamma Distribution

Here, we will examine the gamma distribution. Add a *TestGamma* method to the *Program.cs* file:

```
static void TestGamma()
{
    int nBins = 20;
    int nPoints = 2000;
    double xmin = 0;
    double xmax = 15;

    double[] rand = RandomGenerators.NextGamma(2, 0.5, nPoints);
    ArrayList aList = RandomGenerators.HistogramData(rand, xmin, xmax, nBins);
    double[] xdata = new double[nBins];
    double[] ydata = new double[nBins];

    double[] ydistribution = new double[nBins];
    for (int i = 0; i < nBins; i++)
    {
        xdata[i] = xmin + (i + 0.5) * (xmax - xmin) / nBins;
        ydata[i] = (double)aList[i];
        ydistribution[i] = DistributionFunctions.Gamma(xdata[i], 2, 0.5);
    }
    double normalizeFactor = RandomGenerators.ArrayMax(ydata) /
                        RandomGenerators.ArrayMax(ydistribution);
    Console.WriteLine("");
    for (int i = 0; i < nBins; i++)
    {
        Console.WriteLine(" x = {0:n1}, Random data = {1:n0},
                        density distribution = {2:n0}",
                        xdata[i], ydata[i],
                        Math.Round(ydistribution[i] * normalizeFactor, 0));
    }
}
```

Here, we create a random array with the gamma distribution, and then compare the histogram of the random data with the theoretical probability density function of the gamma distribution. Note that the probability density function is normalized so that its maximum overlaps with the maximum of the random distribution, which allows us to compare them directly. Again, the number of random data points cannot be set too small. Otherwise, the random distribution will deviate dramatically from the theoretical probability function.

Running this example generates the results shown in Figure 7-11. These results are also plotted in Figure 7-12. You can see that the results from the gamma random generator are very close to the theoretical gamma distribution function.

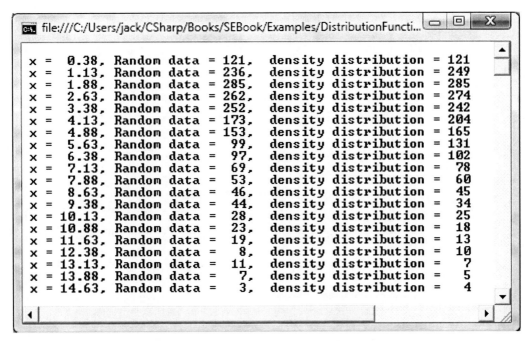

Figure 7-11 Results of the gamma distribution.

Beta Distribution

The beta distribution describes a family of continuous probability distributions that are unique because they are nonzero only in the interval [0, 1].

Probability Density Function

The probability density function is the same as the incomplete beta function:

$$f(x;\alpha,\beta) = \frac{x^{\alpha-1}(1-x)^{\beta-1}}{B(\alpha,\beta)}$$

where $B(.)$ is the beta function.

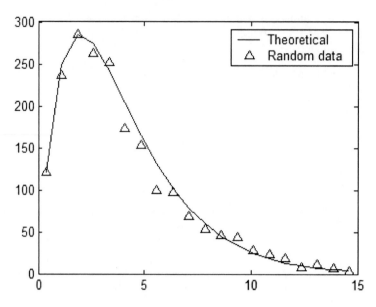

Figure 7-12 Gamma distributions.

This function can be easily implemented in C# by adding the following methods to the *DistributionFunctions* class:

```
public static double Beta(double x, double alpha, double beta)
{
    double b = SpecialFunctions.Beta(alpha, beta);
    return Math.Pow(x, alpha - 1) * Math.Pow(1 - x, beta - 1) / b;
}

public static double[] Beta(double[] x, double alpha, double beta)
{
    double[] y = new double[x.Length];
    for (int i = 0; i < x.Length; i++)
    {
        y[i] = Beta(x[i], alpha, beta);
    }
    return y;
}
```

Beta Random Number Generator

A beta random distribution can be constructed using gamma random distribution. Suppose

$$Gamma(a, 1) = -\sum_{i=1}^{a} U_i(0, 1), \qquad Gamma(b, 1) = -\sum_{j=1}^{b} U_j(0, 1)$$

where both parameters, a and b, are positive integers, and $U(0, 1)$ is the uniform random generator defined in the range of [0, 1]. Therefore, the beta random distribution can be written in the form:

$$Beta(a, b) = \frac{Gamma(a, 1)}{Gamma(a, 1) + Gamma(b, 1)}$$

The random number generator from the beta distribution can be implemented using the above relationship by adding the following methods to the *RandomGenerators* class:

```
public static double NextBeta(int a, int b)
{
    double gamma1 = NextGamma(a, 1);
    double gamma2 = NextGamma(b, 1);
    return gamma1 / (gamma1 + gamma2);
}

public static double[] NextBeta(int a, int b, int nLength)
{
    double[] array = new double[nLength];
    for (int i = 0; i < nLength; i++)
    {
        array[i] = NextBeta(a, b);
    }
    return array;
}
```

Testing the Beta Distribution

Here, we will examine the beta distribution. Add a TestBeta method to the Program.cs file:

```
static void TestBeta()
{
    int nBins = 10;
    int nPoints = 2000;
    double xmin = 0;
    double xmax = 1;

    double[] rand = RandomGenerators.NextBeta(2, 5, nPoints);
    ArrayList aList = RandomGenerators.HistogramData(rand, xmin, xmax, nBins);
    double[] xdata = new double[nBins];
    double[] ydata = new double[nBins];

    double[] ydistribution = new double[nBins];
    for (int i = 0; i < nBins; i++)
    {
        xdata[i] = xmin + (i + 0.5) * (xmax - xmin) / nBins;
        ydata[i] = (double)aList[i];
        ydistribution[i] = DistributionFunctions.Beta(xdata[i], 2, 5);
    }
    double normalizeFactor = RandomGenerators.ArrayMax(ydata) /
                             RandomGenerators.ArrayMax(ydistribution);
    Console.WriteLine("");
    for (int i = 0; i < nBins; i++)
    {
        Console.WriteLine(" x = {0:n2}, Random data = {1:n0},
                          density distribution = {2:n0}",
                          xdata[i], ydata[i],
```

```
                                    Math.Round(ydistribution[i] * normalizeFactor, 0));
        }
    }
```

Here, we create the random array with the beta distribution, and compare the histogram of the random data with the theoretical probability density function of the beta distribution. Note that the probability density function is normalized so that its maximum overlaps with the maximum of the random distribution, which allows us to compare them directly.

Running this example generates the results shown in Figure 7-13. These results are also plotted in Figure 7-14. You can see that the results from the beta random generator are very close to the theoretical beta distribution function.

```
file:///C:/Users/jack/CSharp/Books/SEBook/Examples/DistributionFu...

x = 0.05, Random data = 235, density distribution = 258
x = 0.15, Random data = 439, density distribution = 496
x = 0.25, Random data = 501, density distribution = 501
x = 0.35, Random data = 347, density distribution = 396
x = 0.45, Random data = 255, density distribution = 261
x = 0.55, Random data = 141, density distribution = 143
x = 0.65, Random data =  60, density distribution =  62
x = 0.75, Random data =  16, density distribution =  19
x = 0.85, Random data =   6, density distribution =   3
x = 0.95, Random data =   0, density distribution =   0
```

Figure 7-13 Results of the beta distribution.

Poisson Distribution

In probability theory and statistics, the Poisson distribution is a discrete probability distribution. It describes the probability of a number of events occurring in a fixed time period. The Poisson distribution is not limited to a time interval domain. It can also be used for the number of events in other specified intervals such as distance, area, or volume.

The Poisson and exponential distributions are related. If the number of counts follows the Poisson distribution, then the interval between individual counts follows the exponential distribution.

Probability Density Function

The probability density function of the Poisson distribution can be expressed as

$$f(x; \lambda) = \frac{\lambda^x}{x!} e^{-\lambda}$$

where x is a positive integer, representing the number of occurrences of an event, $x!$ is the factorial of x, and λ is a positive real number, equal to the expected number of occurrences during a given interval.

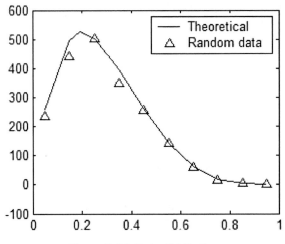

Figure 7-14 Beta distributions.

This function can be easily implemented in C# by adding the following methods to the *DistributionFunctions* class:

```
public static double Poisson(int x, double lambda)
{
    double y = Math.Exp(-lambda) * Math.Pow(lambda, x);
    return y / SpecialFunctions.Gamma(x + 1);
}

public static double[] Poisson(int[] x, double lambda)
{
    double[] y = new double[x.Length];
    for (int i = 0; i < x.Length; i++)
    {
        y[i] = Poisson(x[i], lambda);
    }
    return y;
}
```

Here the x! is calculated using the gamma function $\Gamma(x + 1)$ implemented in the *SpecialFunctions* class. You can use the Poisson method to compute the probability density function of Poisson distributions.

Poisson Random Number Generator

The following code for generating Poisson random numbers is based on a rejection method, which is a powerful technique for generating random numbers whose distribution function is known and computable.

Add the following public methods to the *RandomGenerators* class:

```
public static double NextPoisson(double lambda)
{
    int i = 0;
```

```
        double f = Math.Exp(-lambda);
        double p = f;
        double u = rand.NextDouble();
        while (f <= u)
        {
            p *= (lambda / (i + 1.0));
            f += p;
            i++;
        }
        return i;
    }

    public static double[] NextPoisson(double lambda, int nLength)
    {
        double[] array = new double[nLength];
        for (int i = 0; i < nLength; i++)
        {
            array[i] = NextPoisson(lambda);
        }
        return array;
    }
```

Testing the Poisson Distribution

Here, we will examine the Poisson distribution. Add a *TestPoisson* method to the *Program.cs* file:

```
static void TestPoisson()
{
    int nBins = 15;
    int nPoints = 2000;

    double[] rand = RandomGenerators.NextPoisson(4.0, nPoints);
    ArrayList aList = RandomGenerators.HistogramData(rand, nBins);
    double[] xdata = new double[nBins];
    double[] ydata = new double[nBins];

    double[] ydistribution = new double[nBins];
    for (int i = 0; i < nBins; i++)
    {
        xdata[i] = i;
        ydata[i] = (double)aList[i];
        ydistribution[i] = DistributionFunctions.Poisson(i, 4.0);
    }
    double normalizeFactor = RandomGenerators.ArrayMax(ydata) /
                        RandomGenerators.ArrayMax(ydistribution);
    Console.WriteLine("");
    for (int i = 0; i < nBins; i++)
    {
        Console.WriteLine(" x = {0:n2}, Random data = {1:n0},
                        density distribution = {2:n0}",
                        xdata[i], ydata[i],
                        Math.Round(ydistribution[i] * normalizeFactor, 0));
    }
}
```

Here, we create the random array with the Poisson distribution, and compare the histogram of the random data with the theoretical probability density function of the Poisson distribution. Note that the probability density function is normalized so that its maximum overlaps with the maximum of the random distribution, which allows us to compare them directly.

Running this example generates the results shown in Figure 7-15. These results are also plotted in Figure 7-16. You can see that the results from the Poisson random generator are very close to the theoretical Poisson distribution function.

```
file:///C:/Users/jack/CSharp/Books/SEBook/Examples/Distribution...

x =   0,  Random data =   38,   density distribution =   39
x =   1,  Random data =  156,   density distribution =  157
x =   2,  Random data =  293,   density distribution =  314
x =   3,  Random data =  378,   density distribution =  419
x =   4,  Random data =  419,   density distribution =  419
x =   5,  Random data =  299,   density distribution =  335
x =   6,  Random data =  208,   density distribution =  223
x =   7,  Random data =  111,   density distribution =  128
x =   8,  Random data =   58,   density distribution =   64
x =   9,  Random data =   28,   density distribution =   28
x =  10,  Random data =    8,   density distribution =   11
x =  11,  Random data =    3,   density distribution =    4
x =  12,  Random data =    1,   density distribution =    1
x =  13,  Random data =    0,   density distribution =    0
x =  14,  Random data =    0,   density distribution =    0
```

Figure 7-15 Results of the Poisson distribution.

Binomial Distribution

The binomial distribution is another discrete probability distribution. It describes the number of successes in a sequence of n independent yes-or-no experiments, each of which yields success with probability p. This distribution is the basis for the popular binomial test of statistical significance.

Probability Density Function

The probability density function of the binomial distribution is given by the following formula:

$$f(x; n, p) = \frac{n!}{x!(n-x)!} p^x (1-p)^{n-x}$$

where n is an integer, $x = 0, 1, 2, \ldots n$, and $0 \leq p \leq 1$.

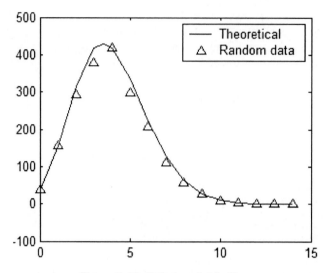

Figure 7-16 Poission distributions.

Add the following methods to the *DistributionFunctions* class:

```
public static double Binomial(int x, int n, double p)
{
    return SpecialFunctions.Gamma(n + 1) * Math.Pow(p, x) *
        Math.Pow(1 - p, n - x) /SpecialFunctions.Gamma(x + 1) /
        SpecialFunctions.Gamma(n - x + 1);
}

public static double[] Binomial(int[] x, int n, double p)
{
    double[] y = new double[x.Length];
    for (int i = 0; i < x.Length; i++)
    {
        y[i] = Binomial(x[i], n, p);
    }
    return y;
}
```

You can use the *Binomial* method to compute the probability density function of binomial distributions.

Binomial Random Number Generator

The binomial random numbers can be generated using the C# built-in uniform random generator. Add the following public methods to the *RandomGenerators* class:

```
public static double NextBinomial(int n, double p)
{
    double result = 0.0;
    for (int i = 0; i < n; i++)
    {
        if (rand.NextDouble() < p)
        {
```

```
                    result++;
                }
            }
        return result;
    }

    public static double[] NextBinomial(int n, double p, int nLength)
    {
        double[] array = new double[nLength];
        for (int i = 0; i < nLength; i++)
        {
            array[i] = NextBinomial(n, p);
        }
        return array;
    }
```

Testing the Binomial Distribution

Here, we will examine the binomial distribution. Add a *TestBinomial* method to the *Program.cs* file:

```
static void TestBinomial()
{
    int nBins = 21;
    int nPoints = 2000;

    double[] rand = RandomGenerators.NextBinomial(20, 0.5, nPoints);
    ArrayList aList = RandomGenerators.HistogramData(rand, nBins);
    double[] xdata = new double[nBins];
    double[] ydata = new double[nBins];

    double[] ydistribution = new double[nBins];
    for (int i = 0; i < nBins; i++)
    {
        xdata[i] = i;
        ydata[i] = (double)aList[i];
        ydistribution[i] = DistributionFunctions.Binomial(i, 20, 0.5);
    }
    double normalizeFactor = RandomGenerators.ArrayMax(ydata) /
                        RandomGenerators.ArrayMax(ydistribution);
    Console.WriteLine("");
    for (int i = 0; i < nBins; i++)
    {
        Console.WriteLine(" x = {0:n0}, Random data = {1:n0},
                        density distribution = {2:n0}",
                        xdata[i], ydata[i],
                        Math.Round(ydistribution[i] * normalizeFactor, 0));
    }
}
```

Here, we create the random array with the binomial distribution and compare the histogram of the random data with the theoretical probability density function of the binomial distribution. Note that the probability density function is normalized so that its maximum overlaps with the maximum of the random distribution, which allows us to compare them directly.

Running this example generates the results shown in Figure 7-17. These results are also plotted in Figure 7-18. You can see that the results from the binomial random generator are very close to the theoretical binomial distribution function.

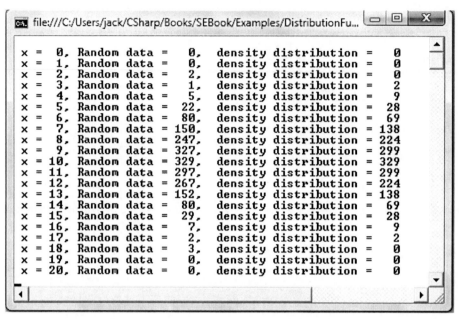

```
file:///C:/Users/jack/CSharp/Books/SEBook/Examples/DistributionFu...

x =   0,  Random data =    0,  density distribution =   0
x =   1,  Random data =    0,  density distribution =   0
x =   2,  Random data =    2,  density distribution =   0
x =   3,  Random data =    1,  density distribution =   2
x =   4,  Random data =    5,  density distribution =   9
x =   5,  Random data =   22,  density distribution =  28
x =   6,  Random data =   80,  density distribution =  69
x =   7,  Random data =  150,  density distribution = 138
x =   8,  Random data =  247,  density distribution = 224
x =   9,  Random data =  327,  density distribution = 299
x =  10,  Random data =  329,  density distribution = 329
x =  11,  Random data =  297,  density distribution = 299
x =  12,  Random data =  267,  density distribution = 224
x =  13,  Random data =  152,  density distribution = 138
x =  14,  Random data =   80,  density distribution =  69
x =  15,  Random data =   29,  density distribution =  28
x =  16,  Random data =    7,  density distribution =   9
x =  17,  Random data =    2,  density distribution =   2
x =  18,  Random data =    3,  density distribution =   0
x =  19,  Random data =    0,  density distribution =   0
x =  20,  Random data =    0,  density distribution =   0
```

Figure 7-17 Results of the binomial distribution.

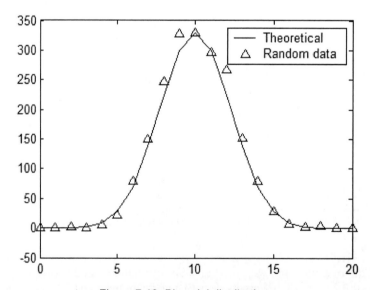

Figure 7-18 Binomial distributions.

Chapter 8

Interpolation

In numerical analysis, interpolation is a method of constructing new data points within the range of a discrete set of known data. In science and engineering, when we have a number of data points obtained by sampling and experimentation, we try to construct a function that closely fits those data points. This usually called curve fitting or regression analysis. Interpolation can be regarded as a special case of curve fitting, in which the function must go exactly through every single data point. This means an implicit assumption that the given data points in interpolation are accurate and distinct.

Suppose that a function $y = f(x)$ is given only at discrete points such as (x_0, y_0), (x_1, y_1), ..., (x_N, y_N). The purpose of interpolation is to find the y value at any other point of x. In this chapter, we will present some popular methods for interpolating the function of $y = f(x)$. Most of the interpolation methods are based on the Taylor series expansion of a function $f(x)$ about a specific value of x_0. Each method manipulates the Taylor series differently to yield an interpolation algorithm.

At the end of this chapter, I will also present a bilinear interpolation for two-dimensional functions $z = f(x, y)$. This interpolation method plays an important role in computer vision and image processing.

Linear Interpolation

Linear interpolation is a simple method of curve fitting based on linear polynomials. It is widely used in numerical ananlysis and computer graphics.

Algorithm

As shown in Figure 8-1, if two known points are given by (x_0, y_0) and (x_1, y_1), the linear interpolation is the straight line between these two points. For any value of x in the range (x_0, x_1), the corresponding y value along the straight line can be found by the following relation:

$$y = y_0 + (x - x_0) \frac{y_1 - y_0}{x_1 - x_0}$$

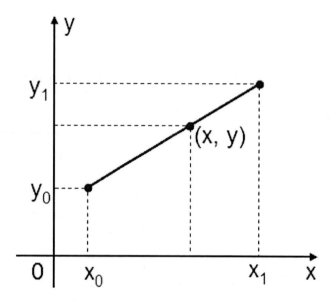

Figure 8-1 Linear interpolation: given two points (x_0, y_0) and (x_1, y_1), the line between these two points is the result of the linear interpolation.

The linear interpolation on a set of data points (x_0, y_0), (x_1, y_1),..., (x_n, y_n) is defined as the concatenation of linear interpolants between each pair of data points. This results in a continuous curve with a discontinuous derivative.

Implementation

Here, we will implement the linear interpolation. Start with a new C# Console application and name it *InterpolationTest*. Add a new class, *Interpolation*, to the project and change its namespace to *XuMath*.

Add a public static method, *Linear*, to the *Interpolation* class:

```
public static double Linear(double[] xarray, double[] yarray, double x)
{
    double y = double.NaN;
    for (int i = 0; i < xarray.Length - 1; i++)
    {
        if (x >= xarray[i] && x < xarray[i + 1])
        {
            y = yarray[i] + (x - xarray[i]) * (yarray[i + 1] - yarray[i]) /
                (xarray[i + 1] - xarray[i]);
        }
    }
    return y;
}

public static double[] Linear(double[] xarray, double[] yarray, double[] x)
{
    double[] y = new double[x.Length];
    for (int i = 0; i < x.Length; i++)
```

```
            y[i] = Linear(xarray, yarray, x[i]);
        return y;
    }
```

Here, we implement an overloaded method, *Linear*. The *xarray* and *yarray* are the sets of *x* and *y* data, which present a set of given data points. The *Linear* method will find a single *y* value or a *y* array at the input *x* (a double value or double array), depending on the input variable *x*.

Testing the Linear Interpolation

You can perform the linear interpolation using the *Linear* method. To display the results more conveniently, we first add the *VectorR* and *MatrixR.cs* classes from the *RealMatrixTest* project of Chapter 3 to the current project. Then add a method, *TestLinear*, to the *Program.cs* file:

```
static void TestLinear()
{
    double[] xarray = new double[] { 0, 2, 4, 6, 8 };
    double[] yarray = new double[] { 0, 4, 16, 36, 64 };
    double[] x = new double[] { 1, 3, 5, 7 };
    double[] y = Interpolation.Linear(xarray,yarray,x);
    VectorR vx = new VectorR(x);
    VectorR vy = new VectorR(y);
    Console.WriteLine(" x = " + vx.ToString());
    Console.WriteLine(" y = " + vy.ToString());
}
```

Here, we first create a set of data points by defining the *xarray* and *yarray*, and then compute the *y* values at the *x* values specified by a double array *x*. Finally, we convert the *x* and *y* arrays to corresponding *VectorR* objects to display the results.

Running this example generates following results:

```
x = (1, 3, 5, 7)
y = (2, 10, 26, 50)
```

From the input data points described by *xarray* and *yarray*, we see that the given data can be expressed in terms of an analytic function $y = x^2$. The results from the *Linear* method derivates slightly from this analytic function. This means that the linear interpolation only provides approximate results.

Lagrange Interpolation

The Lagrange interpolation is a well known, classical technique for performing interpolation. Sometimes, the Lagrange interpolation is also called the polynomial interpolation. In the first order approximation, it reduces to the linear interpolation, which has been discussed in the previous section.

Algorithm

For a given set of $n + 1$ data points (x_0, y_0), (x_1, y_1),..., (x_n, y_n), where no two x_i are the same, the interpolation polynomial in the Lagrange form is a linear combination:

$$y = f(x) = \sum_{i=0}^{n} l_i(x) f(x_i)$$

where

$$l_i(x) = \prod_{j=0, j \neq i}^{n} \frac{x - x_j}{x_i - x_j} = \frac{(x - x_0)}{(x_i - x_0)} \cdots \frac{(x - x_{i-1})}{(x_i - x_{i-1})} \frac{(x - x_{i+1})}{(x_i - x_{i+1})} \cdots \frac{(x - x_n)}{(x_i - x_n)}$$

From the numerator of the above definition, we see that $l_i(x)$ is an nth order polynomial with zeros at all of the sample points except the nth.

Implementation

Here, we will implement the Lagrange interpolation. Add a new public static method, *Lagrangian*, to the *Interpolation* class:

```
public static double Lagrangian(double[] xarray, double[] yarray, double x)
{
    double y = 0.0;
    double product = yarray[0];
    for (int i = 0; i < xarray.Length; i++)
    {
        product = yarray[i];
        for (int j = 0; j < xarray.Length; j++)
        {
            if (i != j)
            {
                product *= (x - xarray[j]) / (xarray[i] - xarray[j]);
            }
        }
        y += product;
    }
    return y;
}

public static double[] Lagrangian(double[] xarray, double[] yarray, double[] x)
{
    double[] y = new double[x.Length];
    for (int i = 0; i < x.Length; i++)
        y[i] = Lagrangian(xarray, yarray, x[i]);
    return y;
}
```

Here, we implement an overloaded method, *Lagrangian*. The *xarray* and *yarray* are the sets of *x* and *y* data, which present a set of given data points. The *Lagrangian* method will find a single *y* value or a *y* array at the input *x* (a double value or double array), depending on the input variable *x*.

Testing the Lagrange Interpolation

You can perform the Lagrange interpolation using the *Lagrangian* method. Add a method, *TestLagrangian*, to the *Program.cs* file:

```
static void TestLagrangian()
{
    double[] xarray = new double[5] { 1, 2, 3, 4, 5 };
    double[] yarray = new double[5] { 1, 4, 9, 16, 25 };
```

```
        double[] x = new double[3] { 2.5, 3.5, 1.5 };
        double[] y = Interpolation.Lagrangian(xarray, yarray, x);
        VectorR vx = new VectorR(x);
        VectorR vy = new VectorR(y);
        Console.WriteLine(" x = " + vx.ToString());
        Console.WriteLine(" y=" + vy.ToString());
    }
```

Here, we first create a set of data points by defining the *xarray* and *yarray*. We then compute the *y* values at the *x* values specified by a double array. Finally, we convert the *x* and *y* arrays to corresponding *VectorR* objects to display the results.

Running this example generates the following results:

```
x = (2.5, 3.5, 1.5)
y = (6.25, 12.25, 2.25)
```

Barycentric Interpolation

The Lagrange interpolation discussed above requires recomputation for all of the terms for each distinct *x* value, meaning that this method can only be applied for small numbers *n* of nodes. Several shotcomings are associated with the Lagrange interpolation:

- Each evaluation of *f*(*x*) requires $O(n^2)$ additions and multiplications.

- Adding a new data point (x_{n+1}, y_{n+1}) requires a new computation from scratch.

- The computation is numerically unstable.

In order to improve the Lagrange method, we can simply rearrange the terms in the equation of the Lagrange interpolation by defining weight functions that do not depend on the interpolated value of *x*.

Algorithm

Let's introduce the quantity

$$l(x) = (x - x_0)(x - x_1)\cdots(x - x_n)$$

We can then rearrange the Lagrange basis polynomial as

$$l_i(x) = \frac{l(x)}{x - x_i} \frac{1}{\prod_{j=0, j \neq i}^{n} (x_i - x_j)}$$

Defining the barycentric weight functions:

$$w_i = \frac{1}{\prod_{j=0, j \neq i}^{n} (x_i - x_j)}$$

We can simply write

$$l_i(x) = l(x) \frac{w_i}{x - x_i}$$

which is usually referred to as the barycentric interpolation formula. The advantage of this representation is that the interpolation polynomial may now be evaluated as

$$y = f(x) = l(x) \sum_{i=0}^{n} \frac{w_i}{x - x_i} f(x_i)$$

Now, if the weight function w_i has been pre-computed, the barycentric interpolation requires only $O(n)$ operations as opposed to $O(n^2)$ for evaluating the Lagrange basis polynomials $l_i(x)$ individually.

Implementation

Here, we will implement the barycentric interpolation. Add a new public static method, *Barycentric*, to the *Interpolation* class:

```
public static double Barycentric(double[] xarray, double[] yarray, double x)
{
    double product;
    double dx;
    double c1 = 0;
    double c2 = 0;
    int n = xarray.Length;
    double[] wt = new double[n];

    for (int i = 0; i < n; i++)
    {
        product = 1;
        for (int j = 0; j < n; j++)
        {
            if (i != j)
            {
                product *= (xarray[i] - xarray[j]);
                wt[i] = 1.0 / product;
            }
        }
    }

    for (int i = 0; i < n; i++)
    {
        dx = wt[i] / (x - xarray[i]);
        c1 += yarray[i] * dx;
        c2 += dx;
    }
    return c1 / c2;
}

public static double[] Barycentric(double[] xarray, double[] yarray, double[] x)
{
    double[] y = new double[x.Length];
    for (int i = 0; i < x.Length; i++)
        y[i] = Barycentric(xarray, yarray, x[i]);
    return y;
}
```

Here, we implement an overloaded method, *Barycentric*. The *xarray* and *yarray* are the sets of x and y data, which present a set of given data points. The *Barycentric* method returns a single y value or a y array at the input x (a double value or double array), depending on the input variable x.

Testing the Barycentric Interpolation

You can perform the interpolation using the *Barycentric* method. Add a method, *TestBarycentric*, to the *Program.cs* file:

```
static void TestBarycentric()
{
    double[] xarray = new double[] { 0, 2, 4, 6, 8 };
    double[] yarray = new double[] { 0, 4, 16, 36, 64 };
    double[] x = new double[] { 1, 3, 5, 7 };
    double[] y = Interpolation.Barycentric(xarray, yarray, x);
    VectorR vx = new VectorR(x);
    VectorR vy = new VectorR(y);
    Console.WriteLine(" x = " + vx.ToString());
    Console.WriteLine(" y=" + vy.ToString());
}
```

Here, we first create a set of data points by defining the *xarray* and *yarray*. We then compute the y values at the x values specified by a double array. Finally, we convert the x and y arrays to corresponding *VectorR* objects to display the results.

Running this example generates following results:

```
x = (1, 3, 5, 7)
y = (1, 9, 25, 49)
```

From the input data points described by *xarray* and *yarray*, we see that the given data can be expressed in terms of an analytic function $y = x^2$. The *Barycentric* method gives the exact results expected from the analytic function.

Newton Divided Difference Interpolation

The Newton divided difference interpolation is the interpolation polynomial approximation for a given set of data points in the Newton form. It uses the Taylor expansion to perform the interpolation. The divided differences are used to approximately calculate the various derivatives.

Algorithm

Let's first examine the linear and quadratic approximations. Given two data point (x_0, y_0) and (x_1, y_1), the linear interpolation can be expressed in the form

$$f_1(x) = c_0 + c_1(x - x_0) = f(x_0) + \frac{f(x_1) - f(x_0)}{x_1 - x_0}(x - x_0)$$

For a given set of three data points (x_0, y_0), (x_1, y_1), and (x_2, y_2), we can fit a quadratic interpolant through the data:

$$f_2(x) = c_0 + c_1(x - x_0) + c_2(x - x_0)(x - x_1)$$

$$= f(x_0) + \frac{f(x_1) - f(x_0)}{x_1 - x_0}(x - x_0) + \frac{\dfrac{f(x_2) - f(x_1)}{x_2 - x_1} - \dfrac{f(x_1) - f(x_0)}{x_1 - x_0}}{x_2 - x_0}(x - x_0)(x - x_1)$$

In this two cases, we see how the linear and quadratic interpolations are derived by the Newton divided difference polynomial method.

We can rewrite the quadratic polynomial interpolant formula by introducing the notations:

$$f[x_0] = f(x_0)$$

$$f[x_1, x_0] = \frac{f(x_1) - f(x_0)}{x_1 - x_0}$$

$$f[x_2, x_1, x_0] = \frac{f[x_2, x_1] - f[x_1, x_0]}{x_2 - x_0}$$

where $f[x_0]$, $f[x_1, x_0]$, and $f[x_2, x_1, x_0]$ are called bracketed functions, as their variables are enclosed in square brackets. Using these notations, we can then write the quadratic interpolant formula as

$$f_2(x) = f[x_0] + f[x_1, x_0](x - x_0) + f[x_2, x_1, x_0](x - x_0)(x - x_1)$$

This leads to the general form of the Newton divided difference polynomial for a given set of $(n + 1)$ data points, $(x_0, y_0), (x_1, y_1), \ldots, (x_n, y_n)$, which can be written as

$$f_n(x) = c_0 + c_1(x - x_0) + \cdots + c_n(x - x_0)(x - x_1) \cdots (x - x_{n-1})$$

where

$$c_0 = f[x_0]$$
$$c_1 = f[x_1, x_0]$$
$$\vdots$$
$$c_n = f[x_n, x_{n-1}, \cdots, x_0]$$

From the above definition, we see that the Newton divided differences are calculated recursively.

The divided differences can also be written in the form of a table:

$$x_0 \quad f[x_0]$$
$$f[x_1, x_0]$$
$$x_1 \quad f[x_1] \qquad f[x_2, x_1, x_0]$$
$$f[x_2, x_1] \qquad\qquad f[x_3, x_2, x_1, x_0]$$
$$x_2 \quad f[x_2] \qquad f[x_3, x_2, x_1] \qquad\qquad \ddots$$
$$f[x_3, x_2] \qquad\qquad \vdots \qquad\qquad f[x_n, x_{n-1}, \cdots, x_0]$$
$$\qquad\qquad\qquad \ddots$$
$$x_3 \quad f[x_3] \qquad\qquad \vdots \qquad f[x_{n-3}, \cdots, x_n]$$
$$\vdots \quad \vdots \qquad \vdots \qquad f[x_{n-2}, x_{n-1}, x_n]$$
$$f[x_{n-1}, x_n]$$
$$x_n \quad f[x_n]$$

The above difference table is an upper triangular matrix. You can store this kind of matrix in a one-dimensional array to save space.

Implementation

Here, we will implement the Newton divided difference interpolation. Add a new public static method, *NewtonDividedDifference*, to the *Interpolation* class:

```
public static double NewtonDividedDifference(double[] xarray,
        double[] yarray, double x)
{
    double y;
    int n = xarray.Length;
    double[] temp = new double[n];
    for (int i = 0; i < n; i++)
    {
        temp[i] = yarray[i];
    }

    for (int i = 0; i < n - 1; i++)
    {
        for (int j = n - 1; j > i; j--)
        {
            temp[j] = (temp[j - 1] - temp[j]) / (xarray[j - 1 - i] - xarray[j]);
        }
    }

    y = temp[n - 1];
    for (int i = n - 2; i >= 0; i--)
    {
        y = temp[i] + (x - xarray[i]) * y;
    }
    return y;
}

public static double[] NewtonDividedDifference(double[] xarray,
        double[] yarray, double[] x)
```

```
{
    double[] y = new double[x.Length];
    for (int i = 0; i < x.Length; i++)
        y[i] = NewtonDividedDifference(xarray, yarray, x[i]);
    return y;
}
```

Here, we implement an overloaded method, *NewtonDividedDifference*. The *xarray* and *yarray* are the sets of *x* and *y* data, which present a set of given data points. This method returns a single *y* value or a *y* array at the input *x* (a double value or double array), depending on the input variable *x*.

Testing the Newton Divided Difference Interpolation

You can perform an interpolation using the Newton divided difference method. There are two data arrays, *xarray* and *yarray*, which represent the census years from 1950 to 1990 and corresponding United States population in millions of people:

```
xarray = {1950, 1960, 1970, 1980, 1990 };
yarray = {150.697, 179.323, 203.212, 226.505, 249.633 };
```

Now, you can use the Newton divided difference interpolation to interpolate within the census data to estimate the population in 1955, 1965, 1975, and 1985.

Add a method, *TestNewtonDividedDifference*, to the *Program.cs* file:

```
static void TestNewtonDividedDifference()
{
    double[] xarray = new double[] {1950, 1960, 1970, 1980, 1990 };
    double[] yarray =
            new double[] {150.697, 179.323, 203.212, 226.505, 249.633 };
    double[] x = new double[] { 1955, 1965, 1975, 1985 };
    double[] y = Interpolation.NewtonDividedDifference(xarray, yarray, x);
    VectorR vx = new VectorR(x);
    VectorR vy = new VectorR(y);
    Console.WriteLine(" x = " + vx.ToString());
    Console.WriteLine(" y=" + vy.ToString());
}
```

Here, we first create a set of data points by defining the *xarray* and *yarray*. We then compute the *y* values at the *x* values specified by a double array. Finally, we convert the *x* and *y* arrays to corresponding *VectorR* objects to display the results.

Running this example generates following results:

```
x = (1955, 1965, 1975, 1985)
y = (166.006, 191.514, 214.819, 238.208)
```

You can examine how good the interpolation is by plotting the results on a chart, as shown in Figure 8-2.

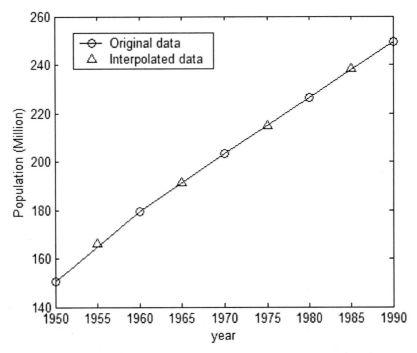

Figure 8-2 Results by the Newton divided difference interpolation.

Cubic Spline Interpolation

In numerical analysis, the spline interpolation is a form of interpolation where the interpolant is a special type of piecewise polynomial called a spline. This method provides a great deal of smoothness. for interpolations with significantly varying data.

The cubic spline interpolation uses weight coefficients on the cubic polynomials to interpolate the data. These coefficients bend the line so that it passes through each of the data points without any erratic behavior or breaks in continuity.

Algorithm

The basic idea of the cubic spline is to fit a piecewise function of the form

$$S(x) = \begin{cases} s_1(x) & \text{if} \quad x_1 \le x < x_2 \\ s_2(x) & \text{if} \quad x_2 \le x < x_3 \\ \quad \vdots \\ s_{n-1}(x) & \text{if} \quad x_{n-1} \le x < x_n \end{cases}$$

where s_i is a third degree polynomial defined by

$$s_i(x) = a_i(x - x_i)^3 + b_i(x - x_i)^2 + c_i(x - x_i) + d_i, \quad \text{for } i = 1, 2, \cdots, n-1$$

The first and second derivatives of these $n-1$ equations are fundamental to this process and they are given by

$$s_i'(x) = 3a_i(x-x_i)^2 + 2b_i(x-x_i) + c_i$$
$$s_i''(x) = 6a_i(x-x_i) + 2b_i$$

We require that the piecewise function $S(x)$ interpolate all data points and that its first and second derivatives must be continuous in the interval $[x_1, x_n]$. Using these properties, we can compute the coefficients:

$$a_i = \frac{M_{i+1} - M_i}{6h}$$

$$b_i = \frac{M_i}{2}$$

$$c_i = \frac{y_{i+1} - y_i}{h} - \left(\frac{M_{i+1} + 2M_i}{6}\right)h$$

$$d_i = y_i$$

where:

$$M_i = s_i''(x_i), \quad y_i = S(x_i) = s_i(x_i)$$

Implementation

In order to implement the cubic spline interpolation, we need first calculate the second derivatives using the following two private static methods, *SecondDerivative* and *Tridiagonal*:

```
private static double[] SecondDerivatives(double[] xarray, double[] yarray)
{
    int n = xarray.Length;
    double[] c1 = new double[n];
    double[] c2 = new double[n];
    double[] c3 = new double[n];
    double[] dx = new double[n];
    double[] derivative = new double[n];

    for (int i = 1; i < n; i++)
    {
        dx[i] = xarray[i] - xarray[i - 1];
        derivative[i] = (yarray[i] - yarray[i - 1]) / dx[i];
    }

    for (int i = 1; i < n - 1; i++)
    {
        c2[i - 1] = 2;
        c3[i - 1] = dx[i + 1] / (dx[i] + dx[i + 1]);
        c1[i - 1] = 1 - c3[i - 1];
        derivative[i - 1] = 6 * (derivative[i + 1] - derivative[i]) /
                            (dx[i] + dx[i + 1]);
    }
```

```
        derivative = Tridiagonal(n - 2, c1, c2, c3, derivative);
        return derivative;
}

private static double[] Tridiagonal(int n, double[] c1, double[] c2,
                                    double[] c3, double[] derivative)
{
    double tol = 1.0e-12;
    bool isSingular = (c2[0] < tol) ? true : false;

    for (int i = 1; i < n && !isSingular; i++)
    {
        c1[i] = c1[i] / c2[i - 1];
        c2[i] = c2[i] - c1[i] * c3[i - 1];
        isSingular = (c2[i] < tol) ? true : false;
        derivative[i] = derivative[i] - c1[i] * derivative[i - 1];
    }

    if (!isSingular)
    {
        derivative[n - 1] = derivative[n - 1] / c2[n - 1];
        for (int i = n - 2; i >= 0; i--)
        {
            derivative[i] = (derivative[i] - c3[i] * derivative[i + 1]) / c2[i];
        }
        return derivative;
    }
    else
        return null;
}
```

Once we obtain the derivatives, we can then use them to implement the cubic spline interpolation. Add a new public static method, *Spline*, to the *Interpolation* class:

```
public static double Spline(double[] xarray, double[] yarray, double x)
{
    double d1, d2;
    double y = double.NaN;
    int n = xarray.Length;
    double[]dx=new double[n];
    double[] derivative = SecondDerivatives(xarray, yarray);

    for (int i = 1; i < n; i++)
    {
        dx[i] = xarray[i] - xarray[i - 1];
    }

    for (int i = 1; i < n - 1; i++)
    {
        if (x >= xarray[i] && x < xarray[i + 1])
        {
            d1 = x - xarray[i];
            d2 = xarray[i + 1] - x;
            y = derivative[i - 1] * d2 * d2 * d2 / (6.0 * dx[i + 1]) +
                derivative[i] * d1 * d1 * d1 / (6.0 * dx[i + 1]) +
                (yarray[i + 1] / dx[i + 1] - derivative[i] * dx[i + 1] / 6.0) * d1 +
```

```
                    (yarray[i] / dx[i + 1] - derivative[i - 1] * dx[i + 1] / 6.0) * d2;
            }
        }
        return y;
    }

    public static double[] Spline(double[] xarray, double[] yarray, double[] x)
    {
        double[] y = new double[x.Length];
        for (int i = 0; i < x.Length; i++)
            y[i] = Spline(xarray, yarray, x[i]);
        return y;
    }
```

Here, we implement an overloaded method, *Spline*. The *xarray* and *yarray* are the sets of *x* and *y* data, which present a set of given data points. This method returns a single *y* value or a *y* array at the input *x* (a double value or double array), depending on the input variable *x*.

Testing the Spline Interpolation

You can perform the interpolation using the *Spline* method. Add a method, *TestSpline*, to the *Program.cs* file:

```
static void TestSpline()
{
    double[] xarray = new double[] { 0, 2, 4, 6, 8 };
    double[] yarray = new double[] { 0, 4, 16, 36, 64 };
    double[] x = new double[] { 3, 5, 7 };
    double[] y = Interpolation.Spline(xarray, yarray, x);
    VectorR vx = new VectorR(x);
    VectorR vy = new VectorR(y);
    Console.WriteLine(" x = " + vx.ToString());
    Console.WriteLine(" y=" + vy.ToString());
}
```

Here, we first create a set of data points by defining the *xarray* and *yarray*. We then compute the *y* values at the *x* values specified by a double array. Finally, to display the results, we convert the *x* and *y* arrays to corresponding *VectorR* objects.

Running this example generates following results:

```
x = (3, 5, 7)
y = (8.92857142857143, 24.9285714285714, 46.8571428571429)
```

Bilinear Interpolation

In this section, we will discuss the bilinear interpolation, which is an extension of the linear interpolation for interpolating functions of two variables, $z = f(x, y)$, on a regular grid. The key step for the bilinear interpolation is to perform a linear interpolation first in one direction, and then again in the other direction.

The bilinear interpolation plays an important role in computer vision and image processing. It is a texture mapping method that produces a reasonably realistic image, also known as bilinear filtering or bilinear texture mapping. The bilinear algorithm is used to map a screen pixel location to a corresponding point on the texture map. A weighted average of the attributes, such as color and alpha value, of the four surrounding texels is calculated and applied to the screen pixel. This process is repeated for each pixel that forms the object being textured. I presented an example on how to create a color shading effect using bilinear interpolation in my other book – *"Practical C# Charts and Graphics"*. If you are interested in color mapping based on the bilinear algorithm, please read that book.

Algorithm

The bilinear interpolation uses four vertex values surrounding each rectangular unit cell to obtain any value inside the unit cell. Suppose you want to get the value at (x, y), and the vertices of the unit cell are located at (x_0, y_0), (x_0, y_1), (x_1, y_0), and (x_1, y_1), where they have the given function values z_{00}, z_{01}, z_{10}, and z_{11}, respectively, as shown in Figure 8-3.

The linear interpolation on the top row of neighbors, between (x_0, y_0) and (x_1, y_0), estimates the value $f(x, y_0)$ at (x, y_0) as

$$f(x, y_0) = \frac{x_1 - x}{x_1 - x_0} z_{00} + \frac{x - x_0}{x_1 - x_0} z_{10}$$

Likewise, the linear interpolation on the bottom row of neighbors, between (x_0, y_1) and (x_1, y_1), estimates the value $f(x, y_1)$ as

$$f(x, y_1) = \frac{x_1 - x}{x_1 - x_0} z_{01} + \frac{x - x_0}{x_1 - x_0} z_{11}$$

Finally, the linear interpolation between (x, y_0) and (x, y_1) estimates the value $f(x, y)$ at point (x, y) as

$$f(x, y) = \frac{y_1 - y}{y_1 - y_0} f(x, y_0) + \frac{y - y_0}{y_1 - y_0} f(x, y_1)$$

By substituting the expressions for $f(x, y_0)$ and $f(x, y_1)$ into the above equation, we obtain:

$$f(x, y) = \frac{y_1 - y}{y_1 - y_0} \left(\frac{x_1 - x}{x_1 - x_0} z_{00} + \frac{x - x_0}{x_1 - x_0} z_{10} \right) + \frac{y - y_0}{y_1 - y_0} \left(\frac{x_1 - x}{x_1 - x_0} z_{01} + \frac{x - x_0}{x_1 - x_0} z_{11} \right)$$

You can see that the equation for the function value $f(x, y)$ at point (x, y) is a polynomial involving powers of x and y no greater than 1, and with four coefficients :

$$f(x, y) = a_0 + a_1 x + a_2 y + a_3 xy$$

Because these four coefficients are determined by four values (z_{00}, z_{01}, z_{10}, z_{11}), they are usually uniquely determined by the given data. This implies that the comparable procedure of first interpolating along columns (in the y-direction) and then interpolating the results in the x-direction will give the same result, because it, too, will have a similar formula with a unique solution.

Note that the term "bilinear" derives from the process of the linear interpolation (twice in one direction, then once in the perpendicular direction), not from the formula for $f(x, y)$. The formula involves a term with xy, which is not linear.

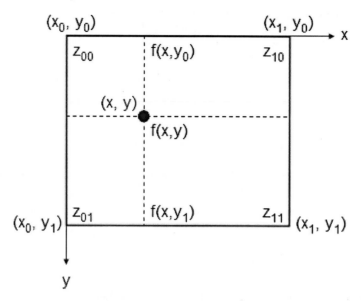

Figure 8-3 Coordinates used for the bilinear interpolation.

Implementation

Using the formula derived in the previous section, we can implement the bilinear interpolation in C#. Add a public static method, *Bilinear*, to the *Interpolation* class:

```
public static double Bilinear(double[] xarray, double[] yarray,
                              double[,] zarray, double x, double y)
{
    double z = double.NaN;
    for (int i = 0; i < xarray.Length - 1; i++)
    {
        for (int j = 0; j < yarray.Length - 1; j++)
        {
            if (x >= xarray[i] && x < xarray[i + 1] &&
                y >= yarray[j] && y < yarray[j + 1])
            {
                z = zarray[i, j] * (xarray[i + 1] - x) * (yarray[j + 1] - y) /
                    (xarray[i + 1] - xarray[i]) / (yarray[j + 1] - yarray[j]) +
                    zarray[i + 1, j] * (x - xarray[i]) * (yarray[j + 1] - y) /
                    (xarray[i + 1] - xarray[i]) / (yarray[j + 1] - yarray[j]) +
                    zarray[i, j + 1] * (xarray[i + 1] - x) * (y - yarray[j]) /
                    (xarray[i + 1] - xarray[i]) / (yarray[j + 1] - yarray[j]) +
                    zarray[i + 1, j + 1] * (x - xarray[i]) * (y - yarray[j]) /
                    (xarray[i + 1] - xarray[i]) / (yarray[j + 1] - yarray[j]);
            }
        }
    }
}
```

```
    return z;
}
```

Here, the *xarray* and *yarray* are the sets of *x* and *y* data, which present a set of grid points. The two-dimensional array, *zarray*, consists of a set of given data values at the grid points specified by *xarray* and *yarray*. The double variables *x* and *y* are the position where we want to compute the function value.

Testing the Bilinear Interpolation

We can perform interpolation by using the bilinear interpolation implemented in the previous section. We will perform a two-dimensional data mapping for a grid with four points: (0, 0), (0, 1), (1, 0), and (1, 1), which have the values 0, 10, 10, and 5, respectively, as shown in Figure 8-4. We want to compute any point inside this grid.

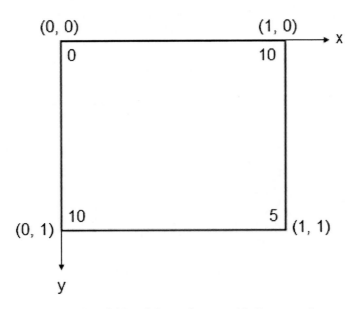

Figure 8-4 Grid and data values used in the example.

Add a method, *TestBilinear*, to the *Program.cs* file:

```
static void TestBilinear()
{
    double[] xarray = new double[] { 0, 1 };
    double[] yarray = new double[] { 0, 1 };
    double[,] zarray = new double[,] { { 0, 10 }, { 10, 5 } };
    double[] x = new double[9];
    double[] y = new double[9];
    MatrixR z = new MatrixR(9, 9);
    for (int i = 0; i < 9; i++)
    {
        x[i] = (i + 1.0) / 10.0;
        y[i] = (i + 1.0) / 10.0;
    }
```

```
VectorR vx = new VectorR(x);
VectorR vy = new VectorR(y);

for (int i = 0; i < 9; i++)
{
    for (int j = 0; j < 9; j++)
    {
        z[i, j] =
              Interpolation.Bilinear(xarray, yarray, zarray, x[i], y[j]);
    }
}
Console.WriteLine("x = " + vx.ToString());
Console.WriteLine("y = " + vy.ToString());
Console.WriteLine("\nResults z = \n" + z.ToString());
}
```

Here, we first create a set of data points by defining *xarray*, *yarray*, and *zarray*. We then compute the *z* values at point (*x*, *y*) specified by double arrays *x* and *y*. Finally, we convert the *x* and *y* arrays to *VectorR* objects, to display the results.

Running this example generates the results shown in Figure 8-5.

```
x = (0.1, 0.2, 0.3, 0.4, 0.5, 0.6, 0.7, 0.8, 0.9)
y = (0.1, 0.2, 0.3, 0.4, 0.5, 0.6, 0.7, 0.8, 0.9)

Results z =
(1.85, 2.7, 3.55, 4.4, 5.25, 6.1, 6.95, 7.8, 8.65
 2.7, 3.4, 4.1, 4.8, 5.5, 6.2, 6.9, 7.6, 8.3
 3.55, 4.1, 4.65, 5.2, 5.75, 6.3, 6.85, 7.4, 7.95
 4.4, 4.8, 5.2, 5.6, 6, 6.4, 6.8, 7.2, 7.6
 5.25, 5.5, 5.75, 6, 6.25, 6.5, 6.75, 7, 7.25
 6.1, 6.2, 6.3, 6.4, 6.5, 6.6, 6.7, 6.8, 6.9
 6.95, 6.9, 6.85, 6.8, 6.75, 6.7, 6.65, 6.6, 6.55
 7.8, 7.6, 7.4, 7.2, 7, 6.8, 6.6, 6.4, 6.2
 8.65, 8.3, 7.95, 7.6, 7.25, 6.9, 6.55, 6.2, 5.85)
```

Figure 8-5 Results from the bilinear interpolation.

Chapter 9
Curve Fitting

In science and engineering, the data obtained from experiments usually contain a significant amount of random noise due to measurement errors. The purpose of curve fitting is to find a smooth curve that, one average, fits the data points. This curve should have a simple form with a low-order polynomial, so it does not reproduce the random errors.

There is a distinction between interpolation and curve fitting. Interpolation, as discussed in the previous chapter, can be regarded as a special case of curve fitting in which the function must go exactly through the data points. This means an implicit assumption that the given data points in interpolation are accurate and distinct. Curve fitting is applied to data that contain noise, usually because of measurement errors. It tries to find the best fit to a set of given data. Thus, the curve does not necessarily pass through all of the given data points.

Least Squares Fit

The most popular curve fitting technique is the least squares method, which is usually used to solve overdetermined systems. It is often applied in statistics, particularly regression analysis. The best fit, in the least-squares sense, is the instance of the model in which the sum of squared residuals has the least value. The residual in the least squares is the difference between the observed data value and the value provided by the model.

Suppose that a given data set consists of n data points (x_i, y_i), $i = 0, 1,..., n-1$, where x_i is an independent variable and y_i is a dependent variable whose value is obtained by observation. The model function can be defined as

$$f(x; \mathbf{a}) = f(x; a_0, a_1, \cdots, a_m)$$

This function is to be fitted to the data set with $n + 1$ data points. The above model function contains $m + 1$ variable parameters $a_0, a_1, ..., a_m$, where $m < n$. We wish to find those parameter values for which the model best fits the data.

The form of the model function is determined beforehand, usually based on the theory associated with the experiment from which the data is obtained. The curve fitting process involves two steps: choosing the form of the model function, followed by computation of the parameters that produce the best fit to the data.

The least squares method minimizes the sum of squared residuals

$$S(\mathbf{a}) = \sum_{i=0}^{n} r_i^2, \quad r_i = y_i - f(x_i; \mathbf{a})$$

with respect to each parameter a_i. The terms in the above equation are called residuals, and they represent the discrepancy between the data points and the fitting function. Therefore, the optimal values of the parameters are given by the conditions:

$$\frac{\partial S(\mathbf{a})}{\partial a_i} = 2\sum_{i=0}^{n} r_i \frac{\partial r_i}{\partial a_i} = -2\sum_{i=0}^{n} \frac{\partial f(x_i; \mathbf{a})}{\partial a_i} r_i = 0, \quad i = 0, 1, \cdots, m$$

The above equations are generally nonlinear in a_i and may be difficult to solve. These gradient equations apply to all least squares problems. Each particular problem requires particular expressions for the model function and its partial derivatives. For example, the model function is often chosen as a linear combination of the specified functions $f_i(x)$:

$$f(x; \mathbf{a}) = \sum_{i=0}^{m} a_i f_i(x)$$

In this case, the gradient equation becomes linear. If the model function is a polynomial, we have $f_0(x) = 1, f_1(x) = x, f_2(x) = x^2$, etc.

Straight Line Fit

The simplest linear regression is the straight line fit, which attempts to fit a straight line using the least squares technique. It examines the correlation between an independent variable x and a dependent variable y. In this case, the model function has the following simple form:

$$f(x; \mathbf{a}) = a + bx$$

The sum function of the linear regression to be minimized becomes

$$S(a,b) = \sum_{i=0}^{n} [y_i - f(x_i; \mathbf{a})]^2 = \sum_{i=0}^{n} (y_i - a - bx_i)^2$$

The corresponding gradient equation becomes

$$\frac{\partial S}{\partial a} = -2\sum_{i=0}^{n} (y_i - a - bx_i) = 0$$

$$\frac{\partial S}{\partial b} = -2\sum_{i=0}^{n} (y_i - a - bx_i)x_i = 0$$

We can find solution for parameters a and b by solving the above two equations:

$$a = y_m - x_m b$$

$$b = \frac{\sum_{i=0}^{n} y_i (x_i - x_m)}{\sum_{i=0}^{n} x_i (x_i - x_m)}$$

where x_m and y_m are the mean values of the x and y data:

$$x_m = \frac{1}{n+1}\sum_{i=0}^{n} x_i, \quad y_m = \frac{1}{n+1}\sum_{i=0}^{n} y_i$$

The standard deviation σ can be expressed by

$$\sigma = \sqrt{\frac{S}{n-m}}$$

Implementation

Here, we will implement the straight line fit. Start with a new C# Console application and name it *CurveFittingTest*. Add a new class, *CurveFitting*, to the project and change its namespace to *XuMath*.

Using the algorithm developed in the previous section, we can implement a linear regression. Add a new public static method, *StraightLineFit*, to the *CurveFitting* class:

```
public static double[] StraightLineFit(double[] xarray, double[] yarray)
{
    int n = xarray.Length;
    double xm = 0.0;
    double ym = 0.0;
    double b1 = 0.0;
    double b2 = 0.0;
    double a = 0.0;
    double b = 0.0;
    double s = 0.0;
    double sigma = 0.0

    for (int i = 0; i < n; i++)
    {
        xm += xarray[i] / n;
        ym += yarray[i] / n;
    }

    for (int i = 0; i < n; i++)
    {
        b1 += yarray[i] * (xarray[i] - xm);
        b2 += xarray[i] * (xarray[i] - xm);
    }
    b = b1 / b2;
    a = ym - xm * b;

    for (int i = 0; i < n; i++)
    {
        s += (yarray[i] - a - b * xarray[i]) * (yarray[i] - a - b * xarray[i]);
    }
    sigma = math.Sqrt(s /(n - 2));
    return new double[] {a, b, sigma };
}
```

Here, the *xarray* and *yarray* are the sets of x and y data, which present a set of given data points. This method returns the coefficients a and b of the model function, and the standard deviation sigma.

Testing the Straight Line Fit

You can perform the straight line fit using the *StraightLineFit* method. Add a method, *TestStraightLineFit*, to the *Program.cs* file:

```
using System;
using XuMath;

namespace CurveFittingTest
{
    class Program
    {
        static void Main(string[] args)
        {
            TestLinearRegression();
            Console.ReadLine();
        }

        private static void TestStraightLineFit()
        {
            double[] xarray = new double[] { 0.0, 1.0, 2.0, 3.0, 4.0, 5.0 };
            double[] yarray = new double[] { 1.9, 2.7, 3.3, 4.4, 5.5, 6.5 };
            double[] results = CurveFitting.StraightLineFit(xarray, yarray);
            VectorR v = new VectorR(results);
            Console.WriteLine(v.ToString());
        }
    }
}
```

Here, we provide the input data points using two double arrays *xarray* and *yarray*. Running this example generates the following results:

```
(1.729, 0.929, 0.191)
```

Therefore, the regression line is given by

$$f(x) = 1.729 + 0.929x$$

and the standard deviation is 0.191.

Figure 9-1 shows the results of the straight line fitting to the original data.

Linear Regression

Let's first consider the least squares fit of the linear form:

$$f(x; \boldsymbol{a}) = \sum_{i=0}^{m} a_i f_i(x)$$

where $f_i(x)$ is a predetermined function of x, called a basis function. In this case, the sum of residuals is given by

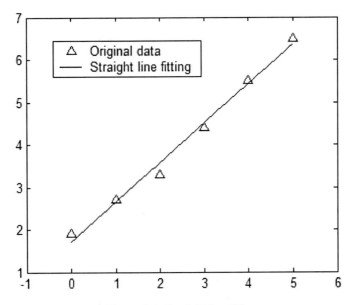

Figure 9-1 Straight line fitting.

$$S(\mathbf{a}) = \sum_{i=0}^{n}\left[y_i - \sum_{j=0}^{m} a_j f_j(x_i)\right]^2$$

The corresponding gradient equation from $\partial S / \partial a_k = 0$ reduces to

$$\sum_{j=0}^{m}\sum_{i=0}^{n} f_j(x_i) f_k(x_i)\, a_j = \sum_{i=0}^{n} f_k(x_i) y_i, \quad k = 0,1,\cdots,m$$

The above equation can be rewritten in a matrix form:

$$\mathbf{A\alpha} = \mathbf{\beta}, \quad A_{jk} = \sum_{i=0}^{n} f_j(x_i) f_k(x_i), \quad \beta_k = \sum_{i=0}^{n} f_k(x_i) y_i$$

The above matrix form is also called the normal equations of the least-squares fit, which can be solved with the methods discussed in Chapter 4.

Implementation

Using the algorithum developed in the previous section, we can implement the linear regression method. Add a new public static method, *LinearRegression*, to the *CurveFitting* class:

```
public delegate double ModelFunction(double x);

public static VectorR LinearRegression(double[] xarray, double[] yarray,
                ModelFunction[] f, out double sigma)
{
    int m = f.Length;
```

```csharp
    MatrixR A = new MatrixR(m, m);
    VectorR b = new VectorR(m);
    int n = xarray.Length;

    for (int k = 0; k < m; k++)
    {
        b[k] = 0.0;
        for (int i = 0; i < n; i++)
        {
            b[k] += f[k](xarray[i]) * yarray[i];
        }
    }

    for (int j = 0; j < m; j++)
    {
        for (int k = 0; k < m; k++)
        {
            A[j, k] = 0.0;
            for (int i = 0; i < n; i++)
            {
                A[j, k] += f[j](xarray[i]) * f[k](xarray[i]);
            }

        }
    }

    LinearSystem ls = new LinearSystem();
    VectorR coef = ls.GaussJordan(A, b);

    // Calculate the standard deviation:
    double s = 0.0;

    for (int i = 0; i < n; i++)
    {
        double s1 = 0.0;
        for (int j = 0; j < m; j++)
        {
            s1 += coef[j] * f[j](xarray[i]);
        }
        s += (yarray[i] - s1) * (yarray[i] - s1);
    }
    sigma = Math.Sqrt(s / (n - m));
    return coef;
}
```

Notice that we first define a delegate function that takes a double variable x as its input parameter. Then we implement a public static method, *LinearRegression*, that returns a *VectorR* object with its components as the coefficients of a basis function.

Inside the *LinearRegression* method, we specify the coefficient matrix A and vector b. The normal equations are then solved by the *GaussJordan* method in the *LinearSystem* class presented in Chapter 4. So you need to add the *LinearSystem.cs* class from the *LinearSystemTest* project in Chapter 4 to the current project in order to solve the normal equations. Following the solution, the standard deviation is also computed. Since C# can only return a single variable, the coefficients of the polynomial (a

VectorR object) in this case, the standard deviation is specified using the *out* prefix in the *PolynomialFit* method:

```
public static VectorR LinearRegression(double[] xarray, double[] yarray,
                      ModelFunction[] f, out double sigma)
{
    ⋮
}
```

In this way, you can obtain multiple output results from a single C# method.

Testing the Linear Regression

You can perform the linear regression using the *LinearRegression* method implemented in the previous section. Add a method, *TestLinearRegression*, to the *Program.cs* file:

```
private static void TestLinearRegression()
{
    double[] xarray = new double[] { 0, 1, 2, 3, 4, 5 };
    double[] yarray = new double[] { 2, 1, 4, 4, 3, 2 };

    // First order polynomial (m = 1):
    CurveFitting.ModelFunction[] f = new CurveFitting.ModelFunction[] { f0, f1};
    double sigma = 0.0;
    VectorR results =
            CurveFitting.LinearRegression(xarray, yarray,f, out sigma);
    Console.WriteLine("Order of polynomial m = 1" + ",
            Standard deviation = " + sigma.ToString());
    Console.WriteLine("Ceofficients = " + results.ToString() +"\n");

    //Second order polynomial (m = 2):
    f = new CurveFitting.ModelFunction[] { f0, f1, f2 };
    results = CurveFitting.LinearRegression(xarray, yarray, f, out sigma);
    Console.WriteLine("Order of polynomial m = 2" + ",
            Standard deviation = " + sigma.ToString());
    Console.WriteLine("Ceofficients = " + results.ToString() + "\n");

    //Third order polynomial (m = 3):
    f = new CurveFitting.ModelFunction[] { f0, f1, f2, f3 };
    results = CurveFitting.LinearRegression(xarray, yarray, f, out sigma);
    Console.WriteLine("Order of polynomial m = 3" + ",
            Standard deviation = " + sigma.ToString());
    Console.WriteLine("Ceofficients = " + results.ToString() + "\n");
}
```

In this method, we fit a polynomial of different order *m* (from 1 to 3) to the data points. We also need to define the delegate functions:

```
private static double f0(double x)
{
    return 1.0;
}

private static double f1(double x)
{
```

```
        return x;
    }

    private static double f2(double x)
    {
        return x * x;
    }

    private static double f3(double x)
    {
        return x * x * x;
    }
```

Running this application generates the results shown in Figure 9-4.

```
Order of polynomial m = 1, Standard deviation = 1.30566531115823
Ceofficients = (2.23809523809524, 0.171428571428571)

Order of polynomial m = 2, Standard deviation = 1.12122382116278
Ceofficients = (1.28571428571429, 1.6, -0.285714285714285)

Order of polynomial m = 3, Standard deviation = 1.15813204828717
Ceofficients = (1.67460317460324, -0.175925925926153, 0.686507936508051, -0.1296
29629629644)
```

Figure 9-2 Polynomial curve fitting.

From these results we see that the quadratic polynomial with $m = 2$:

$$f(x) = 1.2857 + 1.6x - 0.2857x^2$$

produces the smallest standard deviation, which can be considered as the "best" fit to the data. Note that the standard deviation is not a reliable measure of the integrity of the fit. It is always a good idea to plot the data points and $f(x)$ before you make your final determination. Here, we plot the data points and the polynomials with $m = 2$ and 3 in Figure 9-3. From this figure, it is hard to tell which polynomial provides a better fit for the data.

The LinearRegression method is very general. You can use any basis functions $f_j(x)$ that you like.

Polynomial Fit

As mentioned previously, the polynomial fit is a special case of the linear least squares method. In this case, the basis function becomes

$$f_j(x) = x^j, \quad j = 0, 1, \cdots, m$$

Thus, the matrix and vector in the normal equation become

$$A_{jk} = \sum_{i=0}^{n} x_i^{j+k}, \quad \beta_k = \sum_{i=0}^{n} x_i^k y_i$$

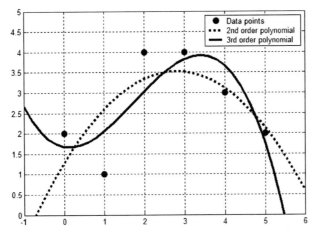

Figure 9-3 Curve fitting using the linear regression technique.

Implementation

Using the algorithum developed in the previous section, we can implement the polynomial curve fitting method. This method is very similar to the *LinearRegression* method presented previously, where you simply replace the basis functions $f_j(x)$ with x^j. Add a new public static method, *PolynomialFit*, to the *CurveFitting* class:

```
public static VectorR PolynomialFit(double[] xarray, double[] yarray,
                                    int m, out double sigma)
{
    m++;
    MatrixR A = new MatrixR(m, m);
    VectorR b = new VectorR(m);
    int n = xarray.Length;

    for (int k = 0; k < m; k++)
    {
        b[k] = 0.0;
        for (int i = 0; i < n; i++)
        {
            b[k] += Math.Pow(xarray[i], k) * yarray[i];
        }
    }

    for (int j = 0; j < m; j++)
    {
        for (int k = 0; k < m; k++)
        {
            A[j, k] = 0.0;
            for (int i = 0; i < n; i++)
            {
                A[j, k] += Math.Pow(xarray[i], j + k);
            }
        }
    }
```

```
        LinearSystem ls = new LinearSystem();
        VectorR coef = ls.GaussJordan(A, b);

        // Calculate the standard deviation:
        double s = 0.0;

        for (int i = 0; i < n; i++)
        {
            double s1 = 0.0;
            for (int j = 0; j < m; j++)
            {
                s1 += coef[j] * Math.Pow(xarray[i], j);
            }
            s += (yarray[i] - s1) * (yarray[i] - s1);
        }
        sigma = Math.Sqrt(s / (n - m));

        return coef;
    }
```

Here, we don't need to define the delegate function, since the basis functions are predetermined as polynomials. The method *PolynomialFit* sets up and solves the normal equation for the coeffients of a polynomial of degree *m*. It returns the coefficients of the polynomial.

Testing the Polynomial Fit

You can perform the polynomial curve fitting using the *Polynomial* method. We will calculate coefficients for the same polynomial used in examining the *LinearRegression* method to see if we can obtain the same results. Add a method, *TestPolynomial*, to the *Program.cs* file:

```
private static void TestPolynomialFit()
{
    double[] xarray = new double[] { 0, 1, 2, 3, 4, 5 };
    double[] yarray = new double[] { 2, 1, 4, 4, 3, 2 };
    for (int m = 1; m < 4; m++)
    {
        double sigma = 0.0;
        VectorR results =
                CurveFitting.PolynomialFit(xarray, yarray, m, out sigma);
        Console.WriteLine("\nOrder of polynomial m = " + m.ToString() +
                        ", Standard deviation = " + sigma.ToString());
        Console.WriteLine("Ceofficients = " + results.ToString());
    }
}
```

In this method, we fit a polynomial of different order *m* (from 1 to 3) to the data points. Running this application generates the results shown in Figure 9-4.

```
Order of polynomial m = 1, Standard deviation = 1.30566531115823
Coefficients = <2.23809523809524, 0.171428571428571>

Order of polynomial m = 2, Standard deviation = 1.12122382116278
Coefficients = <1.28571428571429, 1.6, -0.285714285714285>

Order of polynomial m = 3, Standard deviation = 1.15813204828717
Coefficients = <1.67460317460324, -0.175925925926153, 0.686507936508051, -0.1296
29629629644>
```

Figure 9-4 Polynomial curve fitting.

You can see that results from the *PolynomialFit* method, as expected, are exactly the same as those from the *LinearRegression* method.

Weighted Linear Regression

As mentioned previously, the linear regression is a useful technique for representing observed data by a model function, which is formulated as a least squares minimization problem. For the case of the linear least squares, the resulting analysis requires the solution of a set of simultaneous equations through the Gauss-Jordan method. One important assumption in linear regression is that all errors have the same significance.

However, there are occasions when the confidence in the accuracy of data varies from point to point. For example, there may be a drift in the precision of the measurements, and some errors may be more or less important than others. In order to take these factors into account, we can introduce a weight factor to each data point and minimize the sum of the squares of the weighted residuals

$$S(\alpha) = \sum_{i=0}^{n} r_i^2 = \sum_{i=0}^{n} w_i^2 [y_i - f(x_i)]^2$$

This procedure forces the ftting function $f(x)$ closer to the data points that have higher weights.

For the simplest linear regression, i.e., the straight line fit, the fitting function is given by $f(x) = a + bx$, the above equation becomes

$$S(a,b) = \sum_{i=0}^{n} w_i^2 (y_i - a - bx_i)^2$$

From the minimization conditions $\partial S / \partial a = 0$ and $\partial S / \partial b = 0$, we can determine the coefficients a and b:

$$a = y_w - bx_w, \quad b = \frac{\sum_{i=0}^{n} w_i^2 y_i (x_i - x_w)}{\sum_{i=0}^{n} w_i^2 x_i (x_i - x_w)}$$

where x_w and y_w are weighted averages:

$$x_w = \frac{\sum_{i=0}^{n} w_i^2 x_i}{\sum_{i=0}^{n} w_i^2}, \quad y_w = \frac{\sum_{i=0}^{n} w_i^2 y_i}{\sum_{i=0}^{n} w_i^2}$$

Implementation

Using the algorithum developed in the previous section, we can implement the weighted linear regression. Add a new public static method, *WeightedLinearRegression*, to the *CurveFitting* class:

```
public static double[] WeightedLinearRegression(double[] xarray,
                        double[] yarray, double[] warray)
{
    int n = xarray.Length;
    double xw = 0.0;
    double yw = 0.0;
    double b1 = 0.0;
    double b2 = 0.0;
    double a = 0.0;
    double b = 0.0;

    for (int i = 0; i < n; i++)
    {
        xw += xarray[i] / n;
        yw += yarray[i] / n;
    }

    for (int i = 0; i < n; i++)
    {
        b1 += warray[i] * warray[i] * yarray[i] * (xarray[i] - xw);
        b2 += warray[i] * warray[i] * xarray[i] * (xarray[i] - xw);
    }
    b = b1 / b2;
    a = yw - xw * b;

    return new double[] { a, b };
}
```

Here, the *xarray* and *yarray* are the sets of *x* and *y* data, which present a set of given data points. The *warray* is the weights for each data point. This method returns the coefficients *a* and *b* of the model function.

Exponential Function Fit

We can use the weighted linear regression to fit various exponential functions to a given set of data points. For example, the model function $f(x)=ae^{bx}$ used in the least squares technique usually leads to nonlinear dependence on coefficients *a* and *b*. However, if we use log *y* rather than *y*, the problem is transformed into a linear regression. In this case, the fit function becomes

$$F(x) = \ln f(c) = \ln a + bx$$

Note that the least-squares fit to the logarithm of the data is different from the least-squares fit to the original data. The residuals of the logarithmic fit are given by

$$R_i = \ln y_i - F(x_i) = \ln y_i - \ln a - bx_i$$

while the residuals used in fitting the original data are

$$r_i = y_i - f(x_i) = y_i - ae^{bx_i}$$

We can eliminate this discrepancy by weighting the logarithmic fit. From the above equation, we have

$$\ln(r_i - y_i) = \ln(ae^{bx_i}) = \ln a + bx_i$$

So that the residuals of the logarithmic fit can be rewritten in the form:

$$R_i = \ln y_i - \ln(r_i - y_i) = \ln\left(1 - \frac{r_i}{y_i}\right)$$

In the limit of $r_i \ll y_i$, we can use the approximation for $R_i = r_i / y_i$. It can be seen that by minimizing $\sum_{i=0}^{n} R_i^2$, we have to introduce the weight factor $1 / y_i$. This can be done if we apply the weights $w_i = y_i$ when fitting $F(x)$ to data points $(x_i, \ln y_i)$. Thus, minimizing

$$S = \sum_{i=0}^{n} y_i^2 R_i^2$$

will be a good approximation to minimizing $\sum_{i=0}^{n} r_i^2$.

Suppose we have the following set of data:

$x = 1, 2, 3, 4, 5, 6, 7, 8, 9, 10$

$y = 1.9398, 2.9836, 5.9890, 10.2, 20.7414, 23.232, 69.5855, 82.5836, 98.1779\ 339.3256$

we want to fit this data with an exponential function $y = ae^{bx}$. We first need to perform a logarithmic transformation $\log y = \log a + bx$, and then use the *WeightedLinearRegression* method to calculate the coefficients $\log a$ and b. Add a static method, *TestWeightedLinearRegression*, to the *Program.cs* class:

```
private static void TestWeightedLinearRegression()
{
    double[] x =  new double[] { 1, 2, 3, 4, 5, 6, 7, 8, 9, 10 };
    double[] y = new double[] { 1.9398, 2.9836, 5.9890, 10.2000, 20.7414,
                                23.2320, 69.5855, 82.5836, 98.1779, 339.3256 };
    double[] ylog = new double[] { 0.6626, 1.0931, 1.7899, 2.3224, 3.0321,
                                3.1455, 4.2426, 4.4138, 4.5868, 5.8270 };
    double[] results = CurveFitting.WeightedLinearRegression(x, ylog, y);
    VectorR v = new VectorR(results);
    Console.WriteLine(v.ToString());
}
```

In this case, the input parameters become: *xarray* = x, *yarray* = ylog, and *warray* = y. Running this example generates the following results:

$$(-0.0686983921044781, 0.578232434928087)$$

The first number is the parameter ($\log a$) and the second is the parameter b, which gives

$$\ln y \approx -0.0687 + 0.5782x$$

$$y \approx 0.9336 e^{0.5782x}$$

We can check how good the fit results are by plotting them graphically (see Figure 9-5). You can see from Figure 9-5 that the weighted linear regression indeed gives a reasonably good fit.

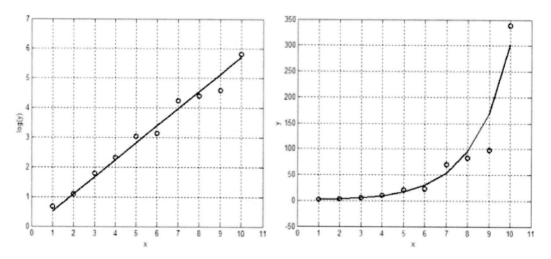

Figure 9-5 Results of the weighted linear regression. x ~ log y (left) and x ~ y (right).

Simple Moving Average

A moving averages is often used in analyzing time series data. It is widely applied in finance, particularly in technical analysis. It can also be used as a generic smoothing operation, in which case the data need not be a time series.

A moving average series can be calculated for any time series. In finance it is most often applied to stock prices, returns, or trading volumes. Moving averages are used to smooth out short-term fluctuations, thus highlighting longer-term trends or cycles. The threshold between short-term and long-term depends on the application, and the parameters of the moving average will be set accordingly.

A moving average smooths data by replacing each data point with the average of the neighboring data points defined within the time span. This processs is equivalent to the lowpass filters used in signal processing.

In this section, we first discuss the simple moving average. A simple moving average (SMA) is the mean value of the previous n data points. For example, a 5-day simple moving average of an opening price is the mean of the previous 5 days' opening prices. If those prices are $p_0, p_{-1}, p_{-2}, p_{-3}, p_{-4}$ then the moving average is described by

$$SMA = \frac{p_0 + p_{-1} + p_{-2} + p_{-3} + p_{-4}}{5}$$

In general, for a n-day moving average, we have

$$SMA = \frac{p_0 + p_{-1} + \cdots + p_{-n+1}}{n} = \frac{1}{n}\sum_{i=0}^{n-1} p_{-i}$$

When calculating successive values, a new value comes into the sum and an old value drops out, meaning a full summation each time is unnecessary

$$SMA_{today} = SMA_{yesterday} - \frac{P_{-n+1}}{n} + \frac{P_1}{n}$$

Technical analysis uses various popular values for n, like 10 days, 40 days, or 100 days. The period selected depends on the kind of movement you are concentrating on, such as short, intermediate, or long term. In any case moving average levels are interpreted as support in a rising market, or resistance in a falling market.

In general, a moving average lags behind the latest data point, simply from the nature of the smoothing. An SMA can lag to an undesirable extent, and can be disproportionately influenced by old data points dropping out of the average. To address this, extra weight is given to more recent data points, as in weighted and exponential moving averages, which will be discussed in following sections.

One characteristic of the SMA is that if the data have periodic fluctuations, then applying an SMA of those periods will eliminate that variation.

Implementation

Using the algorithum developed in the previous section, we can implement the simple moving average in C#. Add a new public static method, *SimpeMovingAverage*, to the *CurveFitting* class:

```
public static VectorR SimpleMovingAverage(double[] data, int n)
{
    int m = data.Length;
    double[] sma = new double[m - n + 1];
    if (m > n)
    {
        double sum = 0.0;
        for (int i = 0; i < n; i++)
        {
            sum += data[i];
        }

        sma[0] = sum / n;

        for (int i = 1; i <= m - n; i++)
        {
            sma[i] = sma[i - 1] + (data[n + i - 1] - data[i - 1]) / n;
        }
    }
    return new VectorR(sma);
}
```

In this method, the simple moving average is calculated by adding the prices for a number of periods and dividing by the number of periods. In the input, you simply provide the data and the number of periods.

Testing the Simple Moving Average

You can perform the simple moving average using the *SimpleMovingAverage* method. Here, we consider a sample closing data of a stock for 20 days, and use it to calculate a five-day simple moving average.

Add a method, *TestSimpleMovingAverage*, to the *Program.cs* file:

```
private static void TestSimpleMovingAverage()
{
    double[] data = new double[] {45.375, 45.500, 45.000, 43.625, 43.375,
                                  43.125, 43.125, 44.250, 43.500, 44.375,
                                  45.875, 46.750, 47.625, 48.000, 49.125,
                                  48.750, 46.125, 46.750, 46.625, 46.000 };
    VectorR sma = CurveFitting.SimpleMovingAverage(data, 5);
    Console.WriteLine(sma.ToString());
}
```

Running this application produces the result:

```
SMA = (44.575, 44.125, 43.650, 43.500, 43.475, 43.675, 44.225, 44.950,
       45.625, 46.525, 47.475, 48.050, 47.925, 47.750, 47.475, 46.850)
```

Note that there is no simple moving average data for the first four days, since there are not enough data to make the calculation until the fifth day.

For comparison, we plot the data and the five-day simple moving average in Figure 9-6. In practice, longer time periods are typically used for simple moving averages.

Weighted Moving Average

A weighted average is any average that has multiplying factors that give different weights to different data points. But in technical analysis, a weighted moving average (WMA) specifically means weights which decrease arithmetically. In an n-day WMA the latest day has weight n, the second latest $n - 1$, etc., down to zero:

$$WMA = \frac{np_0 + (n-1)p_{-1} + \cdots + 2p_{-n+2} + p_{-n+1}}{n + (n-1) + \cdots + 2 + 1} = \frac{2}{n(n+1)} \sum_{i=0}^{n-1} (n-i)p_{-i}$$

when calculating the WMA across successive values, please note the difference between the numerators of WMA_{+1} and WMA is $np_{+1} - p_0 - \cdots - p_{-n+1}$. If we denote $p_0^{sum} = p_0 + p_{-1} + \cdots + p_{-n+1}$, then we have

$$p_{+1}^{sum} = p_0^{sum} + p_{+1} - p_{-n+1}$$

$$Numerator_{+1} = Numerator_0 + np_{+1} - p_0^{sum}$$

$$WMA_{+1} = \frac{2}{n(n+1)} Numerator_{+1}$$

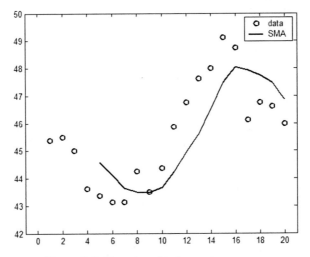

Figure 9-6 Five-day simple moving average.

Implementation

Using the algorithm developed in the previous section, we can implement the weighted moving average in C#. Add a new public static method, *WeightedMovingAverage*, to the *CurveFitting* class:

```
public static VectorR WeightedMovingAverage(double[] data, int n)
{
    int m = data.Length;
    double[] wma = new double[m - n + 1];
    double psum = 0.0;
    double numerator = 0.0;
    double[] numerator1 = new double[m - n  + 1];
    double[] psum1 = new double[m - n + 1];

    if (m > n)
    {
        for (int i = 0; i < n; i++)
        {
            psum += data[i];
            numerator += (i + 1) * data[i];
        }
        psum1[0] = psum;
        numerator1[0] = numerator;
        wma[0] = 2 * numerator / n / (n + 1);

        for (int i = 1; i <= m - n; i++)
        {
            numerator1[i] =
                numerator1[i - 1] + n * data[i + n - 1] - psum1[i - 1];
            psum1[i] = psum1[i - 1] + data[i + n - 1] - data[i - 1];
            wma[i] = 2 * numerator1[i] / n / (n + 1);
        }
    }
}
```

```
            return new VectorR(wma);
    }
```

In this method, we see that each period's price is multipled by a given weight. The products of the calculation are summed and divided by the total of the weights.

Testing the Weighted Moving Average

You can perform the weighted moving average using the *WeightedMovingAverage* method. Here, we will use the same sample closing data of a stock for 20 days that was used in the previous example, to calculate a five-day weighted moving average.

Add a method, *TestWeightedMovingAverage*, to the *Program.cs* file:

```
private static void TestWeightedMovingAverage()
{
    double[] data = new double[] {45.375, 45.500, 45.000, 43.625, 43.375,
                          43.125, 43.125, 44.250, 43.500, 44.375,
                          45.875, 46.750, 47.625, 48.000, 49.125,
                          48.750, 46.125, 46.750, 46.625, 46.000 };
    VectorR sma = CurveFitting.WeightedMovingAverage(data, 5);
    Console.WriteLine(sma.ToString());
}
```

Running this application produces the result:

```
WMA = (44.183, 43.700, 43.367, 43.567, 43.567, 43.867, 44.600, 45.442,
       46.333, 47.125, 47.992, 48.417, 47.775, 47.383  47.008, 46.517)
```

Note that there is no weighted moving average data for the first four days, since there are not enough data to make the calculation until the fifth day.

For comparison, we plot the data, the five-day weighted, and simple moving averages in Figure 9-7. You can see that the weighted moving average is closer to data than the simple moving average is.

Exponential Moving Average

One drawback of both the simpe and weighted moving average is that they include data for only the number of periods the moving average covers. For example, a five-day simple or weighted moving average only uses five days' worth of data. Data prior to those five days are not included in the calculation of the moving average.

In some situations, however, the prior data is an important reflection of prices and should be included in a moving average calculation. This can be achieved by using an exponential moving average.

An exponential moving average (EMA) uses weight factors that decrease exponentially. The weight for each older data point decreases exponentially, giving much more importance to recent observations while still not discarding older observations entirely.

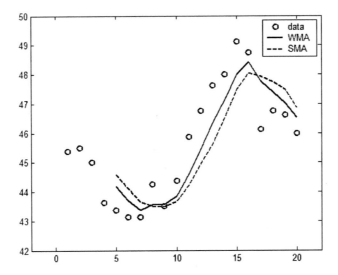

Figure 9-7 Results of the simple and weighted moving averages.

The degree of weight decrease is expressed as a constant smoothing factor α, a number between 0 and 1. α may be expressed as a percentage, so a smoothing factor of 10% is equivalent to $\alpha = 0.1$. Alternatively, α may be expressed in terms of n time periods, where $\alpha = 2/(n+1)$. For example, $n = 19$ is equivalent to $\alpha = 0.1$.

The observation at a time period t is designated Y_t, and the value of the EMA at any time period t is designated S_t. S_1 is undefined. S_2 may be initialized in a number of different ways, most commonly by setting S_2 to Y_1, though other techniques exist, such as setting S_2 to an average of the first 4 or 5 observations. The prominence of the S_2 initialization's effect on the resultant moving average depends on α; smaller α values make the choice of S_2 relatively more important than larger α values, since a higher α discounts older observations faster.

The formula for calculating the EMA at time periods $t \geq 2$ is

$$S_t = \alpha Y_{t-1} + (1-\alpha)S_{t-1}$$

This formula can also be expressed in technical analysis terms as follows, showing how the EMA steps towards the latest data point:

$$EMA_{today} = EMA_{yesterday} + \alpha\left(p_0 - EMA_{yesterday}\right)$$

where p_0 is the current price. Expanding out $EMA_{yesterday}$ each time results in the following power series, showing how the weighting factor on each data point p_1, p_2, etc, decrease exponentially:

$$EMA = \frac{p_{-1} + (1-\alpha)p_{-2} + (1-\alpha)^2 p_{-3} + \cdots}{1 + (1-\alpha) + (1-\alpha)^2 + \cdots} = \frac{\sum_{i=0}^{\infty}(1-\alpha)^i p_{-i-1}}{\sum_{i=0}^{\infty}(1-\alpha)^i} = \alpha \sum_{i=0}^{\infty}(1-\alpha)^i p_{-i-1}$$

Theoretically, this is an infinite sum, but because $1-\alpha$ is less than one, the terms become smaller and smaller, and can be ignored when they are small enough.

The n periods in an n-day EMA only specify the α factor. n is not a stopping point for the calculation as it is in an SMA or WMA. The first n data points in an EMA represent about 86% of the total weight in the calculation. As an approximation, we set $\alpha \approx 2/(n+1)$.

Implementation

Using the algorithum developed in the previous section, we can implement the exponential moving average in C#. Add a new public static method, *ExponentialMovingAverage*, to the *CurveFitting* class:

```
public static VectorR ExponentialMovingAverage(double[] data, int n)
{
    int m = data.Length;
    double[] ema = new double[m - n + 1];
    double psum = 0.0;
    double alpha = 2.0 / n;

    if (m > n)
    {
        for (int i = 0; i < n; i++)
        {
            psum += data[i];
        }
        ema[0] = psum / n + alpha * (data[n - 1] - psum / n);

        for (int i = 1; i <= m - n; i++)
        {
            ema[i] = ema[i - 1] + alpha * (data[i + n - 1] - ema[i - 1]);
        }
    }
    return new VectorR(ema);
}
```

In this method, before calculating the exponential moving average, we must have a beginning moving average number. To start, here we use a n-day simple moving average for the previous day's exponential moving average. Each day we make the following calculation: the previous day's exponential moving average is subtracted from the current day's price. That difference is multiplied by the exponential factor α to arrive at a number that we add to the previous day's exponential moving average, resulting in the current day's exponential moving average.

Testing the Exponential Moving Average

You can calculate the exponential moving average using the *ExponentialMovingAverage* method. Here, we will use the same sample closing data of a stock for 20 days that we used in previous examples, to calculate a five-day exponential moving average.

Add a method, *TestExponentialMovingAverage*, to the *Program.cs* file:

```
private static void TestExponentialMovingAverage()
{
    double[] data = new double[] {45.375, 45.500, 45.000, 43.625, 43.375,
                                  43.125, 43.125, 44.250, 43.500, 44.375,
                                  45.875, 46.750, 47.625, 48.000, 49.125,
                                  48.750, 46.125, 46.750, 46.625, 46.000 };
```

```
VectorR sma = CurveFitting.ExponentialMovingAverage(data, 5);
Console.WriteLine(sma.ToString());
}
```

Running this application produces the result:

```
EMA = (44.095, 43.707, 43.474, 43.785, 43.671, 43.952, 44.721, 45.533,
       46.370, 47.022, 47.863, 48.218, 47.381, 47.128  46.927, 46.556)
```

Note that there is no exponential moving average data for the first four days, since there are not enough data to make the calculation until the fifth day.

For comparison, we plot the data, the five-day exponential, weighted, and simple moving averages in Figure 9-8.

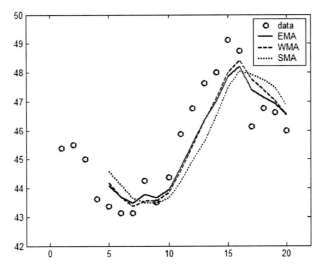

Figure 9-8 Results of the exponential, weighted, and simple moving averages.

Chapter 10
Optimization

Optimization is often used to find the values of variables that yield a minimum or maximum function value. It usually only considers the problem of minimization because maximization of $f(x)$ can be achieved by simply minimizing $-f(x)$.

In this chapter, we will present a variety of the optimization methods that can be applied to real-world scientific and engineering problems. Note that no optimization methods are absolutely reliable, i.e., one method may work on one problem and fail on another. As a rule of thumb, you may gain computational efficiency by using sophisticated methods, but not necessarily reliability.

As with interpolation, the algorithms for optimization are also iterative procedures that require starting values for the variables. If the function $f(x)$ has several local minima, the initial guess determines which of these will be computed. There is no guaranteed way to find the global optimal point. One good way is to make several computer runs using different starting points, and then pick the best result.

In this chapter, I will also present several popular methods for optimizing functions with multiple variable, including the simplex, simulated annealing, and the more sophisticated methods, such as genetic and evolution algorithms. In particular, simulated annealing, genetic, and evolution methods can deal with highly nonlinear models, chaotic, noisy data, and constraints. They represent more robust and general optimization techniques.

Bisection Method

The bisection method for finding the minimum works in a similar way to the bisection method for finding a root of a nonlinear function, which was discussed in Chapter 5. The method starts with an interval that contains the minimum and then divides that interval into two parts to zoom in on the minimum location.

For a give inverval $[x_a, x_b]$, you can use the sign of $f'(x_a) f'(x_b)$ to determine whether there are mimina in the interval. If $f'(x_a) f'(x_b) < 0$, as shown in Figure 10-1, $f(x)$ has at least one minimum between x_a and x_b.

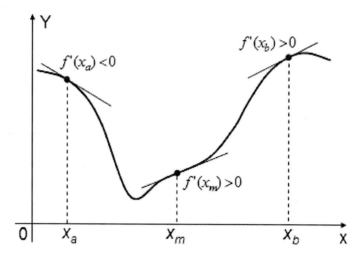

Figure 10-1 Illustration of the bisection method.

In this case, we can always find the midpoint, x_m, between x_a and x_b. This gives us two new intervals, (x_a, x_m) and (x_m, x_b). To find out which interval, (x_a, x_m) or (x_m, x_b), contains the minimum, we can find the sign of $f'(x_a) f'(x_m)$, and if $f'(x_a) f'(x_m) < 0$, then the minimum is in the interval (x_a, x_m). Otherwise, it is between x_m and x_b. As we repeat the process, the width of the interval that contains the minimum becomes smaller and smaller, and eventually you can reach the minimum of the function of $f(x)$.

Implementation

The steps to apply the bisection method to find the minimum of the function $f(x)$ are listed below:

- Chooese x_a and x_b as two guesses for the minimum such that $f'(x_a) f'(x_b) < 0$.

- Set $x_m = (x_a + x_b)/2$ as the mid-point between x_a and x_b.

- Check the sign of $f'(x_a) f'(x_m)$. If this quantity is greater than zero, then set $x_a = x_m$; otherwise set $x_b = x_m$.

- Repeat the above steps until the specified accuracy is reached.

Start with a new C# Console application and name it *OptimizationTest*. Add a new class, *Optimization*, to the project and change its namespace to *XuMath*. Add a public static method, *Bisection*, to the *Optimization* class:

```
using System;
using System.Collections;

namespace XuMath
{
    public class Optimization
    {
        public delegate double Function(double x);
        public static double Bisection(Function f, double xa, double xb,
            double tolerance)
        {
```

```
       double xm, fa, fb, fm;
       fa = f(xa);
       fb = f(xb);
       do
       {
           xm = 0.5 * (xa + xb);
           fm = f(xm);

           if (Derivative(f, xm) * Derivative(f, xa) > 0)
           {
               xa = xm;
               fm = f(xa);
           }
           else
           {
               xb = xm;
               fm = f(xb);
           }

       }
       while (Math.Abs(xb - xa) > tolerance);
       return xm;
   }

   private static double Derivative(Function f, double x)
   {
       double dx = (Math.Abs(x) > 1) ? 0.01 * x : 0.01;
       return (f(x + dx) - f(x - dx)) / (2.0 * dx);
   }
}
}
```

Here, we first introduce a delegate function Function(double *x*), which is a user-supplied nonlinear function. The *Bisection* method takes this user-supplied function *f*, the starting interval [x_a, x_b], and accuracy control parameter *tolerance* as its input parameters.

Testing the Bisection Method

Now, we can use the bisection method to find the minimum of a nonlinear function. To test it out, we will try to find the minimum of the following function in the interval [0, 1]:

$$f(x) = 1.6x^3 + 3x^2 - 2x$$

Add a new static method, *TestBisection*, to the *Program.cs* file:

```
using System;
using XuMath;

namespace OptimizationTest
{
    class Program
    {
        static void Main(string[] args)
        {
            TestBisection();
```

```
            Console.ReadLine();
        }

        static void TestBisection()
        {
            double result = Optimization.Bisection(f, 0.0, 1.0, 1.0e-5);
            Console.WriteLine("x = " + result.ToString() + ",
                        f(x) = " + f(result).ToString());
        }

        static double f(double x)
        {
            return 1.6 * x * x * x + 3 * x * x - 2 * x;
        }
    }
}
```

Running this example produces the following results

```
x    =  0.273475646972656
f(x) = -0.289859784079365
```

Golden Search Method

The golden search, also called the golden section search, works similarly to the bisection method in finding the minimum in an interval. Suppose that the minimum of $f(x)$ is in the interval $[x_a, x_b]$ of length w. The golden search method uses an interval reduction factor that is based on the Fibonacci numbers, instead of just selecting the middle point of the interval.

To divide the interval using the golden section method, we evaluate the function at $x_1 = x_b - gw$ and $x_2 = x_a + gw$, as shown in Figure 10-2 (top). The constant g will be determined later. If $f_1 > f_2$ as indicated in the figure, the minimum lies in the range of x_a and x_2; otherwise it is located in (x_1, x_b).

Assuming that $f_1 > f_2$, we reset $x_a \rightarrow x_1$ and $x_2 \rightarrow x_1$, which yields a new interval $[x_a, x_b]$ with a length of $w' = gw$, as illustrated in the bottom of Figure 10-2. To carry out the next sectioning operation, we evaluate the function at $x_2 = x_a + gw'$ and repeat the process.

The procedure works only if the same constant g locates x_1 and x_2 in both cases. From the top figure, we have

$$x_2 - x_1 = 2g(w - 1)$$

On the other hand, we have the following relation from the bottom:

$$x_1 - x_a = w'(1 - g)$$

Note that the above two equation describe the same distance. Equating the two and using the relation $w' = gw$, we get the golden ratio:

$$g = \frac{\sqrt{5} - 1}{2} = 0.618033989...$$

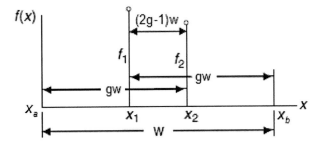

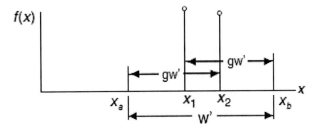

Figure 10-2 Illustration of the golden section method.

Note that each sectioning step decreases the interval containing the minimum by the factor of *g*, which does not seem as efficient as the factor of 0.5 in the bisection method. However, the golden search method achieves this reduction with a single function evaluation, whereas two evaluations would be needed if we were using the bisection method.

Implementation

The algorithm of the golden section search method is similar to the algorithm of the bisection method. For a given interval [x_a, x_b] that contains the minimum value for the function *f(x)*, and the tolerance level, the golden search can be implemented by adding a public static method, *GoldenSearch*, to the *Optimization* class:

```
public static double GoldenSearch(Function f, double xa, double xb,
                                  double tolerance)
{
    double x1, x2, f1, f2;
    double g = 1.0 - (Math.Sqrt(5.0) - 1.0) / 2.0;
    x1 = xa + g * (xb - xa);
    x2 = xb - g * (xb - xa);
    f1 = f(x1);
    f2 = f(x2);
    do
    {
        if (f1 < f2)
        {
            xb = x2;
            x2 = x1;
            x1 = xa + g * (xb - xa);
            f2 = f1;
```

```
            f1 = f(x1);
        }
        else
        {
            xa = x1;
            x1 = x2;
            x2 = xb - g * (xb - xa);
            f1 = f2;
            f2 = f(x2);
        }
    }
    while (Math.Abs(xb - xa) > tolerance);
    return 0.5 * (xa + xb);
}
```

Testing the Golden Search Method

Now, we can use the golden search method to find the minimum of a nonlinear function. To test it out, we will find the minimum of the same function used in testing the bisection method previously.

Add a new static method, *TestGoldenSearch*, to the *Program.cs* file:

```
static void TestGoldenSearch()
{
    double result = Optimization.GoldenSearch(f, 0.0, 1.0, 1.0e-5);
    Console.WriteLine("x = " + result.ToString() + ",
                        f(x) = " + f(result).ToString());
}
```

Running this example produces the following results

```
x    =  0.273494020639686
f(x) = -0.289859785549557
```

which gives the same accuracy within the tolerance as that from the bisection method.

Newton Method

You can also use the Newton's root seeking method to find the minimum, maximum, or saddle point of a function, because the derivative of the targeted function is zero at these points. In this method, the minimum is not bracketed and only one initial guess value of the solution is needed to get the iterative process started to find the minimum of a nonlinear function. Thus, unlike bracketing-based methods such as the bisection method and the golden search, which always give the convergent result, the Newton method may fail to converge if the initial value is too far from the true mimimum.

The Newton-Raphson method uses the following iteration relation:

$$x_{n+1} = x_n - \frac{f'(x_n)}{f''(x_n)}$$

where $f'(x)$ and $f''(x)$ are the first and second derivatives of function $f(x)$ respectively.

Starting with an initial value, x_n, we can find the next guess, x_{n+1}, by using the above iterative relation. This process can be repeated until the solution within a given tolerance is found.

Implementation

For a given minimized function $f(x)$, the initial guess for the minimum x, and the tolerance, we can implement the Newton method by adding a public static method, *Newton*, to the *Optimization* class:

```
public static double Newton(Function f, double x, double tolerance)
{
    double dx;
    double fm, f0, fp;
    double d1, d2;
    do
    {
        dx = (Math.Abs(x) > 1) ? 0.01 * x : 0.01;
        f0 = f(x);
        fm = f(x - dx);
        fp = f(x + dx);
        d1 = (fp - fm) / (2.0 * dx);
        d2 = (fp - 2.0 * f0 + fm) / dx / dx;
        x -= d1 / d2;
    }
    while (Math.Abs(d1 / d2) > tolerance);
    return x;
}
```

Testing the Newton Method

Now, we can use the Newton method to find the minimum of a nonlinear function. To test it out, we will find the minimum of the same function we used to test the bisection method previously.

Add a new static method, *TestNewton*, to the *Program.cs* file:

```
static void TestNewton()
{
    double result = Optimization.Newton(f, 0.0, 1.0e-5);
    Console.WriteLine("x = " + result.ToString() + ",
                      f(x) = " + f(result).ToString());
}
```

Running this example produces the following results

```
x    =  0.273475560790101
f(x) = -0.289859784065608
```

which gives the same accuracy within the tolerance as that from the bisection method.

Brent Method

The Brent method is based on the quadratic or parabolic interpolation technique. If the function is smooth and parabolic near to the minimum, then a parabola fitted through any three points should lead to the minimum, or at least very close to it.

The formula for a location x that is the minimum of a parabola through three points $(x_a, f(x_a))$, $(x_b, f(x_b))$, and $(x_c, f(x_c))$ is given by

$$x = x_b - \frac{1}{2} \frac{(x_b - x_a)^2 \left[f(x_b) - f(x_c) \right] - (x_b - x_c)^2 \left[f(x_b - f(x_a) \right]}{(x_b - x_a) \left[f(x_b) - f(x_c) \right] - (x_b - x_c) \left[f(x_b - f(x_a) \right]}$$

This formula can be easily derived. It fails when the three points are collinear, in which case the denominator is zero.

The Brent method is an optimization scheme based on the parabolic interpolation. At any particular stage, the method keeps track of six function points, x_a, x_b, x_u, x_v, x_w, and x, defined as follows: x_a and x_b define the interval $[x_a, x_b]$ that contains the minimum; x is the point with the very least function value found so far; x_w is the point with the second least function value; x_v is the previous value of x_w; x_u is the point at which the function was evaluated most recently. We also need the point x_m, the midpoint between x_a and x_b.

The Brent method attempts to fit through the points x, x_v, and x_w, which requires the parabolic step to fall within the interval $[x_a, x_b]$ and imply a movement from the best current value x that is less than half the movement of the step before the last. These requirements ensure that the parabolic steps converges, rather than bounces around in a nonconvergent limit cycle. In the worst possible scenario, where the parabolic steps do not converge, the method will switch to the golden section search method.

Implementation

For a given minimized function $f(x)$, the initial interval $[x_a, x_b]$ that contains the minimum, and the tolerance, we can implemented the Brent method by adding a public static method, *Brent*, to the *Optimization* class:

```
public static double Brent(Function f, double xa, double xb, double tolerance)
{
    double x1 = 0;
    double x2 = 0;
    double bx = 0;
    double xd = 0;
    double xe = 0;
    double xtemp = 0;
    double fu = 0;
    double fv = 0;
    double fw = 0;
    double fx = 0;
    double p = 0;
    double q = 0;
    double r = 0;
    double xu = 0;
    double xv = 0;
    double xw = 0;
    double x = 0;
    double xm = 0;
    double tao = 0.5 * (3.0 - Math.Sqrt(5));

    bx = 0.5 * (xa + xb);
    if (xa < xb)
    {
        x1 = xa;
    }
    else
```

```
{
    x1 = xb;
}
if (xa > xb)
{
    x2 = xa;
}
else
{
    x2 = xb;
}
xv = bx;
xw = xv;
x = xv;
xe = 0.0;
fx = f(x);
fv = fx;
fw = fx;

do
{
    xm = 0.5 * (x1 + x2);
    if (Math.Abs(xe) > tolerance)
    {
        r = (x - xw) * (fx - fv);
        q = (x - xv) * (fx - fw);
        p = (x - xv) * q - (x - xw) * r;
        q = 2 * (q - r);
        if (q > 0)
        {
            p = -p;
        }
        q = Math.Abs(q);
        xtemp = xe;
        xe = xd;
        if (!(Math.Abs(p) >= Math.Abs(0.5 * q * xtemp) |
            p <= q * (x1 - x) | p >= q * (x2 - x)))
        {
            xd = p / q;
            xu = x + xd;
            if (xu - x1 < tolerance * 2 | x2 - xu < tolerance * 2)
            {
                xd = Math.Sign(xm - x) * tolerance;
            }
        }
        else
        {
            if (x >= xm)
            {
                xe = x1 - x;
            }
            else
            {
                xe = x2 - x;
            }
```

```
                xd = tao * xe;
        }
    }
    else
    {
        if (x >= xm)
        {
            xe = x1 - x;
        }
        else
        {
            xe = x2 - x;
        }
        xd = tao * xe;
    }
    if (Math.Abs(xd) >= tolerance)
    {
        xu = x + xd;
    }
    else
    {
        xu = x + Math.Sign(xd) * tolerance;
    }
    fu = f(xu);
    if (fu <= fx)
    {
        if (xu >= x)
        {
            x1 = x;
        }
        else
        {
            x2 = x;
        }
        xv = xw;
        fv = fw;
        xw = x;
        fw = fx;
        x = xu;
        fx = fu;
    }
    else
    {
        if (xu < x)
        {
            x1 = xu;
        }
        else
        {
            x2 = xu;
        }
        if (fu <= fw | xw == x)
        {
            xv = xw;
            fv = fw;
```

```
                    xw = xu;
                    fw = fu;
                }
                else
                {
                    if (fu <= fv | xv == x | xv == 2)
                    {
                        xv = xu;
                        fv = fu;
                    }
                }
            }
        }
        while (Math.Abs(x - xm) > tolerance * 2 - 0.5 * (x2 - x1));
        return x;
    }
```

Testing the Brent Method

Now, we can use the Brent method to find the minimum of a nonlinear function. To test it out, we will find the minimum of the same function we used to test the bisection method previously.

Add a new static method, *TestBrent*, to the *Program.cs* file:

```
static void TestBrent()
{
    double result = Optimization.Brent(f, 0.0, 1.0, 1.0e-5);
    Console.WriteLine("x = " + result.ToString() + ",
                    f(x) = " + f(result).ToString());
}
```

Running this example produces the following results

```
x    =  0.273494354133259
f(x) = -0.289859785549336
```

which gives the same accuracy within the tolerance as that from the bisection method.

Newton Method for Multi-Variable Functions

The optimization methods discussed so far are applicable only for functions with a single variable. This means that the functions are defined in one-dimensional space. Here, we will look at optimization for functions with multiple variables. It is possible to extend a single-variable optimization method to work with multiple variables.

In this section, we will extend the Newton method to find the minimum of a function with multiple variables. The basic idea is simple:

- Start with a initial array, which represents initial points in n-dimensional space.
- For each variable, i.e., x_i, we minimize the multi-variable function $f(x)$, where x is a n-dimensional vector, using the Newton method.
- Loop over all variables.

The minimization along a line with a single variable can be accomplished with any one-dimensional optimization algorithm (it is not limited to the Newton method).

Implementation

For a given multi-variable function $f(x)$, the initial guess for the minimum x, and the tolerance, we can implement the Newton method by adding a public static method, *MultiNewton*, to the Optimization class:

```
public delegate double MultiFunction(VectorR x);

public static VectorR multiNewton(MultiFunction f, double[] xarray,
                                   double tolerance)
{
    for (int i = 0; i < xarray.Length; i++)
    {
        double dx, fm, f0, fp;
        double d1, d2;
        double x = xarray[i];
        do
        {
            dx = (Math.Abs(x) > 1) ? 0.01 * x : 0.01;
            xarray[i] = x - dx;
            fm = f(new VectorR(xarray));
            xarray[i] = x + dx;
            fp = f(new VectorR(xarray));
            xarray[i] = x;
            f0 = f(new VectorR(xarray));
            d1 = (fp - fm) / (2.0 * dx);
            d2 = (fp + fm - 2.0 * f0) / dx / dx;
            x -= d1 / d2;
            xarray[i] = x;
        }
        while (Math.Abs(d1 / d2) > tolerance);
    }
    return new VectorR(xarray);
}
```

Here, we first define a delegate function called *MultiFunction*, which represents a function with multiple variables. The *MultiNewton* method takes the *Multifunction*, initial array, and tolerance as input parameters. You can see that the *MultiNewton* method performs repeated sequential optimization on variables, one at a time. At the end of each cycle the method examines for convergence.

Testing the MultiNewton Method

Now, we can use the *MultiNewton* method to find the minimum of a function with multiple variables. As an example, we will find the minimum of a function given by

$$f(x, y) = 1.5(x - 0.5)^2 + 3.4(y + 1.2)^2 + 2.5$$

Add a new static method, *TestMultiNewton*, to the *Program.cs* file:

```
static void TestMultiNewton()
{
    double[] xarray = new double[] { 0, 0 };
    VectorR result = Optimization.multiNewton(f1, xarray, 1.0e-5);
    Console.WriteLine("x = " + result.ToString());
    Console.WriteLine("f1(x) = " + f1(result).ToString());
}

static double f1(VectorR x)
{
    return 1.5 * (x[0] - 0.5) * (x[0] - 0.5) +
           3.4 * (x[1] + 1.2) * (x[1] + 1.2) + 2.5;
}
```

Running this example produces the following result

```
x     = (0.5, -1.2)
f1(x) = 2.5
```

Note that the *MultiNewton* method is simple and works well for simple multi-variable functions. However, it may not work for some nonlinear functions. In this case, you need more sophiscated optimization methods, such as simplex, simulated annealing, genetic, and evolution techniques.

Simplex Method

The simplex method is an optimization algorithm, which is used to find the global minimum of any multi-variable function. This method is especially popular in the fields of chemistry, chemical engineering, and medicine.

The simplex method attempts to minimize a nonlinear function with n variables by using only function values, without any deriative information. The method thus falls in the general class of direct search methods. At each step, it maintains a nondegenerate simplex – a geometric figure in n dimensions of nonzero valume that is the convex hull of $n + 1$ vertices.

Each iteration of the simplex method begins with a simplex, specified by its $n + 1$ vertices and the associated function values. One or more test points are computed, along with their function values. The iterations are terminated when no more significant improvement of the minimum is observed on moving from one simplex to the other.

You can see that the simplex method obtains the minimum for a function with n variable by examining the function values at $n + 1$ points. The method locates the points with the best and worst function values, and then attempts to replace the worst-value point with a better point. This replacement process involves expanding and contracting the simplex near the worst-value point to determine a better replancement point. The iterations of the method shrink the multidimensional simplex around the minimum point, as shown in Figure 10-3.

The simplex algorithm consists of few key rules. The first rule is to reject the trial with the least favorable response value in the current simplex. A new set of control variable levels is calculated, by reflection into the control variable space opposite the undesirable result. This new trial replaces the least favorable trial in the simplex. This leads to a new least favorable response in the simplex that, in turn, leads to another new trial, and so on. At each step you move away from the least favorable conditions. In this way the simplex will move steadily towards more favorable conditions.

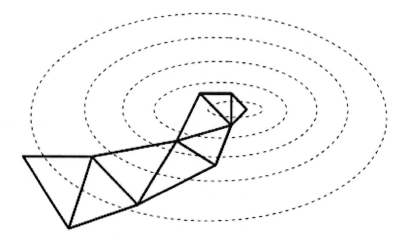

Figure 10-3 Illustration of the simplex method. The iterations of the method guadually approach
the minimum at the center of the contours.

The second key rule is never to return to control variable levels that have just been rejected. The calculated reflection in the control variables can also produce a least favorable result. Without this second rule the simplex would just oscillate between the two control variable levels. This problem is nicely avoided by choosing the second least favorable condition and moving away from it.

The simplex method occasionally may get stuck in a nut. The standard approach to handle this situation is to restart the algorithm with a new simplex starting at the current best value.

Implementation

For a given function $f(x)$ with n variable, the $(n + 1) \times n$ matrix x that contains $n + 1$ vertices used to construct the initial simplex, and the maximum number of iterations, we can implemented the simplex algorithm by adding a public static method, *Simplex*, to the *Optimization* class:

```
public static VectorR Simplex(MultiFunction f, MatrixR x, int MaxIterations)
{
    double reflect = 1.0;
    double expand = 2.0;
    double contract = 0.5;
    bool flag;

    VectorR Y = new VectorR(x.GetRows());
    int nv = x.GetRows - 1;

    int iw, ib;
    double y1, y2, x0;
    VectorR x1 = new VectorR(nv);
    VectorR x2 = new VectorR(nv);
    VectorR centroid = new VectorR(nv);

    // calculate Y using x1:
    for (int i = 0; i <= nv; i++)
```

```
{
    for (int j = 0; j < nv; j++)
    {
        x1[j] = x[i, j];
    }
    Y[i] = f(x1);
}

int iteration = 0;
do
{
    iteration++;

    // find worst and best points:
    iw = 0;
    ib = 0;

    for (int i = 1; i <= nv; i++)
    {
        if (Y[i] < Y[ib])
            ib = i;
        else if (Y[i] > Y[iw])
            iw = i;
    }

    // calculate centriod:
    for (int i = 0; i < nv; i++)
    {
        centroid[i] = 0;
        for (int j = 0; j <= nv; j++)
        {
            if (j != iw)
            centroid[i] += x[j, i];
        }
        centroid[i] /= nv;
    }

    // calculate reflected points:
    for (int i = 0; i < nv; i++)
    {
        x1[i] = (1 + reflect) * centroid[i] - reflect * x[iw, i];
    }
    y1 = f(x1);

    if (y1 < Y[ib])
    {
        // calculate expended points:
        for (int i = 0; i < nv; i++)
        {
            x2[i] = (1 + expand) * x1[i] - expand * centroid[i];
        }
        y2 = f(x2);
        if (y2 < Y[ib])
        {
            for (int i = 0; i < nv; i++)
```

```
                    {
                        x[iw, i] = x2[i];
                    }
                }
                else
                {
                    for (int i = 0; i < nv; i++)
                    {
                        x[iw, i] = x1[i];
                    }
                }
            }
            else
            {
                flag = true;
                for (int i = 0; i <= nv; i++)
                {
                    if (i != iw && y1 <= Y[i])
                    {
                        flag = false;
                        break;
                    }
                }
                if (flag)
                {
                    if (y1 < Y[iw])
                    {
                        for (int i = 0; i < nv; i++)
                        {
                            x[iw, i] = x1[i];
                        }
                        Y[iw] = y1;
                    }

                    // calculate contracted points:
                    for (int i = 0; i < nv; i++)
                    {
                        x2[i] = contract * x[iw, i] + (1 - contract) * centroid[i];
                    }
                    y2 = f(x2);
                    if (y2 > Y[iw])
                    {
                        for (int i = 0; i < nv; i++)
                        {
                            x2[i] = x[ib, i];
                        }
                        for (int j = 0; j <= nv; j++)
                        {
                            for (int i = 0; i < nv; i++)
                            {
                                x[j, i] = 0.5 * (x2[i] + x[j, i]);
                            }
                        }
                    }
                    else
```

```
        {
            for (int i = 0; i < nv; i++)
            {
                x[iw, i] = x2[i];
            }
        }
    }
    else
    {
        for (int i = 0; i < nv; i++)
        {
            x[iw, i] = x1[i];
        }
    }

    // calculate Y[] using x1:
    for (int i = 0; i <= nv; i++)
    {
        for (int j = 0; j < nv; j++)
        {
            x1[j] = x[i, j];
        }
        Y[i] = f(x1);
    }
}

// find the best points:
ib = 0;
for (int i = 1; i <= nv; i++)
{
    if (Y[i] < Y[ib])
    {
        ib = i;
    }
}
if (ib != 0)
{
    for (int i = 0; i < nv; i++)
    {
        x0 = x[0, i];
        x[0, i] = x[ib, i];
        x[ib, i] = x0;
    }
    y1 = Y[0];
    Y[0] = Y[ib];
    Y[ib] = y1;
}
}
while (iteration < MaxIterations);
return x.GetRowVector(0);
}
```

Inside this method, we first specify several factors used in the computation, including reflection, contraction, and expansion factors. You can change these parameters for different applications, or move them to the input parameters that can be specified by the user directly. The reflection factor is a positive value that can be unity. The expansion factor should be greater than 1 (usually is set to 2). The extraction parameter is a value between 0 and 1 (usually 0.5).

Testing the Simplex Method

In this section, I will show you how to find the minimum of a multi-variable function using the simplex method. First, let's minimize a two-dimensional function:

$$f(x, y) = x^2 - 4x + y^2 - y - xy$$

In order to use the simplex method, we need first construct a 3×2 vertex matrix using the following guesses for initial vertices:

$$\mathbf{V}_1 = (0, 0), \quad \mathbf{V}_2 = (0, 1), \quad \mathbf{V}_3 = (1, 0)$$

Then the corresponding vertex matrix can be constructed using the above vertices:

$$\begin{pmatrix} 0 & 0 \\ 0 & 1 \\ 1 & 0 \end{pmatrix}$$

With the above information, we can now find the minimum of the function using the simplex method. Add a new *TestSimplex* method to the *Program.cs* file:

```
static void TestSimplex()
{
    MatrixR x = new MatrixR(3, 2);

    x[0, 0] = 0;
    x[0, 1] = 0;
    x[1, 0] = 1;
    x[1, 1] = 0;
    x[2, 0] = 0;
    x[2, 1] = 1;

    VectorR result = Optimization.Simplex(f2, x, 100);
    Console.WriteLine("x = " + result.ToString());
    Console.WriteLine("f(x) = " + f2(result));
}

static double f2(VectorR x)
{
    return x[0] * x[0] - 4 * x[0] + x[1] * x[1] - x[1] - x[0] * x[1];
}
```

Running this example generates the result:

$$x = (2.99999997164077, 1.99999997619986)$$
$$f(x) = -7$$

which is very close to the exact minimum of $f(x) = -7$ at the point (3, 2). You can change the parameter of the number of the maximum iterations to obtain results with different accuracy. The key to use the simplex method is correcly construct the initial vertex matrix.

Simulated Annealing Method

Simulated annealing is a random-search technique which exploits an analogy between the way in which a metal cools and freezes into a minimum energy crystalline structure (the annealing process). In an annealing process, a melt, initially at high temperature and disordered system is slowly cooled so that the system at any time is approximately in thermodynamic equilibrium state. As cooling proceeds, the system becomes more ordered and approaches a frozen ground state at zero temperature.

Simulated annealing searches for the global minimum in a more general system and forms the basis of an optimisation technique for combinatorial and other problems. By analogy, the generalization of this simulated annealing method to combinatorial problems is straightforward. The current state of the thermodynamic system is analogous to the current solution to the combinatorial problem; the energy equation for the thermodynamic system is analogous to the model function; and the ground state is analogous to the global minimum. The major difficulty in implementation of the algorithm is that there is no obvious analogy for the temperature with respect to a free parameter in the combinatorial problem. Furthermore, avoidance of entrainment in local minima (quenching) is dependent on the "annealing schedule", the choice of initial temperature, how many iterations are performed at each temperature, and how much the temperature is decremented at each step as cooling proceeds.

Simulated annealing can deal with highly nonlinear models, chaotic, noisy data, and functions with constraints. It is a robust and general technique. Its main advantages over other local search methods are its flexibility and its ability to approach global optimality.

Algorithm

The algorithm of the simulated annealing method is listed below:

- Choose a random configuration array X_i, select an initial system temperature, and specify a cooling (i.e., annealing) schedule.

- Evaluate the energy $f(X_i)$ using the model function that is to be minimized.

- Perturb the configuration array X_i using a random permutation technique and a random normal distribution to obtain a neighboring configurion array X_{i+1}.

- Evaluate energy $f(X_{i+1})$ using the model function.

- If $f(X_{i+1}) < f(X_i)$, then X_{i+1} is the new current solution.

- If $f(X_{i+1}) > f(X_i)$, then accept X_{i+1} as the new current solution with a probability $e^{-\Delta/T}$ where $\Delta = f(X_{i+1}) - f(X_i)$ and T is the system temperature.

- Repeat computation with new configuration array X_i ($i = 1, 2, 3,\dots$) until the system is in thermodynamic equilibrium state (i.e., very little change in $f(X_i)$ with different configurations).

- Reduce the system temperature according to the cooling schedule.

- Terminate the computation until the system temperature reaches the minimum temperature specified by the user.

There are two basic algorithms in the simulated annealing method, the random configuration and cooling schedule. For a randomly generated configuration, the algorithm can produce a new solution from an old one. Here, we require that the algorithm can create a new configuration which slightly differs from the old one in one element. To this end, we need to add the *RandomGenerators.cs* from the *DistributionFunctionsTest* project in Chapter 7 to the current project. Open the *RandomGenerators.cs* file and add a new public static method, *RandomPermutation*:

```
public static int[] RandomPermutation(int n)
{
    ArrayList numbers = new ArrayList();
    int[] permutation = new int[n];

    // create a list that holds the numbser 0, 1, 2 ... n
    for (int i = 0; i < n; i++)
    {
        numbers.Add(i);
    }

    // for each entry in the permutation list,
    // grab the number from a random position in the number list
    for (int i = 0; i < n; i++)
    {
        int n1 = rand.Next(numbers.Count);
        permutation[i] = (int)numbers[n1];
        numbers.RemoveAt(n1);
    }
    return permutation;
}
```

This method generates a random integer array with *n* elements. For example, the following code snippet

```
int[] x = RandomGenerators.RandomPermutation(10);
foreach (int i in x)
    Console.WriteLine(i.ToString());
```

Generates a random integer array = (7, 6, 1, 8, 0, 2, 3, 9, 4, 5).

We can now use this *RandomPermutation* method to create a random perturbation to a configuration array. Add a new method, RandomPerturbation, to the Optimization class:

```
public static double[] RandomPerturbation(double[] xarray)
{
    int[] r = RandomGenerators.RandomPermutation(xarray.Length);
    for (int i = 0; i < xarray.Length; i++)
    {
        if (r[i] == xarray.Length - 1)
        {
            xarray[i] += RandomGenerators.NextNormal(0,1) / 100.0;
        }
    }
    return xarray;
}
```

This method creates a new configuration double array which is slightly different from the input double array in one element that is randomly selected. The modified value to this element is created using a random normal distribution function (see Chapter 7). For example, the following code snippet will create a new double array with random modification from an input double array:

```
double[] x = new double[] { 1, 2, 3, 4, 5, 6 };
double[] y = Optimization.RandomPerturbation(x);
Console.WriteLine(new VectorR(y).ToString());
```

This gives the results (1, 2, 2.97605087089402, 4, 5, 6). Here only the third element is modified slightly. The element to be modified is selected according to the random permutation algorithm, while the modified value is determined from the random normal distribution.

Another key algorithm to simulated annealing is the cooling schedule. Various cooling schedules can be used, including linear, step-wise, and exponential coolings, depending on your application requirements. Usually the exponential cooling works better. So in our implementation of the simulated annealing method, the exponential cooling schedule will be used.

Implementation

Here, we will implement a simple version of the simulated annealing. Add a now public static method, *Anneal*, to the *Optimization* class:

```
public static VectorR Anneal(MultiFunction f, double[] xarray, double Tmin,
                             int nEquilibrium)
{
    double T0 = 1.0;
    double T = T0;

    double e0 = f(new VectorR(xarray));
    double e1;
    double de;
    double[] xcurrent;
    Random rand = new Random();

    int j = 0;
    do
    {
        j++;

        int i = 0;
        while (i < nEquilibrium)
        {
            i++;

            xcurrent = RandomPerturbation((double[])xarray.Clone());
            e1 = f(new VectorR(xcurrent));
            de = e1 - e0;

            if (de < 0)
            {
                xarray = xcurrent;
                e0 = e1;
            }
            else
            {
                if (rand.NextDouble() < Math.Exp(-de / T))
                {
```

```
                    xarray = xcurrent;
                    e0 = e1;
                }
            }
        }

        T *= Math.Pow(0.9, j);
    }
    while (T > Tmin);
    return new VectorR(xarray);
}
```

This method takes the multiple-variable function, initial solution array, minimum system temperature, and the maximum number of iterations to reach the thermodynamic equilibrium state as input parameters. Here we set the initial system temperature $T_0 = 1$. In fact, the solution is not very sensitive to the initial system temperature. Note how we generate the new configuration array using the RandomPermutation method:

```
xcurrent = RandomPerturbation((double[])xarray.Clone());
```

Here we use *xarray.clone()* instead of *xarray* itself, which is required to obtain a correct solution. This means that we only need to manipulate *xarray*'s elements in the perturbation process not its reference.

Testing the Simulated Annealing Method

In this section, I will show you how to find the minimum of a multi-variable function using the simplex method. First, let's minimize the so-called "six-hump camelback" function:

$$f(x, y) = (4 - 2.1x^2 + x^{4/3})x^2 + xy + 4(y^2 - 1)y^2$$

which has several local minima in the range $-3 < x < 3$ and $-2 < y < 2$. It has two global minima, namely $f(-0.0883, 0.7126) = (0.0883, -0.7126) = -1.0313$.

Add a new method, *TestAnneal*, to the *Program.cs* file:

```
static void TestAnneal()
{
    double[] xarray = new double[] { 0, 1};
    VectorR result = Optimization.Anneal(f3, xarray, 1e-15, 200);
    Console.WriteLine(result.ToString());
    Console.WriteLine(" f3 = " + f3(result).ToString());
}

static double f3(VectorR x)
{
    return (4.0 - 2.1 * x[0] * x[0] + Math.Pow(Math.Abs(x[0]), 4.0 / 3.0)) *
            x[0] * x[0] + x[0] * x[1] + 4.0 * (x[1] * x[1] - 1) * x[1] * x[1];
}
```

Here, the initial array is set to (0, 1). The minimum system temperature is 1e-15, and the maximum number of iterations is 200. Running this example generates the result:

$$\mathbf{x} = (-0.0883602586099963, 0.712593257529641), \quad f(\mathbf{x}) = -1.03131283862155$$

which reasonably approximates the global minimum. Note that the result from different runs may not be exactly the same due to random feature of the simulated annealing method.

In practice, you may need to adjust the input parameters such as minimum system temperature and the maximum number of iterationc needed to reach the equilibrium state. The initial configuration (*xarray*) is also critical to the simulated annealing method. You may try to use different initial configurations in order to find the true global minimum.

Let's consider another example, namely the peaks function, which is used extensively in Matlab:

$$f(x,y) = 3(1-x)^2 e^{-x^2-(y+1)^2} - 10\left(\frac{1}{5}x - x^3 - y^5\right)e^{-x^2-y^2} - \frac{1}{3}e^{-(x+1)^2-y^2}$$

This function has several local minima and maxima. Add a new *TestPeaks* method to the *Program.cs* file:

```
static void TestPeaks()
{
    double[] xarray = new double[] { 0, -2 };
    VectorR result = Optimization.Anneal(Peaks, xarray, 1e-15, 200);
    Console.WriteLine(result.ToString());
    Console.WriteLine(" Peaks = " + Peaks(result).ToString());
}

static double Peaks(VectorR x)
{
    double z = 3 * (1 - x[0]) * (1 - x[0]) * Math.Exp(-x[0] * x[0] -
               (x[1] + 1) * (x[1] + 1)) - 10 * (x[0] / 5 - Math.Pow(x[0], 3) -
               Math.Pow(x[1], 5)) * Math.Exp(-x[0] * x[0] - x[1] * x[1])
               - 1 / 3 * Math.Exp(-(x[0] + 1) * (x[0] + 1) - x[1] * x[1]);
    return z;
}
```

In this example, we use the same set of parameters as those in the previous example, except for the ininial configuration array. Here (0, –2) is used for the initial guess of *xarray*.

Running this example generates the result:

$$x = (0.228870418146029, -1.62602258504236), \quad f(x) = -6.54589199712958$$

You can also find the global maximum of the Peaks function. Add a new methods, *Peaks*1 and *TestPeaks*1, to the *Program.cs* file. The *Peaks*1 is defined by simply returning the negative value of the *Peaks* function, then we can find the minimum for the *–f(x,y)* using the simulated annealing method. In this case, we set the initial configuration array *xarray* = (0, 2). Re-running the example produces the results:

$$x = (-0.010572917640621, 1.58032753593398), \quad f(x) = 8.11649000854207$$

Differential Evolution

It is clear from the implementation of the simulated annealing method that stochastic methods are efficient technique for finding the global minimum of a function with multiple variables. Among stochastic methods, the evolutionary algorithm is a very promising direction. Several independently developed but strongly related implementations of evolutionary algorithms are genetic algorithm, evolution strategies, and evolutionary programming.

Evolutionary algorithms are an optimization technique inspired by the biological processes of genetic and evolution. Both binary- and real-coded representations have been used in the development of the evolutionary algorithms. There are several advantages of using the evolutionary algorithms over the traditional optimization methods, include:

- Optimize with continuous or discrete variables.

- Usually do not require derivative information.

- Simultaneously search from a wide sampling of the cost surface.

- Deal with a large number of variables.

- Are well suited for parallel computing.

- Work with numerically generated data, experimental data, or analytical functions.

In this section, I will present one of the evolutionary algorithms, the differential evolution (DE) algorithm. DE is a more efficient genetic type of algorithm for the real-valued function optimization. I will show you how to implement the DE algorithm and how to use it for minimizing real-valued function with multiple variables.

Algorithm

The DE algorithm is a parallel direct search method which utilizes n_p n-dimensional parameter vectors

$$\mathbf{x}_{i,g}, \quad i = 0, 1, 2, \cdots, n_p - 1$$

as a population for each generation g (or each iteration of the optimization). The parameter n_p does not change during the minimization process. The initial population is chosen randomly and should try to cover the entire parameter space uniformly. The DE method generates new parameter vectors by adding the weighted difference between two population vectors to a third vector. If the resulting population yields a lower objective function value than a predetermined population member, then the newly generated population will replace the vector with which it was compared in the following generation; otherwise, the old population is retained.

The above principle may be extended when it comes to the practical variant of DE. For example, an existing population can be perturbed by adding more than one weighted difference vector to it. In some cases, it is also worthwhile to mix the parameters of the old population with those of the perturbed one before comparing the objective function value. Here, we will implement several variants of DE which have proven to be useful.

Implementation

In this section, we will inplement the DE method. Add a new public method, *DifferentialEvolution*, to the *Optimization* class. Here is the code list of this method:

```
private static double minCost = -100;
private static int nVaraibles = 2;
private static int nPopulations = 10;
private static int maxIterations = 200;
```

```
private static double stepSize = 0.8;
private static double crossover = 0.5;
private static VectorR xmin = new VectorR(new double[] { -2, -2 });
private static VectorR xmax = new VectorR(new double[] { 2, 2 });
private static int strategy = 2;
private static int refresh = 10;

public static double MinCost
{
    get { return minCost; }
    set { minCost = value; }
}

public static int NVariables
{
    get { return nVaraibles; }
    set { nVaraibles = value; }
}

public static int NPopulations
{
    get { return nPopulations; }
    set { nPopulations = value; }
}

public static int MaxIterations
{
    get { return maxIterations; }
    set { maxIterations = value; }
}

public static double StepSize
{
    get { return stepSize; }
    set { stepSize = value; }
}

public static double CrossOver
{
    get { return crossover; }
    set { crossover = value; }
}

public static VectorR Xmin
{
    get { return xmin; }
    set { xmin = value; }
}

public static VectorR Xmax
{
    get { return xmax; }
    set { xmax = value; }
}
```

```
public static int Strategy
{
    get { return strategy; }
    set { strategy = value; }
}
public static int Refresh
{
    get { return refresh; }
    set { refresh = value; }
}

public static void DifferentialEvolution(MultiFunction f,
                    out VectorR bestMember, out double bestValue)
{
    int n = NVariables;
    int np = NPopulations * n;
    bestMember = new VectorR(n);
    bestValue = 0.0;

    if (np < 5)
        np = 5;
    if (CrossOver < 0 || CrossOver > 1)
        CrossOver = 0.5;
    if (MaxIterations <= 0)
        MaxIterations = 200;

    MatrixR population = new MatrixR(np, n);

    Random rand = new Random();
    for (int i = 0; i < np; i++)
    {
        for (int j = 0; j < n; j++)
        {
            population[i, j] = Xmin[j] +
                        rand.NextDouble() * (Xmax[j] - Xmin[j]);
        }
    }

    MatrixR population0 = new MatrixR(np, n);
    VectorR values = new VectorR(np);
    VectorR bestMemit = new VectorR(n);

    int ibest = 0;
    values[0] = f(population.GetRowVector(ibest));
    bestValue = values[0];
    for (int i = 1; i < np; i++)
    {
        values[i] = f(population.GetRowVector(i));
        if (values[i] < bestValue)
        {
            ibest = i;
            bestValue = values[i];
        }
    }
```

```
bestMemit = population.GetRowVector(ibest);
double bestvalueit = bestValue;
bestMember = bestMemit;

MatrixR pm1 = new MatrixR(np, n);
MatrixR pm2 = new MatrixR(np, n);
MatrixR pm3 = new MatrixR(np, n);
MatrixR pm4 = new MatrixR(np, n);
MatrixR pm5 = new MatrixR(np, n);
MatrixR bm = new MatrixR(np, n);
MatrixR ui = new MatrixR(np, n);
MatrixR mui = new MatrixR(np, n);
MatrixR mpo = new MatrixR(np, n);
int[] rot = new int[np];
int[] rotn = new int[n];

for (int i = 0; i < np; i++)
    rot[i] = i;
for (int i = 0; i < n; i++)
    rotn[i] = i;

int[] rt = new int[np];
int[] rtn = new int[n];
int[] a1 = new int[np];
int[] a2 = new int[np];
int[] a3 = new int[np];
int[] a4 = new int[np];
int[] a5 = new int[np];
int[] ind = new int[4];

int iterations = 1;
do
{
    population0 = population.Clone();
    ind = RandomGenerators.RandomPermutation(4);
    a1 = RandomGenerators.RandomPermutation(np);

    for (int i = 0; i < np; i++)
    {
        rt[i] = (rot[i] + ind[0]) % np;
        a2[i] = a1[rt[i]];
    }
    for (int i = 0; i < np; i++)
    {
        rt[i] = (rot[i] + ind[1]) % np;
        a3[i] = a2[rt[i]];
    }
    for (int i = 0; i < np; i++)
    {
        rt[i] = (rot[i] + ind[2]) % np;
        a4[i] = a3[rt[i]];
    }
    for (int i = 0; i < np; i++)
    {
        rt[i] = (rot[i] + ind[3]) % np;
```

```
        a5[i] = a4[rt[i]];
}

double[,] randArray = new double[np, n];
for (int i = 0; i < np; i++)
{
    for (int j = 0; j < n; j++)
        randArray[i, j] = rand.NextDouble();
}

for (int i = 0; i < np; i++)
{
    for (int j = 0; j < n; j++)
    {
        pm1[i, j] = population0[a1[i], j];
        pm2[i, j] = population0[a2[i], j];
        pm3[i, j] = population0[a3[i], j];
        pm4[i, j] = population0[a4[i], j];
        pm5[i, j] = population0[a5[i], j];
        bm[i, j] = bestMemit[j];
        mui[i, j] = 0;

        if (randArray[i, j] < CrossOver)
            mui[i, j] = 1;
    }
}

int st = Strategy;
if (Strategy > 5)
    st = Strategy - 5;
else
{
    st = Strategy;
    mui.Transpose();
    mui = MatrixSort(mui);
    for (int i = 0; i < np; i++)
    {
        int nn = (int)Math.Floor((decimal)rand.NextDouble() * n);
        if (nn > 0)
        {
            for (int j = 0; j < n; j++)
            {
                rtn[j] = (rotn[j] + n) % n;
                mui[j, i] = mui[rtn[j], i];
            }
        }
    }
    mui.Transpose();
}

for (int i = 0; i < np; i++)
{
    for (int j = 0; j < n; j++)
    {
        if (mui[i, j] < 0.5)
```

```
                mpo[i, j] = 1;
        }
}

for (int i = 0; i < np; i++)
{
    for (int j = 0; j < n; j++)
    {
        if (st == 1)
        {
            ui[i, j] = bm[i, j] + StepSize * (pm1[i, j] - pm2[i, j]);
            ui[i, j] = population0[i, j] * mpo[i, j] +
                    ui[i, j] * mui[i, j];
        }
        else if (st == 2)
        {
            ui[i, j] = pm3[i, j] + StepSize * (pm1[i, j] - pm2[i, j]);
            ui[i, j] = population0[i, j] * mpo[i, j] +
                     ui[i, j] * mui[i, j];
        }
        else if (st == 3)
        {
            ui[i, j] = population0[i, j] +
                    StepSize * (bm[i, j] - population0[i, j]) +
                    StepSize * (pm1[i, j] - pm2[i, j]);
            ui[i, j] = population0[i, j] * mpo[i, j] +
                    ui[i, j] * mui[i, j];
        }
        else if (st == 4)
        {
            ui[i, j] = bm[i, j] + StepSize * (pm1[i, j] - pm2[i, j] +
                    pm3[i, j] - pm4[i, j]);
            ui[i, j] = population0[i, j] * mpo[i, j] +
                    ui[i, j] * mui[i, j];
        }
        else if (st == 5)
        {
            ui[i, j] = pm5[i, j] + StepSize * (pm1[i, j] - pm2[i, j] +
                    pm3[i, j] - pm4[i, j]);
            ui[i, j] = population0[i, j] * mpo[i, j] +
                    ui[i, j] * mui[i, j];
        }
    }
}

// Select which vectors are allowed to enter the next population:
for (int i = 1; i < np; i++)
{
    double temp = f(ui.GetRowVector(i));
    if (temp <= values[i])
    {
        for (int j = 0; j < n; j++)
        {
            population[i, j] = ui[i, j];
        }
```

```
                    values[i] = temp;
                    if (temp < bestValue)
                    {
                        bestValue = temp;
                        bestMember = ui.GetRowVector(i);
                    }
                }
            }
            bestMemit = bestMember;

            if (Refresh > 0)
            {
                if (iterations % Refresh == 0)
                {
                    Console.WriteLine("\n Iterations = {0}, Best value = {1}",
                                        iterations, bestValue);
                    Console.WriteLine(" Best member = {0}", bestMember);
                }
            }
            iterations++;
        }
        while (iterations < MaxIterations && bestValue > MinCost);
    }
```

In the above code listing, we first define several private fields and corresponding public properties. The *MinCost* property is used to stop the program if the value of the merit function is less than or equal to the *MinCost* value. The vector *XMin* and *Xmax* properties are lower and upper bounds of the initial population respectively. Note that those vectors are not bound constraints. The *StepSize* property has the value that lies in the interval [0, 2], and the *CrossOver* property represents the crossover probability from [0, 1]. The *Refresh* property controls the intermediate output. Namely, the intermediate output will be generated after the "*Refresh*" iterations, and there will be no intermediate results produced if *Refresh* < 1.

The *Strategy* takes different integer value :

- Strategy = 1 – uses the scheme: DE/best/1/exp.
- Strategy = 2 – uses the scheme: DE/rand/1/exp.
- Strategy = 3 – uses the scheme: DE/rand-to-best/1/exp.
- Strategy = 4 – uses the scheme: DE/best/2/exp.
- Strategy = 5 – uses the scheme: DE/rand/2/exp.
- Strategy = 6 – uses the scheme: DE/best/1/bin.
- Strategy = 7 – uses the scheme: DE/rand/1/bin.
- Strategy = 8 – uses the scheme: DE/rand-to-best/2/bin.
- Strategy = 9 – uses the scheme: DE/rand/1/bin.
- else – DE/rand/2/bin

In the above, the different *Strategy* represents different DE variant scheme. Here, the *exp* and *bin* stand for the exponential and binomial crossovers. For example, the DE/best/1 scheme generates the new population vector according to the formula:

$$\mathbf{v}_{i,g+1} = \mathbf{x}_{best,g} + StepSize \cdot (\mathbf{x}_{r1,g} - \mathbf{x}_{r2,g})$$

Where $r1$ and $r2$ are integers in the range $[0, n_p-1]$. The population vector to be perturbed is the best for forming vectors of the current generation.

You can find detailed definitions of different DE schemes in literature, for example, *IEEE Conference on Evolutionary Computation*, Nagoya, 1996, pp. 842-844 by R. Storn and K. Price.

If you want to optimize multi-variable functions with DE method, you can try the following settings for the input parameters first: choose method e.g. DE/rand/1/exp, set the number of parents n_p to 10 times the number of variables, select *StepSize* = 0.8 and crossover constant = 0.9. Make sure that you initialize the population vector by exploiting its full numerical range. For example, if a parameter is allowed to exhibit values in the range $[-10, 10]$, you should pick the initial values from this range instead of unnecessarily restricting diversity. If you experience convergence issues, you usually need to increase the value for n_p.

DE is much more sensitive to the choice of the step size than it is to the choice of the crossover. The crossover parameter is more like a fine tuning element. High values of crossover like *Crossover* = 1 converge faster if convergence occurs. Sometimes, however, you have to go down as much as *Crossover* = 0 to make DE robust enough for a particular problem. If you choose a binomial crossover like, DE/rand/1/bin, *Crossover* is usually higher than in the exponential crossover variant. Note that different problems usually require different settings for n_p, *StepSize*, and *Crossover*. If you still run into convergence problems, you might want to try a different method. Generally, you can use DE/rand/1/... or DE/best/2/... schemes.

Testing Differential Evolution

In this section, I will show you how to find the minimum of a multi-variable function by using the DE algorithm. Here, we want to find the minimum and maximum of the *Peaks* function, the same function we used to test the simulated annealing method.

Add a new method, *Test DifferentialEvolution*, to the *Program.cs* file:

```
static void TestDifferentialEvolution()
{
    VectorR bestmember;
    double bestvalue;
    Optimization.MaxIterations = 201;
    Optimization.MinCost = -50;
    Optimization.Refresh = 50;
    Optimization.DifferentialEvolution(Peaks, out bestmember, out bestvalue);
    Console.WriteLine("\n Minimum = {0}, Location = {1}\n\n\n",
                        bestvalue, bestmember);
    Optimization.DifferentialEvolution(Peaks1, out bestmember, out bestvalue);
    Console.WriteLine("\n Maximum = {0}, Location = {1}",
                        -bestvalue, bestmember);
}
```

Here we use functions, *Peaks* and *Peaks1*, to find the minimum and maximum points of the *Peaks* function. Running this example generates results shown in Figure 10-4.

I should point out here that like other random search methods, the evolutionary algorithms need much more function evalutions than linearized methods, so their convergent speed is slow. They do not necessarily find the true global minimum. You should run the program several times with different parameters to see if you can obtain consistent results.

```
Iterations = 50, Best value = -6.45432367025763
Best member = (0.129897893466939, -1.63475707063207)

Iterations = 100, Best value = -6.49188864736702
Best member = (0.153836811033001, -1.64689389830508)

Iterations = 150, Best value = -6.52608528311864
Best member = (0.232238287922105, -1.66278127656641)

Iterations = 200, Best value = -6.52608528311864
Best member = (0.232238287922105, -1.66278127656641)

Minimum = -6.52608528311864
Location = (0.232238287922105, -1.66278127656641)



Iterations = 50, Best value = -8.02885594970186
Best member = (0.0709256889047689, 1.62800767197542)

Iterations = 100, Best value = -8.10754541562067
Best member = (0.0127772513463988, 1.59747070352556)

Iterations = 150, Best value = -8.10754541562067
Best member = (0.0127772513463988, 1.59747070352556)

Iterations = 200, Best value = -8.10754541562067
Best member = (0.0127772513463988, 1.59747070352556)

Maximum = 8.10754541562067
Location = (0.0127772513463988, 1.59747070352556)
```

Figure 10-4 Results from the Differential Evolution method.

Chapter 11
Numerical Differentiation

Numerical differentiation deals with the following problem: for a given function $y = f(x)$, we want to calculate the derivatives of a smooth function defined on a discrete set of grid points $(x_0, x_1, x_2, \ldots, x_N)$. Assume that the data are the exact values of the function at the data points, and we need to compute derivatives only at the data points. Note that numerical differentiation is applied not only to a smooth function, but also to a set of discrete data points (x_i, y_i), $i = 0, 1, 2, \ldots, N$. In either case, we have access to a finite number of (x, y) data pairs from which we can calculate the derivative.

We will focus on the construction of numerical approximations of the derivatives based on finite difference formalism. There are two approaches to such constructions: using interpolation formulas, or using Taylor series approximations. We can approximate the function locally by polynomial interpolation and then differentiate it. The Taylor series expansion method has the advantage of providing us with information about the error involved in the approximation.

Finite Difference Formulas

The derivation of the finite difference approximations for the derivatives of $f(x)$ is based on forward and backward Taylor series expansions of $f(x)$ about x, such as

$$f(x+h) = f(x) + hf'(x) + \frac{h^2}{2!}f''(x) + \frac{h^3}{3!}f'''(x) + \frac{h^4}{4!}f^{(4)}(x) + \cdots \qquad (11.1)$$

$$f(x-h) = f(x) - hf'(x) + \frac{h^2}{2!}f''(x) - \frac{h^3}{3!}f'''(x) + \frac{h^4}{4!}f^{(4)}(x) - \cdots \qquad (11.2)$$

$$f(x+2h) = f(x) + 2hf'(x) + \frac{(2h)^2}{2!}f''(x) + \frac{(2h)^3}{3!}f'''(x) + \frac{(2h)^4}{4!}f^{(4)}(x) + \cdots \qquad (11.3)$$

$$f(x-2h) = f(x) - 2hf'(x) + \frac{(2h)^2}{2!}f''(x) - \frac{(2h)^3}{3!}f'''(x) + \frac{(2h)^4}{4!}f^{(4)}(x) - \cdots \qquad (11.4)$$

where $h = \Delta x$. From the above equations, we can easily obtain the sums and differences of the series:

$$f(x+h) + f(x-h) = 2f(x) + h^2 f''(x) + \frac{h^4}{12} f^{(4)}(x) + \cdots \tag{11.5}$$

$$f(x+h) - f(x-h) = 2hf'(x) + \frac{h^3}{3} f'''(x) + \cdots \tag{11.6}$$

$$f(x+2h) + f(x-2h) = 2f(x) + 4h^2 f''(x) + \frac{4h^4}{3} f^{(4)}(x) + \cdots \tag{11.7}$$

$$f(x+2h) - f(x-2h) = 4hf'(x) + \frac{8h^3}{3} f'''(x) + \cdots \tag{11.8}$$

Note that the sums contains only even derivatives, whereas the differences contain only odd derivatives. Equations (11.1)-(8) can be regarded as coupled equations that can be solved for various derivatives of $f(x)$. The number of equations involved and the number of terms kept in each equation depend on the order of the derivative and the desired degree of accuracy.

Forward Difference Method

The forward finite difference can be obtained from equations (11.1)-(8). The first four derivatives of the function $f(x)$ are given by the following formulas:

$$f'(x) = \frac{f(x+h) - f(x)}{h}$$

$$f''(x) = \frac{f(x+2h) - 2f(x+h) + f(x)}{h^2}$$

$$f'''(x) = \frac{f(x+3h) - 3f(x+2h) + 3(f+h) - f(x)}{h^3}$$

$$f^{(4)}(x) = \frac{f(x+4h) - 4f(x+3h) + 6f(x+2h) - 4f(x+h) + f(x)}{h^4}$$

The above equations are usually not used to compute derivatives because they have a large truncation error (order of $O(h)$). The common practice is to use expressions of $O(h^2)$. To obtain forward difference formulas of this order, we need to retain more terms in the Taylor series. Below, I only list the results without derivations:

$$f'(x) = \frac{-3f(x) + 4f(x+h) - f(x+2h)}{2h}$$

$$f''(x) = \frac{2f(x) - 5f(x+h) + 4f(x+2h) - f(x+3h)}{h^2}$$

$$f'''(x) = \frac{-5f(x) + 18f(x+h) - 24f(x+2h) + 14f(x+3h) - 3f(x+4h)}{2h^3}$$

$$f^{(4)}(x) = \frac{3f(x) - 14f(x+h) + 26f(x+2h) - 24f(x+3h) + 11f(x+4h) - 2f(x+5h)}{h^4}$$

We will use the above equations to implement the forward differentiation method in C#.

Implementation

Here we will implement the forward numerical derivatives using the equations derived in the previous section. Start with a new C# Console application and name it *DifferentiationTest*. Add a new class, *Differentiation*, to the project and change its namespace to *XuMath*. Here is the code listing of the first four derivatives for a function *f(x)* and array of function values:

```
public delegate double Function(double x);
private const double badResult = double.NaN;

public static double Forward1(Function f, double x, double h)
{
    h = (h == 0) ? 0.01 : h;
    return (-3 * f(x) + 4 * f(x + h) - f(x + 2 * h)) / 2 / h;
}

public static double Forward1(double[] yarray, int yindex, double h)
{
    if (yarray == null || yarray.Length < 3 || yindex < 0 ||
        yindex > yarray.Length - 3 || h == 0)
        return badResult;
    return (-3 * yarray[yindex] + 4 * yarray[yindex + 1] -
            yarray[yindex + 2]) / 2 / h;
}

public static double Forward2(Function f, double x, double h)
{
    h = (h == 0) ? 0.01 : h;
    return (2 * f(x) - 5 * f(x + h) + 4 * f(x + 2 * h) - f(x + 3 * h)) / h / h;
}

public static double Forward2(double[] yarray, int yindex, double h)
{
    if (yarray == null || yarray.Length < 4 || yindex < 0 ||
        yindex > yarray.Length - 4 || h == 0)
        return badResult;
    return (2 * yarray[yindex] - 5 * yarray[yindex + 1] +
            4 * yarray[yindex + 2] - yarray[yindex + 3]) / h / h;
}

public static double Forward3(Function f, double x, double h)
{
    h = (h == 0) ? 0.01 : h;
    return (-5 * f(x) + 18 * f(x + h) - 24 * f(x + 2 * h) +
            14 * f(x + 3 * h) - 3 * f(x + 4 * h)) / 2 / h / h / h;
}

public static double Forward3(double[] yarray, int yindex, double h)
{
```

```
            if (yarray == null || yarray.Length < 5 || yindex < 0 ||
                yindex > yarray.Length - 5 || h == 0)
                return badResult;
            return (-5 * yarray[yindex] + 18 * yarray[yindex + 1] -
                    24 * yarray[yindex + 2] + 14 * yarray[yindex + 3] -
                    3 * yarray[yindex + 4]) / 2 / h / h / h;
        }

        public static double Forward4(Function f, double x, double h)
        {
            h = (h == 0) ? 0.01 : h;
            return (3 * f(x) - 14 * f(x + h) + 26 * f(x + 2 * h) -
                    24 * f(x + 3 * h) + 11 * f(x + 4 * h) -
                    2 * f(x + 5 * h)) / h / h / h / h;
        }

        public static double Forward4(double[] yarray, int yindex, double h)
        {
            if (yarray == null || yarray.Length < 6 || yindex < 0 ||
                yindex > yarray.Length - 6 || h == 0)
                return badResult;
            return (3 * yarray[yindex] - 14 * yarray[yindex + 1] +
                    26 * yarray[yindex + 2] - 24 * yarray[yindex + 3] +
                    11 * yarray[yindex + 4] - 2 * yarray[yindex + 5]) / h / h / h / h;
        }
```

In the above implementation, each derivative has two overloaded methods: one for the function and another for the array of data values. In order to calculate derivatives for a function, you need to pass a delegate function $f(x)$, the value of x at which the derivative is computed, and the increment h as input parameters.

On the other hand, in order to calculate derivatives for an array of data values, each method has input parameters which pass the *yarray* values, the index called *yindex* which specifies the location at which the derivative is computed, and the increment h. Here, the increment parameter is the value used to generate the function values.

Testing the Forward Difference Method

Here I will show you how to calculate the first four derivatives for $f(x) = \sin x$ using the methods implemented in the previous section. Add a new static method, *TestForwardMethod*, to the *Program.cs* file:

```
using System;
using XuMath;

namespace DifferentiationTest
{
    class Program
    {
        static void Main(string[] args)
        {
            TestForwardMethod();
            Console.ReadLine();
        }
```

```
static void TestForwardMethod()
{
    double h = 0.1;
    double[] yarray=new double[10];
    // create yarray:
    for (int i = 0; i < 10; i++)
        yarray[i] = f(i * h);

    // Calculate derivatives for function:
    Console.WriteLine("\n Derivatives for f(x) = sin(x):\n");
    double dy = Differentiation.Forward1(f, 0.3, h);
    Console.WriteLine(" f'(x) = " + dy.ToString());
    dy = Differentiation.Forward2(f, 0.3, h);
    Console.WriteLine(" f''(x) = " + dy.ToString());
    dy = Differentiation.Forward3(f, 0.3, h);
    Console.WriteLine(" f'''(x) = " + dy.ToString());
    dy = Differentiation.Forward4(f, 0.3, h);
    Console.WriteLine(" f''''(x) = " + dy.ToString());

    // Calculate derivatives for array values:
    Console.WriteLine("\n Derivatives for array values:");
    Console.WriteLine("\n yarray =");
    foreach (double y1 in yarray)
        Console.WriteLine(" " + y1.ToString());
    dy = Differentiation.Forward1(yarray, 3, h);
    Console.WriteLine("\n y' = " + dy.ToString());
    dy = Differentiation.Forward2(yarray, 3, h);
    Console.WriteLine(" y'' = " + dy.ToString());
    dy = Differentiation.Forward3(yarray, 3, h);
    Console.WriteLine(" y''' = " + dy.ToString());
    dy = Differentiation.Forward4(yarray, 3, h);
    Console.WriteLine(" y'''' = " + dy.ToString());
}

static double f(double x)
{
    return Math.Sin(x);
}
```

It can seen that we first define the x increment $h = 0.1$ and create the *yarray* of the sine function with this increment. We then calculate the first four derivatives directly for the delegate function at $x = 0.3$. Finally, we calculate derivatives using the data array *yarray*, also at $x = 0.3$, which corresponds to *yindex* = 3.

Running this example generates results shown in Figure 11-1. You can see that the derivatives for the function and for the array values are, as expected, almost identical. The slight difference originates from the numerical round-off error.

```
Derivatives for f(x) = sin(x):

f'(x)  = 0.958436053231903
f''(x) = -0.299161719879693
f'''(x) = -0.971116217218349
f''''(x) = 0.308477068935442

Derivatives for array values:

yarray =
0
0.0998334166468282
0.198669330795061
0.29552020666134
0.389418342308651
0.479425538604203
0.564642473395035
0.644217687237691
0.717356090899523
0.783326909627483

y'   = 0.958436053231901
y''  = -0.299161719879693
y''' = -0.971116217218571
y'''' = 0.308477068944324
```

Figure 11-1 First four derivatives computed using the forward difference method.

Backward Difference Method

The backward finite difference can be also obtained from equations (11.1)–(8). Here are the formulas of the backward difference of the order $O(h^2)$:

$$f'(x) = \frac{3f(x) - 4f(x-h) + f(x-2h)}{2h}$$

$$f''(x) = \frac{2f(x) - 5f(x-h) + 4f(x-2h) - f(x-3h)}{h^2}$$

$$f'''(x) = \frac{5f(x) - 18f(x-h) + 24f(x-2h) - 14f(x-3h) + 3f(x-4h)}{2h^3}$$

$$f^{(4)}(x) = \frac{3f(x) - 14f(x-h) + 26f(x-2h) - 24f(x-3h) + 11f(x-4h) - 2f(x-5h)}{h^4}$$

We will use the above equations to implement the backward differentiation method in C#.

Implementation

Here we will implement the backward numerical derivatives using the above equations. Add the following methods to the *Differentiation* class:

```
public static double Backward1(Function f, double x, double h)
{
    h = (h == 0) ? 0.01 : h;
    return (3 * f(x) - 4 * f(x - h) + f(x - 2 * h)) / 2 / h;
}

public static double Backward1(double[] yarray, int yindex, double h)
{
    if (yarray == null || yarray.Length < 3 || yindex < 0 ||
        yindex < 3 || h == 0)
        return badResult;
    return (3 * yarray[yindex] - 4 * yarray[yindex - 1] +
            yarray[yindex - 2]) / 2 / h;
}

public static double Backward2(Function f, double x, double h)
{
    h = (h == 0) ? 0.01 : h;
    return (2 * f(x) - 5 * f(x - h) + 4 * f(x - 2 * h) - f(x - 3 * h)) / h / h;
}

public static double Backward2(double[] yarray, int yindex, double h)
{
    if (yarray == null || yarray.Length < 4 || yindex < 0 ||
        yindex < 4 || h == 0)
        return badResult;
    return (2 * yarray[yindex] - 5 * yarray[yindex - 1] +
            4 * yarray[yindex - 2] - yarray[yindex - 3]) / h / h;
}

public static double Backward3(Function f, double x, double h)
{
    h = (h == 0) ? 0.01 : h;
    return (5 * f(x) - 18 * f(x - h) + 24 * f(x - 2 * h) -
            14 * f(x - 3 * h) + 3 * f(x - 4 * h)) / 2 / h / h / h;
}

public static double Backward3(double[] yarray, int yindex, double h)
{
    if (yarray == null || yarray.Length < 5 || yindex < 0 ||
        yindex < 5 || h == 0)
        return badResult;
    return (5 * yarray[yindex] - 18 * yarray[yindex - 1] +
            24 * yarray[yindex - 2] - 14 * yarray[yindex - 3] +
            3 * yarray[yindex - 4]) / 2 / h / h / h;
}

public static double Backward4(Function f, double x, double h)
{
    h = (h == 0) ? 0.01 : h;
    return (3 * f(x) - 14 * f(x - h) + 26 * f(x - 2 * h) -
            24 * f(x - 3 * h) + 11 * f(x - 4 * h) -
            2 * f(x - 5 * h)) / h / h / h / h;
}
```

```
public static double Backward4(double[] yarray, int yindex, double h)
{
    if (yarray == null || yarray.Length < 6 || yindex < 0 ||
        yindex < 6 || h == 0)
        return badResult;
    return (3 * yarray[yindex] - 14 * yarray[yindex - 1] +
            26 * yarray[yindex - 2] - 24 * yarray[yindex - 3] +
            11 * yarray[yindex - 4] - 2 * yarray[yindex - 5]) / h / h / h / h;
}
```

In the above implementation, each derivative has two overloaded methods: one for the function and the other for the array of function values. In order to calculate derivatives for a function, you need to pass a delegate function $f(x)$, the value of x at which the derivative is computed, and the increment h as input parameters.

On the other hand, in order to calculate derivatives for an array of function values, each method has input parameters which pass the *yarray* values, the index called *yindex* which specifies the location at which the derivative is computed, and the increment h. Here, the increment parameter is the value used to generate the function values.

Testing the Backward Difference Method

Here I will show you how to calculate the first four derivatives for $f(x) = \sin x$ using the backward difference methods implemented in the previous section. Add a new static method, *TestBackwardMethod*, to the *Program.cs* file:

```
static void TestBackwardMethod()
{
    double h = 0.1;
    double[] yarray = new double[10];
    // create yarray:
    for (int i = 0; i < 10; i++)
    yarray[i] = f(i * h);

    // Calculate derivatives for function at 0.7:
    Console.WriteLine("\n Derivatives for f(x) = sin(x):\n");
    double dy = Differentiation.Backward1(f, 0.7, h);
    Console.WriteLine(" f'(x) = " + dy.ToString());
    dy = Differentiation.Backward2(f, 0.7, h);
    Console.WriteLine(" f''(x) = " + dy.ToString());
    dy = Differentiation.Backward3(f, 0.7, h);
    Console.WriteLine(" f'''(x) = " + dy.ToString());
    dy = Differentiation.Backward4(f, 0.7, h);
    Console.WriteLine(" f''''(x) = " + dy.ToString());

    // Calculate derivatives for array values at yindex = 7:
    Console.WriteLine("\n Derivatives for array values:");
    Console.WriteLine("\n yarray =");
    foreach (double y1 in yarray)
    Console.WriteLine(" " + y1.ToString());
    dy = Differentiation.Backward1(yarray, 7, h);
    Console.WriteLine("\n y' = " + dy.ToString());
```

```
dy = Differentiation.Backward2(yarray, 7, h);
Console.WriteLine(" y'' = " + dy.ToString());
dy = Differentiation.Backward3(yarray, 7, h);
Console.WriteLine(" y''' = " + dy.ToString());
dy = Differentiation.Backward4(yarray, 7, h);
Console.WriteLine(" y'''' = " + dy.ToString());
}
```

We first define the x increment $h = 0.1$ and create the *yarray* of the sine function using this increment. We then calculate the first four derivatives directly for the delegate function at $x = 0.7$. Finally, we calculate derivatives using the data array *yarray*, also at $x = 0.7$, which corresponds to *yindex* = 7. Note that in order to compute the fourth derivative by using the backward difference method, *yindex* must be greater than 6.

Running this example generates the results shown in Figure 11-2. You can see that the derivatives for the function and for the array values, as expected, are almost identical. The slight difference originates from the numerical round-off error.

```
Derivatives for f(x) = sin(x):

f'(x) = 0.767543533685673
f''(x) = -0.649318039163332
f'''(x) = -0.779665379199213
f''''(x) = 0.658341688558606

Derivatives for array values:

yarray =
0
0.0998334166468282
0.198669330795061
0.29552020666134
0.389418342308651
0.479425538604203
0.564642473395035
0.644217687237691
0.717356090899523
0.783326909627483

y'   = 0.767543533685673
y''  = -0.649318039163344
y''' = -0.77966537919949
y''''  = 0.658341688561936
```

Figure 11-2 First four derivatives computed using the backward difference method.

Central Difference Method

The central finite difference can be also obtained from equations (11.1)–(8). It uses the following equations to approximate the first four derivatives:

$$f'(x) = \frac{-f(x-h)+f(x+h)}{2h}$$

$$f''(x) = \frac{f(x-h) - 2f(x) + f(x+h)}{h^2}$$

$$f'''(x) = \frac{-f(x-2h) + 2f(x-h) - 2f(x+h) + f(x+2h)}{2h^3}$$

$$f^{(4)}(x) = \frac{f(x-2h) - 4f(x-h) + 6f(x) - 4f(x+h) + f(x+2h)}{h^4}$$

The central difference privides an $O(h^2)$ approximation. We will use the above equations to implement the central differentiation method in C#.

Implementation

Here we will implement the central numerical derivatives using the above equations. Add the following methods to the *Differentiation* class:

```
public static double Central1(Function f, double x, double h)
{
    h = (h == 0) ? 0.01 : h;
    return (f(x + h) - f(x - h)) / 2 / h;
}

public static double Central1(double[] yarray, int yindex, double h)
{
    if (yarray == null || yarray.Length < 3 || yindex < 1 ||
        yindex > yarray.Length - 2 || h == 0)
        return badResult;
    return (yarray[yindex + 1] - yarray[yindex - 1]) / 2 / h;
}

public static double Central2(Function f, double x, double h)
{
    h = (h == 0) ? 0.01 : h;
    return (f(x - h) - 2 * f(x) + f(x + h)) / h / h;
}

public static double Central2(double[] yarray, int yindex, double h)
{
    if (yarray == null || yarray.Length < 3 || yindex < 1 ||
        yindex > yarray.Length - 2 || h == 0)
        return badResult;
    return (yarray[yindex - 1] - 2 * yarray[yindex] +
            yarray[yindex + 1]) / h / h;
}

public static double Central3(Function f, double x, double h)
{
    h = (h == 0) ? 0.01 : h;
    return (-f(x - 2 * h) + 2 * f(x - h) - 2 * f(x + h) +
            f(x + 2 * h)) / 2 / h / h / h;
}
```

```
public static double Central3(double[] yarray, int yindex, double h)
{
    if (yarray == null || yarray.Length < 5 || yindex < 2 ||
        yindex > yarray.Length - 3 || h == 0)
        return badResult;
    return (-yarray[yindex - 2] + 2 * yarray[yindex - 1] -
        2 * yarray[yindex + 1] + yarray[yindex + 2]) / 2 / h / h / h;
}

public static double Central4(Function f, double x, double h)
{
    h = (h == 0) ? 0.01 : h;
    return (f(x - 2 * h) - 4 * f(x - h) + 6 * f(x) - 4 * f(x + h) +
        f(x + 2 * h)) / h / h / h / h;
}

public static double Central4(double[] yarray, int yindex, double h)
{
    if (yarray == null || yarray.Length < 5 || yindex < 2 ||
        yindex > yarray.Length - 3 || h == 0)
        return badResult;
    return (yarray[yindex - 2] - 4 * yarray[yindex - 1] + 6 * yarray[yindex] -
        4 * yarray[yindex + 1] + yarray[yindex + 2]) / h / h / h / h;
}
```

In the above implementation, each derivative has two overloaded methods: one for the function and the other for the array of function values. In order to calculate derivatives for a function, you need to pass a delegate function $f(x)$, the value of x at which the derivative is computed, and the increment h as input parameters.

On the other hand, in order to calculate derivatives for an array of function values, each method has input parameters which pass the *yarray* values, the index called *yindex* which specifies the location at which the derivative is computed, and the increment h. Here, the increment parameter is the value used to generate the function values.

Testing the Central difference Method

In this section, I will show you how to calculate the first four derivatives for $f(x) = \sin x$ by using the central different methods implemented in the previous section. Add a new static method, *TestCentralMethod*, to the *Program.cs* file:

```
static void TestCentralMethod()
{
    double h = 0.1;
    double[] yarray = new double[10];
    // create yarray:
    for (int i = 0; i < 10; i++)
        yarray[i] = f(i * h);

    // Calculate derivatives for function at 0.5:
    Console.WriteLine("\n Derivatives for f(x) = sin(x):\n");
    double dy = Differentiation.Central1(f, 0.5, h);
    Console.WriteLine(" f'(x) = " + dy.ToString());
    dy = Differentiation.Central2(f, 0.5, h);
```

```
        Console.WriteLine(" f''(x) = " + dy.ToString());
        dy = Differentiation.Central3(f, 0.5, h);
        Console.WriteLine(" f'''(x) = " + dy.ToString());
        dy = Differentiation.Central4(f, 0.5, h);
        Console.WriteLine(" f''''(x) = " + dy.ToString());

        // Calculate derivatives for array values at yindex = 5:
        Console.WriteLine("\n Derivatives for array values:");
        Console.WriteLine("\n yarray =");
        foreach (double y1 in yarray)
            Console.WriteLine(" " + y1.ToString());
        dy = Differentiation.Central1(yarray, 5, h);
        Console.WriteLine("\n y' = " + dy.ToString());
        dy = Differentiation.Central2(yarray, 5, h);
        Console.WriteLine(" y'' = " + dy.ToString());
        dy = Differentiation.Central3(yarray, 5, h);
        Console.WriteLine(" y''' = " + dy.ToString());
        dy = Differentiation.Central4(yarray, 5, h);
        Console.WriteLine(" y'''' = " + dy.ToString());
    }
```

Here, we first define the x increment $h = 0.1$ and create the *yarray* of the sine function using this increment. We then calculate the first four derivatives directly for the delegate function at $x = 0.5$. Finally, we calculate derivatives using the data array *yarray*, also at $x = 0.5$, which corresponds to *yindex* = 5.

Running this example generates the results shown in Figure 11-3. You can see that the derivatives for the function and for the array values are, as expected, almost identical. The slight difference originates from the numerical round-off error.

Extended Centeral Difference Method

The central difference method presented in the previous section provides the first four derivatives with a precision of order $O(h^2)$. It is possible to obtain derivatives with $O(h^4)$ accuracy by modifying the central difference method to include more function values. The following are equations for the extended central difference method:

$$f'(x) = \frac{f(x-2h) - 8f(x-h) + 8f(x+h) - f(x+2h)}{12h}$$

$$f''(x) = \frac{-f(x-2h) + 16f(x-h) - 30f(x) + 16f(x+h) - f(x+2h)}{12h^2}$$

$$f'''(x) = \frac{f(x-3h) - 8f(x-2h) + 13f(x-h) - 13f(x+h) + 8f(x+2h) - f(x+3h)}{8h^3}$$

$$f^{(4)}(x) = \frac{-f(x-3h) + 12f(x-2h) - 39f(x-h) + 56f(x) - 39f(x+h) + 12(x+2h) - f(x+3h)}{6h^4}$$

The extended central difference provides an $O(h^4)$ approximation. We will use the above equations to implement the central differentiation method in C#.

```
Derivatives for f(x) = sin(x):

f'(x) = 0.876120655431924
f''(x) = -0.479026150472006
f'''(x) = -0.875390798209086
f''''(x) = 0.478627095050754

Derivatives for array values:

yarray =
0
0.0998334166468282
0.198669330795061
0.29552020666134
0.389418342308651
0.479425538604203
0.564642473395035
0.644217687237691
0.717356090899523
0.783326909627483

y'   = 0.876120655431925
y''  = -0.479026150471995
y''' = -0.875390798209197
y'''' = 0.478627095047424
```

Figure 11-3 First four derivatives computed using the central difference method.

Implementation

Here we will implement extended central numerical derivatives using the above equations. Add the following methods to the *Differentiation* class:

```
public static double ExtendedCentral1(Function f, double x, double h)
{
    h = (h == 0) ? 0.01 : h;
    return (f(x - 2 * h) - 8 * f(x - h) + 8 * f(x + h) - f(x + 2 * h)) / 12 / h;
}

public static double ExtendedCentral1(double[] yarray, int yindex, double h)
{
    if (yarray == null || yarray.Length < 5 || yindex < 2 ||
        yindex > yarray.Length - 3 || h == 0)
        return badResult;
    return (yarray[yindex - 2] - 8 * yarray[yindex - 1] +
            8 * yarray[yindex + 1] - yarray[yindex + 2]) / 12 / h;
}

public static double ExtendedCentral2(Function f, double x, double h)
{
    h = (h == 0) ? 0.01 : h;
    return (-f(x - 2 * h) + 16 * f(x - h) - 30 * f(x) +
            16 * f(x + h) - f(x + 2 * h)) / 12 / h / h;
}

public static double ExtendedCentral2(double[] yarray, int yindex, double h)
{
```

```
        if (yarray == null || yarray.Length < 5 || yindex < 2 ||
            yindex > yarray.Length - 3 || h == 0)
            return badResult;
        return (-yarray[yindex - 2] + 16 * yarray[yindex - 1] -
            30 * yarray[yindex] + 16 * yarray[yindex + 1] -
            yarray[yindex + 2]) / 12 / h / h;
    }

    public static double ExtendedCentral3(Function f, double x, double h)
    {
        h = (h == 0) ? 0.01 : h;
        return (f(x - 3 * h) - 8 * f(x - 2 * h) + 13 * f(x - h) - 13 * f(x + h) +
            8 * f(x + 2 * h) - f(x + 3 * h)) / 8 / h / h / h;
    }

    public static double ExtendedCentral3(double[] yarray, int yindex, double h)
    {
        if (yarray == null || yarray.Length < 7 || yindex < 3 ||
            yindex > yarray.Length - 4 || h == 0)
            return badResult;
        return (yarray[yindex - 3] - 8 * yarray[yindex - 2] +
            13 * yarray[yindex - 1] - 13 * yarray[yindex + 1] +
            8 * yarray[yindex + 2] - yarray[yindex + 3]) / 8 / h / h / h;
    }

    public static double ExtendedCentral4(Function f, double x, double h)
    {
        h = (h == 0) ? 0.01 : h;
        return (-f(x - 3 * h) + 12 * f(x - 2 * h) - 39 * f(x - h) + 56 * f(x) -
            39 * f(x + h) + 12 * f(x + 2 * h) -
            f(x + 3 * h)) / 6 / h / h / h / h;
    }

    public static double ExtendedCentral4(double[] yarray, int yindex, double h)
    {
        if (yarray == null || yarray.Length < 7 || yindex < 3 ||
            yindex > yarray.Length - 4 || h == 0)
            return badResult;
        return (-yarray[yindex - 3] + 12 * yarray[yindex - 2] -
            39 * yarray[yindex - 1] + 56 * yarray[yindex] -
            39 * yarray[yindex + 1] + 12 * yarray[yindex + 2] -
            yarray[yindex + 3]) / 6 / h / h / h / h;
    }
```

In the above implementation, each derivative has two overloaded methods: one for the function and the other for the array of function values. In order to calculate derivatives for a function, you need to pass a delegate function $f(x)$, the value of x at which the derivative is computed, and the increment h as input parameters.

On the other hand, in order to calculate derivatives for an array of function values, each method has input parameters which pass the *yarray* values, the index called *yindex* which specifies the location at which the derivative is computed, and the increment h. Here, the increment parameter is the value used to generate the function values.

Testing the Extended Central difference Method

Here I will show you how to calculate the first four derivatives for $f(x) = \sin x$ using the extended central different methods that we implemented in the previous section. Add a new static method, *TestExtendedCentralMethod*, to the *Program.cs* file:

```
static void TestExtendedCentralMethod()
{
    double h = 0.1;
    double[] yarray = new double[10];
    // create yarray:
    for (int i = 0; i < 10; i++)
        yarray[i] = f(i * h);

    // Calculate derivatives for function at 0.5:
    Console.WriteLine("\n Derivatives for f(x) = sin(x):\n");
    double dy = Differentiation.ExtendedCentral1(f, 0.5, h);
    Console.WriteLine(" f'(x) = " + dy.ToString());
    dy = Differentiation.ExtendedCentral2(f, 0.5, h);
    Console.WriteLine(" f''(x) = " + dy.ToString());
    dy = Differentiation.ExtendedCentral3(f, 0.5, h);
    Console.WriteLine(" f'''(x) = " + dy.ToString());
    dy = Differentiation.ExtendedCentral4(f, 0.5, h);
    Console.WriteLine(" f''''(x) = " + dy.ToString());

    // Calculate derivatives for array values at yindex = 5:
    Console.WriteLine("\n Derivatives for array values:");
    dy = Differentiation.ExtendedCentral1(yarray, 5, h);
    Console.WriteLine("\n y' = " + dy.ToString());
    dy = Differentiation.ExtendedCentral2(yarray, 5, h);
    Console.WriteLine(" y'' = " + dy.ToString());
    dy = Differentiation.ExtendedCentral3(yarray, 5, h);
    Console.WriteLine(" y''' = " + dy.ToString());
    dy = Differentiation.ExtendedCentral4(yarray, 5, h);
    Console.WriteLine(" y'''' = " + dy.ToString());

    // Analytic results:
    Console.WriteLine("\n Analytic results:");
    dy = Math.Cos(0.5);
    Console.WriteLine("\n y' = " + dy.ToString());
    dy = -Math.Sin(0.5);
    Console.WriteLine(" y'' = " + dy.ToString());
    dy = -Math.Cos(0.5);
    Console.WriteLine(" y''' = " + dy.ToString());
    dy = Math.Sin(0.5);
    Console.WriteLine(" y'''' = " + dy.ToString());
}
```

Here, we first define the x increment $h = 0.1$ and create the *yarray* of the sine function using this increment. We then calculate the first four derivatives directly for the delegate function at $x = 0.5$. Next, we calculate derivatives using the data array *yarray* also at $x = 0.5$, which conrresponds to *yindex* = 5. Finally, we calculate the derivatives analytically, so we can compare them with results from the extended central difference method.

Running this example generates the results shown in Figure 11-4. You can see that the derivatives for the function and for the array values are, as expected, almost identical. By comparing them with the analytic results, it is apparent that the results from the extended central difference method are more accurate than those from the other methods, such as the central difference method.

```
Derivatives for f(x) = sin(x):

f'(x)  = 0.875579640095606
f''(x)  = -0.479425006384552
f'''(x)  = -0.875577452081682
f''''(x)  = 0.479424142338739

Derivatives for array values:

y'  = 0.875579640095607
y''  = -0.479425006384538
y'''  = -0.875577452081682
y''''  = 0.479424142329858

Analytic results:

y'  = 0.875582561890373
y''  = -0.479425538604203
y'''  = -0.875582561890373
y''''  = 0.479425538604203
```

Figure 11-4 Derivatives from the extended central difference method.

Richardson Extrapolation

In numerical simulation, Richardson extrapolation is a sequence acceleration method, often used to improve the rate of convergence. This method can be also applied to finite difference approximations to boost the accuracy of numerical derivatives.

Suppose that $A(h)$ is an estimation of order h^n for A:

$$A = \lim_{h \to 0} A(h)$$

or

$$A = A(h) + a_n h^n + O(h^m), \, a_n \neq 0, \, m > n$$

We can repeat the calculation with $h = th$, so that

$$A = A(th) + a_n (th)^n + O((th)^m)$$

eliminating the coefficient a_n and solving for A, we obtain from the above two equations:

$$A = \frac{t^n A(h) - A(th)}{t^n - 1}$$

which is the Richardson extrapolation formula. It is a common practice to use $t = \frac{1}{2}$, in which case the above equation becomes

$$A = \frac{2^n A(h/2) - A(h)}{2^n - 1}$$

This is an estimate of order h^m for A with $m > n$.

We can apply the Richardson extrapolation to the finite difference methods. In this case, the finite difference approximation for derivatives is in the order of $O(h^2)$. Then, we can set $n = 2$ in the Richardson extrapolation, and the formula reduces to

$$A = \frac{4A(h/2) - A(h)}{3}$$

Implementation

Here we will implement the Richardson extrapolation method to the numerical derivatives based on the finite difference technique. Add the following methods to the *Differentiation* class:

```
public static double Richardson1(Function f, double x, double h, string flag)
{
    double result = badResult;
    if (flag == "Backward")
    {
        result = (4 * Backward1(f, x, h / 2) - Backward1(f, x, h)) / 3;
    }
    else if (flag == "Forward")
    {
        result = (4 * Forward1(f, x, h / 2) - Forward1(f, x, h)) / 3;
    }
    else if (flag == "Central")
    {
        result = (4 * Central1(f, x, h / 2) - Central1(f, x, h)) / 3;
    }
    return result;
}

public static double Richardson1(double[] yarray, int yindex,
                                 double h, string flag)
{
    double result = badResult;
    if (flag == "Backward")
    {
        result = (4 * Backward1(yarray, yindex, h / 2) -
                 Backward1(yarray, yindex, h)) / 3;
    }
    else if (flag == "Forward")
    {
        result = (4 * Forward1(yarray, yindex, h / 2) -
                 Forward1(yarray, yindex, h)) / 3;
    }
    else if (flag == "Central")
    {
```

```
                result = (4 * Central1(yarray, yindex, h / 2) -
                          Central1(yarray, yindex, h)) / 3;
        }
        return result;
}

public static double Richardson2(Function f, double x, double h, string flag)
{
        double result = badResult;
        if (flag == "Backward")
        {
            result = (4 * Backward2(f, x, h / 2) - Backward2(f, x, h)) / 3;
        }
        else if (flag == "Forward")
        {
            result = (4 * Forward2(f, x, h / 2) - Forward2(f, x, h)) / 3;
        }
        else if (flag == "Central")
        {
            result = (4 * Central2(f, x, h / 2) - Central2(f, x, h)) / 3;
        }
        return result;
}

public static double Richardson2(double[] yarray, int yindex,
                                 double h, string flag)
{
        double result = badResult;
        if (flag == "Backward")
        {
            result = (4 * Backward2(yarray, yindex, h / 2) -
                      Backward2(yarray, yindex, h)) / 3;
        }
        else if (flag == "Forward")
        {
            result = (4 * Forward2(yarray, yindex, h / 2) -
                      Forward2(yarray, yindex, h)) / 3;
        }
        else if (flag == "Central")
        {
            result = (4 * Central2(yarray, yindex, h / 2) -
                      Central2(yarray, yindex, h)) / 3;
        }
        return result;
}

public static double Richardson3(Function f, double x, double h, string flag)
{
        double result = badResult;
        if (flag == "Backward")
        {
            result = (4 * Backward3(f, x, h / 2) - Backward3(f, x, h)) / 3;
        }
        else if (flag == "Forward")
        {
```

```
            result = (4 * Forward3(f, x, h / 2) - Forward3(f, x, h)) / 3;
        }
        else if (flag == "Central")
        {
            result = (4 * Central3(f, x, h / 2) - Central3(f, x, h)) / 3;
        }
        return result;
    }

    public static double Richardson3(double[] yarray, int yindex,
                                     double h, string flag)
    {
        double result = badResult;
        if (flag == "Backward")
        {
            result = (4 * Backward3(yarray, yindex, h / 2) -
                        Backward3(yarray, yindex, h)) / 3;
        }
        else if (flag == "Forward")
        {
            result = (4 * Forward3(yarray, yindex, h / 2) -
                        Forward3(yarray, yindex, h)) / 3;
        }
        else if (flag == "Central")
        {
            result = (4 * Central3(yarray, yindex, h / 2) -
                        Central3(yarray, yindex, h)) / 3;
        }
        return result;
    }

    public static double Richardson4(Function f, double x, double h, string flag)
    {
        double result = badResult;
        if (flag == "Backward")
        {
            result = (4 * Backward4(f, x, h / 2) - Backward4(f, x, h)) / 3;
        }
        else if (flag == "Forward")
        {
            result = (4 * Forward4(f, x, h / 2) - Forward4(f, x, h)) / 3;
        }
        else if (flag == "Central")
        {
            result = (4 * Central4(f, x, h / 2) - Central4(f, x, h)) / 3;
        }
        return result;
    }

    public static double Richardson4(double[] yarray, int yindex,
                                     double h, string flag)
    {
        double result = badResult;
        if (flag == "Backward")
        {
```

```
            result = (4 * Backward4(yarray, yindex, h / 2) -
                         Backward4(yarray, yindex, h)) / 3;
        }
        else if (flag == "Forward")
        {
            result = (4 * Forward4(yarray, yindex, h / 2) -
                         Forward4(yarray, yindex, h)) / 3;
        }
        else if (flag == "Central")
        {
            result = (4 * Central4(yarray, yindex, h / 2) -
                         Central4(yarray, yindex, h)) / 3;
        }
        return result;
    }
```

The methods, *Richardson1*, *Richardson2*, *Richardson3*, and *Richardson4*, are used to calculate the first-to-fourth derivatives for a function or a data array of function values. Inside each method, we use the string flag to select a differentiation formalism from the forward, backward, and central difference methods.

Testing the Richardson Extrapolation

Here we will test if Richardson extrapolation improves the accuracy of the derivatives. We will consider the following function:

$$f(x) = x^4 + x + e^{-x}$$

We can easily compute the first four derivatives of the above function analytically. Here, we will calculate the derivatives using the central different methods with and without the Richardson extrapolation. Then we will compare numerical derivatives to the analytic results.

Add a new static method, *TestRichardson*, to the *Program.cs* file:

```
static void TestRichardson()
{
    double h = 0.1;
    double x = 0.5;
    Console.WriteLine("\n Comparison results:\n");
    double exact1 = 4 * x * x * x + 1 - Math.Exp(-x);
    double c1 = Differentiation.Central1(f1, x, h);
    double r1 = Differentiation.Richardson1(f1, x, h, "Central");
    Console.WriteLine(" exact1 = {0:n8}, c1 = {1:n8},
                         r1 = {2:n8}", exact1, c1, r1);
    double exact2 = 12 * x * x + Math.Exp(-x);
    double c2 = Differentiation.Central2(f1, x, h);
    double r2 = Differentiation.Richardson2(f1, x, h, "Central");
    Console.WriteLine(" exact2 = {0:n8}, c2 = {1:n8},
                         r2 = {2:n8}", exact2, c2, r2);
    double exact3 = 24 * x - Math.Exp(-x);
    double c3 = Differentiation.Central3(f1, x, h);
    double r3 = Differentiation.Richardson3(f1, x, h, "Central");
    Console.WriteLine(" exact3 = {0:n8}, c3 = {1:n8},
                         r3 = {2:n8}", exact3, c3, r3);
```

```
        double exact4 = 24 + Math.Exp(-x);
        double c4 = Differentiation.Central4(f1, x, h);
        double r4 = Differentiation.Richardson4(f1, x, h, "Central");
        Console.WriteLine(" exact4 = {0:n8}, c4 = {1:n8},
                        r4 = {2:n8}", exact4, c4, r4);
    }
```

Here, we first set $h = 0.1$ and $x = 0.5$, meaning that we want to calculate derivatives at $x = 0.5$. Inside the *TestRichardson* method, the parameters *exact1*,..., *exact4* represent the exact results for the first four derivatives; *c1*,..., *c4* are the results from the central difference methods; and *r1*,..., *r4* denote the results from the central difference methods with Richardson extrapolation.

Running the example generates results shown in Figure 11-5.

```
Comparison results:

exact1 =  0.89346934, c1 =  0.91245795, r1 =  0.89346947
exact2 =  3.60653066, c2 =  3.62703627, r2 =  3.60653062
exact3 = 11.39346934, c3 = 11.39195150, r3 = 11.39346972
exact4 = 24.60653066, c4 = 24.60754230, r4 = 24.60653047
```

Figure 11-5 Frist four derivatives from different methods.

You can see that the results from the Richardson extrapolation are much closer to the exact results than those from the central difference method. In general, the finite difference based methods provide approximations of $O(h^2)$, while the Richardson extrapolation gives the finite difference approximation of $O(h^4)$.

Derivatives by Interpolation

If $f(x)$ is given as a set of discrete data points generated at uneven intervals of x, the finite difference methods presented in previous sections cannot be applied. For those kinds of data sets, interpolation can be a very effective means of computing derivatives. The idea behind interpolation is to approximate the derivative of $f(x)$ by the derivative of the interpolant.

Here, we will use three-point Lagrange interpolation as an example. In Chapter 8, we discussed the general n-point Lagrange interpolation. The Lagrange interpolation is a well known, classic technique for interpolation. The three-point Lagrange interpolation can be described by the following formula:

$$f(x) = \frac{(x - x_i)(x - x_{i+1})}{(x_{i-1} - x_i)(x_{i-1} - x_{i+1})} f(x_{i-1}) + \frac{(x - x_{i-1})(x - x_{i+1})}{(x_i - x_{i-1})(x_i - x_{i+1})} f(x_i) + \frac{(x - x_{i-1})(x - x_i)}{(x_{i+1} - x_{i-1})(x_{i+1} - x_i)} f(x_{i+1})$$

We can obtain the first and second derivatives by directly differentiating the above equation:

$$f'(x) = \frac{2x - x_i - x_{i+1}}{(x_{i-1} - x_i)(x_{i-1} - x_{i+1})} f(x_{i-1}) + \frac{2x - x_{i-1} - x_{i+1}}{(x_i - x_{i-1})(x_i - x_{i+1})} f(x_i) + \frac{2x - x_{i-1} - x_i)}{(x_{i+1} - x_{i-1})(x_{i+1} - x_i)} f(x_{i+1})$$

$$f''(x) = \frac{2}{(x_{i-1} - x_i)(x_{i-1} - x_{i+1})} f(x_{i-1}) + \frac{2}{(x_i - x_{i-1})(x_i - x_{i+1})} f(x_i) + \frac{2}{(x_{i+1} - x_{i-1})(x_{i+1} - x_i)} f(x_{i+1})$$

The above formulas are applicable to unequally spaced data. You can also use the other interpolations, such as cubic spline interpolation, to compute the derivatives.

Implementation

In this section, we will implement C# methods that can be used to compute the first two derivatives for unequally spaced data by using the three-point Lagrange interpolation. Three methods, *Interpolation0*, *Interpolation1*, and *Interpolation2*, will be added to the *Differentiation* class, which are used to calculate the $f(x)$, $f'(x)$, and $f''(x)$, respectively:

```
public static double Interpolation0(double[] xarray, double[] yarray, double x)
{
    double result = badResult;
    int n = yarray.Length;
    for (int i = 1; i < n - 1; i++)
    {
        if (x > xarray[i - 1] && x < xarray[i + 1])
        {
            result = yarray[i - 1] * (x - xarray[i]) * (x - xarray[i + 1]) /
                     ((xarray[i - 1] - xarray[i]) *
                     (xarray[i - 1] - xarray[i + 1])) +
                     yarray[i] * (x - xarray[i - 1]) * (x - xarray[i + 1]) /
                     ((xarray[i] - xarray[i - 1]) *
                     (xarray[i] - xarray[i + 1])) +
                     yarray[i + 1] * (x - xarray[i - 1]) * (x - xarray[i]) /
                     ((xarray[i + 1] - xarray[i - 1]) *
                     (xarray[i + 1] - xarray[i]));
        }
    }
    return result;
}

public static double Interpolation1(double[] xarray, double[] yarray, double x)
{
    double result = badResult;
    int n = yarray.Length;
    for (int i = 1; i < n - 1; i++)
    {
        if (x > xarray[i - 1] && x < xarray[i + 1])
        {
            result = yarray[i - 1] * (2 * x - xarray[i] - xarray[i + 1]) /
                     ((xarray[i - 1] - xarray[i]) *
                     (xarray[i - 1] - xarray[i + 1])) +
                     yarray[i] * (2 * x - xarray[i - 1] - xarray[i + 1]) /
                     ((xarray[i] - xarray[i - 1]) *
                     (xarray[i] - xarray[i + 1])) +
                     yarray[i + 1] * (2 * x - xarray[i - 1] - xarray[i]) /
                     ((xarray[i + 1] - xarray[i - 1]) *
                     (xarray[i + 1] - xarray[i]));
        }
    }
    return result;
}
```

```
public static double Interpolation2(double[] xarray, double[] yarray, double x)
{
    double result = badResult;
    int n = yarray.Length;
    for (int i = 1; i < n - 1; i++)
    {
        if (x > xarray[i - 1] && x < xarray[i + 1])
        {
            result = yarray[i - 1] * 2 / ((xarray[i - 1] - xarray[i]) *
                     (xarray[i - 1] - xarray[i + 1])) +
                     yarray[i] * 2 / ((xarray[i] - xarray[i - 1]) *
                     (xarray[i] - xarray[i + 1])) +
                     yarray[i + 1] * 2 / ((xarray[i + 1] - xarray[i - 1]) *
                     (xarray[i + 1] - xarray[i]));
        }
    }
    return result;
}
```

The above methods take the double arrays, *xarray* and *yarray*, as input parameters. These data can be either equally or unequally spaced.

Testing Derivatives by Interpolation

Here, I will show you how to calculate derivatives for unequally spaced data by using the methods implemented in the previous section. Add a static method, *TestInterpolation*, to the *Program.cs* file:

```
static void TestInterpolation()
{
    double x = 1.8;
    double[] xa = new double[] { 1.0, 1.1, 1.3, 1.6, 1.7, 2.0 };
    double[] ya =
            new double[] { 1.2772, 1.1414, 0.7880, 0.1435, -0.0729, -0.6215 };
    double y0 = Differentiation.Interpolation0(xa, ya, x);
    double y1 = Differentiation.Interpolation1(xa, ya, x);
    double y2 = Differentiation.Interpolation2(xa, ya, x);
    Console.WriteLine(" y = {0:n6}, y' = {1:n6}, y'' = {2:n6}", y0, y1, y2);
}
```

You can see that the *x* data are unequally spaced. Running this example generates the following results:

$$f(1.8) = -0.272533, \quad f'(1.8) = -1.9125, \quad f''(1.8) = 1.676667$$

In this example, the data points fall on the curve $f(x) = \sin 2x - e^{-x}$, so that the exact values are:

$$f(1.8) = -0.2772, \quad f'(1.8) = -1.9588, \quad f''(1.8) = 1.9354$$

It can be seen that $f(x)$ and $f'(x)$ at $x = 1.8$ from the interpolation methods and exact results differ slightly, but the values of $f''(1.8)$ are much further apart. This is not unexpected, considering the general rule: the higher the order of the derivative, the lower the precision with which it can be computed.

Chapter 12
Numerical Integration

In the previous chapter, various approximations were obtained to derivatives of a function or a data array of function values. In this chapter, we will use a similar process for approximating the integral of a function.

The term numerical quadrature (often abbreviated to quadrature) is more or less a synonym for numerical integration, especially when applied to one-dimensional integrals. Two- and higher-dimensional integration is sometimes known as cubature, although the meaning of quadrature is used for higher dimensional integration as well.

Numerical integration is intrinsically a much more accurate procedure than numerical differentiation. The basic problem considered by numerical integration is computing an approximation to a definite integral

$$I = \int_a^b f(x)dx$$

If $f(x)$ is a smooth well-defined function, integrated over a small number of dimensions, and the limits of integration are bounded, there are many excellent methods to approximate to the integral with arbitrary precision.

The need for a numerical integration arises for two reasons. First, the function to be integrated may be such that the integral is too complicated to evaluate or may be even impossible to obtain analytically. Secondly, the function may be described only by a data array of values, so that a numerical approximation is the only approach available.

Methods of numerical integration can be divided into two groups: Newton–Cotes formulas and Gaussian quadrature. Newton–Cotes formulas are characterized by equally spaced data points, and include well-known methods such as the trapezoidal rule and Simpson's rule. They are most useful if $f(x)$ has already been computed at equal intervals, or can be computed relatively easily. Since Newton–Cotes formulas are based on local interpolation, they require only a piecewise fit to a polynomial.

On the other hand, in Gaussian quadrature the locations of the x-points are chosen to yield the best possible accuracy. Because Gaussian quadrature requires fewer evaluations of the integrand for a given level of precision, it is often used in cases where $f(x)$ is difficult to evaluate. Another advantage of Gaussian quadrature is its ability to handle integrable singularities.

Newton-Cotes Formulas

The Newton-Cotes Formulas are a group of formulas for numerical integration, based on evaluating the integrand at $n + 1$ equally-spaced points. Suppose that the value of a function $f(x)$ is known at equally spaced points x_i, for $i = 0, 1,..., n$. The Newton-Cotes formula is an approximation to the definite integral by the sum

$$I = \int_a^b f(x)\,dx \approx \sum_{i=0}^n w_i f(x_i)$$

where $x_i = x_0 + ih$, with $h = (x_n - x_0)/n$ being the step size. The w_i are called weights, which is derived from the Lagrange basis polynomials. This means that w_i depends only on the x_i and not on the function f. Next we approximate $f(x)$ by the polynomial of degree n that intersects all the data points. Let $L(x)$ be the interpolation polynomial in the Lagrange form for the given data points, then we have

$$I = \int_a^b f(x)\,dx \approx \int_a^b L(x)\,dx = \int_a^b \sum_{i=0}^n f(x_i) l_i(x)\,dx = \sum_{i=0}^n f(x_i) \int_a^b l_i(x)\,dx = \sum_{i=0}^n w_i f(x_i) \tag{12.1}$$

where

$$w_i = \int_a^b l_i(x)\,dx \tag{12.2}$$

Equations (12.1) and (12.2) are the Newton-Cotes formulas.

Trapezoidal Rule

If we set $n = 1$ in Equations (12-1) and (12-2), as illustrated in Figure 12-1, we have

$$l_0 = \frac{x - b}{a - b} = -\frac{x - b}{h}$$

Then we can obtain w_0 from Equation (12-2):

$$w_0 = -\frac{1}{h} \int_a^b (x - b)\,dx = \frac{1}{2h}(b - a)^2 = \frac{h}{2}$$

From the Lagrange interpolation, we also have

$$l_1 = \frac{x - a}{b - a} = \frac{x - a}{h}$$

so that

$$w_1 = \frac{1}{h} \int_a^b (x - a)\,dx = \frac{1}{2h}(b - a)^2 = \frac{h}{2}$$

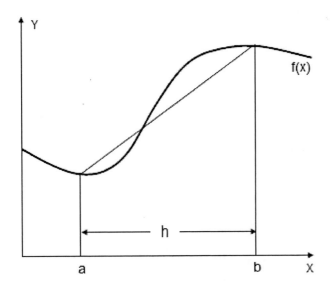

Figure 12-1 trapezoidal rule, n = 1.

Substitution in Equation (12-1) yields

$$I = \frac{h}{2}\left[f(a) + f(b)\right]$$

which is known as the trapezoidal rule. It represents the area of the trapezoid in Figure 12-1.

In practice, the trapezoidal rule is applied in a piecewise fashion – called composite trapezoidal rule. As shown in Figure 12-2, the area under the curve $y = f(x)$ is divided into n strips of width $h = (b-a)/n$. The area of each strip is then approximated to be that of a trapezium. The sum of these trapezoidal areas provides an approximation for the definite integral.

From the trapezoidal rule, we can obtain for the approximation area of the ith strip:

$$I_i = \frac{h}{2}\left[f(x_i) + f(x_{i+1})\right]$$

Hence the total area, representing the integral I, is the summation of each individual strip area:

$$I = \sum_{i=0}^{n-1} I_i = \frac{h}{2}\sum_{i=0}^{n-1}\left[f(x_i) + f(x_{i+1})\right]$$

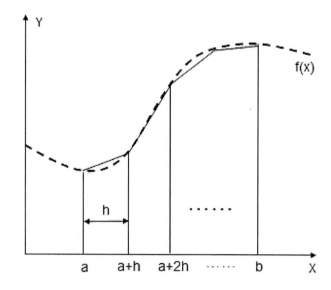

Figure 12-2 Illustration of the trapezoidal rule.

Implementation

Here we will implement the numerical integration using the trapezoidal rule. Start with a new C# Console application and name it *IntegrationTest*. Add a new class, *Integration*, to the project and change its namespace to *XuMath*. Here is code listing of this method:

```
using System;

namespace XuMath
{
    public class Integration
    {
        public delegate double Function(double x);

        public static double Trapezoidal(Function f, double a, double b, int n)
        {
            double sum = 0.0;
            double h = (b - a) / (n - 1);
            for (int i = 0; i < n - 1; i++)
            {
                sum += 0.5 * h * (f(a + i * h) + f(a + (i + 1) * h));
            }
            return sum;
        }

        public static double Trapezoidal(double[] yarray, double h)
        {
            int n = yarray.Length;
            double sum = 0.0;
            for (int i = 0; i < n - 1; i++)
            {
                sum += 0.5 * h * (yarray[i] + yarray[i + 1]);
```

```
            }
            return sum;
        }
    }
}
```

Here, we implement two overloaded methods: one is for the function and the other for the array of function values. In order to calculate integration for a function, you need to pass a delegate function $f(x)$, and the number of strips n.

On the other hand, in order to calculate integration for an array of function values, the method has input parameters which pass the *yarray* values, and the increment h. Here, the increment parameter should be the value used to generate the function values. Note that the data values must be generated in equally spaced x points.

Testing the Trapezoidal Method

Now I will show you how to calculate the integration by using the methods implemented in the previous section. Add a new static method, *TestTrapezoidal*, to the *Program.cs* file:

```
using System;
using XuMath;
namespace IntegrationTest
{
    class Program
    {
        static void Main(string[] args)
        {
            TestTrapezoidal();
            Console.ReadLine();
        }

        static void TestTrapezoidal()
        {

            int n = 101;
            double result;

            result = f1(1) - f1(0);
            Console.WriteLine("\n Analytic result = " + result.ToString());

            result = Integration.Trapezoidal(f, 0, 1, n);
            Console.WriteLine(" Result for function = " + result.ToString());

            double[] ya = new double[n];
            double h = 1.0 / (n - 1);
            for (int i = 0; i < n; i++)
            {
                double x = i * h;
                ya[i] = f(x);
            }
            result = Integration.Trapezoidal(ya, h);
            Console.WriteLine(" Result for data array = " + result.ToString());
        }
```

```
static double f(double x)
{
    return Math.Exp(x) - 3 * x * x;
}

static double f1(double x)
{
    return Math.Exp(x) - x * x * x;
}
    }
}
```

Here, we set $a = 0$, $b = 1$, and $n = 101$, so that the increment $h = (b - a)/(n - 1) = 0.01$. The function $f1$ is the analytic integration of the function f. Running this example generates the result shown in Figure 12-3.

```
Analytic result = 0.718281828459045
Result for function = 0.718246147450418
Result for data array = 0.718246147450418
```

Figure 12-3 Results from the Trapezoidal method.

You can see that the results from both the function and data array are the same, as expected. The trapezoidal method provides 4 digital precision by comparing with the analytic result. You can improve its accuracy by increasing the number of strips n.

Simpson's Rule

The most popular numerical integration method is Simpson's 1/3 rule. This rule can be obtained from the Newton-Cotes formulas (12-1) and (12-2) with $n = 2$. This means that we can replace the integrand $f(x)$ by a quadratic polynomial $p(x)$ which takes the same values as $f(x)$ at the end points a and b and the midpoint $m = (a+b)/2$, as shown in Figure 12-4.

We can use the Lagrange polynomial interpolation to find the expression for this polynomial:

$$p(x) = f(x)\frac{(x-m)(x-b)}{(a-m)(a-b)} + f(m)\frac{(x-a)(x-b)}{(m-a)(m-b)} + f(b)\frac{(x-a)(x-m)}{(b-a)(b-m)}$$

Note from Figure 12-4 that $h = (b-a)/2$. Then we can calculate the integral analytically using $p(x)$:

$$I = \int_a^b f(x)\,dx \approx \int_a^b p(x)\,dx = \frac{h}{3}\left[f(a) + 4f\left(\frac{a+b}{2}\right) + f(b)\right]$$

Using the above equation, we can construct a composite Simpson's rule. If the interval of integration $[a, b]$ is small, then Simpson's rule will provide an adequate approximation to the exact integral. By "small", we really mean that the function being integrated is relatively smooth over the interval $[a, b]$. For such a function, a smooth quadratic interpolant like the one used in Simpson's rule will provide good results.

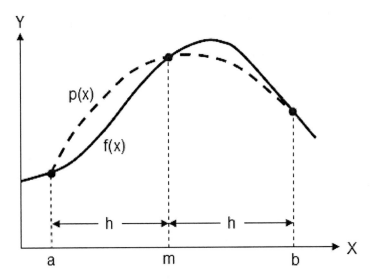

Figure 12-4 Illustration of the Simpson's 1/3 rule.

In practice, the integration range $[a, b]$ can always be divided into n (n must be even) strips of width $h = (b-a)/n,$ as we did for the trapezoidal rule. Applying the Simpson's rule to two adjacent strips, we have

$$\int_{x_i}^{x_{i+2}} f(x)\,dx = \frac{h}{3}\left[f(x_i) + 4f(x_{i+1}) + f(x_{i+2})\right]$$

Then the integral can be obtained by the sum:

$$I = \int_{a}^{b} f(x)\,dx = \sum_{i=0,2,4,\cdots}^{n} \int_{x_i}^{x_{i+2}} f(x)\,dx = \frac{h}{3}\sum_{i=0,2,4,\cdots}^{n}\left[f(x_i) + 4f(x_{i+1}) + f(x_{i+2})\right]$$

Note that Simpson's 1/3 rule requires the number of strips n to be even. If this condition is not satisfied, we can integrate over the first (or last) three strips by using Simpson's 3/8 rule:

$$I = \frac{3h}{8}\left[f(x_0) + 3f(x_1) + 3f(x_2) + f(x_3)\right]$$

and use Simpson's 1/3 rule for the rest of strips.

Implementation

Here we will implement the numerical integration using Simpson's rule. Add a new method, *Simpson*, to the *Intergration* class. Here is the code listing of this method:

```
public static double Simpson(Function f, double a, double b, int n)
{
    if (n < 3)
        return badResult;
    double sum = 0.0;
```

```
        double h = (b - a) / (n - 1);
        if (n % 2 != 0)
        {
            for (int i = 0; i < n - 2; i += 2)
            {
                sum += h * (f(a + i * h) + 4 * f(a + (i + 1) * h) +
                        f(a + (i + 2) * h)) / 3;
            }
        }
        else
        {
            sum = 3 * h * (f(a) + 3 * f(a + h) + 3 * f(a + 2 * h) +
                    f(a + 3 * h)) / 8;
            for (int i = 3; i < n - 2; i += 2)
            {
                sum += h * (f(a + i * h) + 4 * f(a + (i + 1) * h) +
                        f(a + (i + 2) * h)) / 3;
            }
        }
        return sum;
    }

    public static double Simpson(double[] yarray, double h)
    {
        int n = yarray.Length;
        if (n < 3 || h == 0)
            return badResult;

        double sum = 0.0;
        if (n % 2 != 0)
        {
            for (int i = 0; i < n - 2; i += 2)
            {
                sum += h * (yarray[i] + 4 * yarray[i + 1] + yarray[i + 2]) / 3;
            }
        }
        else
        {
            sum = 3 * h * (yarray[0] + 3 * yarray[1] +
                    3 * yarray[2] + yarray[3]) / 8;
            for (int i = 3; i < n - 2; i += 2)
            {
                sum += h * (yarray[i] + 4 * yarray[i + 1] + yarray[i + 2]) / 3;
            }
        }
        return sum;
    }
```

Here, we implement two overloaded methods: one for the function and the other for the array of function values. In order to calculate integration for a function, you need to pass a delegate function $f(x)$, and the number of strips n.

On the other hand, in order to calculate integration for an array of function values, the method has input parameters which pass the *yarray* values, and the increment h. Here, the increment parameter should be the value used to generate the function values. Note that the data values must be generated in equally spaced *x* points.

Testing the Simpson Method

Now I will show you how to calculate the integration by using methods implemented in the previous section. Add a new static method, *TestSimpson*, to the *Program.cs* file:

```
static void TestSimpson()
{
    int n = 101;
    double result;

    result = f1(1) - f1(0);
    Console.WriteLine("\n Analytic result = " + result.ToString());

    result = Integration.Simpson(f, 0, 1, n);
    Console.WriteLine(" Result for function = " + result.ToString());

    double[] ya = new double[n];
    double h = 1.0 / (n - 1);
    for (int i = 0; i < n; i++)
    {
        double x = i * h;
        ya[i] = f(x);
    }
    result = Integration.Simpson(ya, h);
    Console.WriteLine(" Result for data array = " + result.ToString());
}
```

Here, we use the same intgrand function *f* as in the Trapezoidal example. The function *f1* is the analytic integration of function *f*. Running this example generates the result shown in Figure 12-5.

```
Analytic result = 0.718281828459045
Result for function = 0.718281828554504
Result for data array = 0.718281828554504
```

Figure 12-5 Results from the Simpson method.

You can see that the results from both the function and data array are the same, as expected. The Simpson method provides 9 digit precision compared to the analytic result. With the same number of strips (*n* = 101), the Simpson method gives much more accurate results than the trapezoidal method does.

Higher Order Rules

In the previous sections, the trapezoidal and Simpson's 1/3 and 3/8 rules are, respectively, for two-, three- and four-point formulas. Higher order rules are also possible. For example, the five-point formula, also called Bode's rule, is given by

$$\int_{x1}^{x5} f(x)\,dx = \frac{2h}{45}\left[7f(x_1)+32f(x_2)+12f(x_3)+32f(x_4)+7f(x_5)\right]$$

At this point the formulas stop being named aften famous personages. Here I only list formulas up to 9-point rules. For simplicity's sake, the $f(x_i)$ will be denoted by f_i.

6-point rule:

$$\int_{x_1}^{x_6} f(x)\,dx = \frac{5h}{288}\left(19 f_1 + 75 f_2 + 50 f_3 + 50 f_4 + 75 f_5 + 19 f_6\right)$$

7-point rule:

$$\int_{x_1}^{x_7} f(x)\,dx = \frac{h}{140}\left(41 f_1 + 216 f_2 + 27 f_3 + 272 f_4 + 27 f_5 + 216 f_6 + 41 f_7\right)$$

8-point rule:

$$\int_{x_1}^{x_8} f(x)\,dx = \frac{7h}{17280}\left(751 f_1 + 3577 f_2 + 1323 f_3 + 2989 f_4 + 2989 f_5 + 1232 f_6 + 3577 f_7 + 751 f_8\right)$$

9-point rule:

$$\int_{x_1}^{x_9} f(x)\,dx = \frac{4h}{14175}\left(989 f_1 + 5888 f_2 - 928 f_3 + 10496 f_4 - 4540 f_5 + 10496 f_6 - 928 f_7 + 5888 f_8 + 989 f_9\right)$$

The benefit of going to higher orders is obtaining more accurate results. Usually, the error of numerical integration with an n-point rule is proportional to h^{n+1}.

You can easily implement different point rules using the formulas listed here according to the requirements of your application, following the same procedure used to implement the trapezoidal and Simpson methods.

Romberg Integration

Romberg integration combines the trapezoidal rule with Richardson extrapolation. We have used Richardson extrapolation in the previous chapter to compute numerical differentiation. Romberg's method evaluates the integrand at equally spaced points. The integrand must have continuous derivatives, though fair results may be obtained even if only a few derivatives exist.

The method can be defined inductively in this way:

$$R(0,0) = \frac{1}{2}(b-a)\left[f(a) + f(b)\right]$$

$$R(n,0) = \frac{1}{2}R(n-1,0) + h_n \sum_{k=1}^{2^{n-1}} f\left(a + (2k-1)h_n\right)$$

$$R(n,m) = R(n,m-1) + \frac{1}{4^m - 1}\left[R(n,m-1) - R(n-1,m-1)\right]$$

or

$$R(n, m) = \frac{1}{4^m - 1} \left[4^m R(n, m-1) - R(n-1, m-1) \right]$$

where $m, n \geq 1$ and $h_n = (b-a)/2^n$.

The zeroth extrapolation, $R(n,0)$, is equivalent to the trapezoidal rule with $2^{n-1} + 1$ points; the first extrapolation, $R(n,1)$, is equivalent to Simpson's rule with $2^{n-1} + 1$ points.

Note that

$$R(1,1) = \frac{4}{3} R(1,0) - \frac{1}{3} R(0,0)$$

It is convenient to store the results in an array of the form

$$\begin{pmatrix} R(0,0) & \\ R(1,0) & R(1,1) \end{pmatrix}$$

Repeating this process, the terms can be arranged in a tableau:

$$\begin{pmatrix} R(0,0) & & & & \\ R(1,0) & R(1,1) & & & \\ R(2,0) & R(2,1) & R(2,2) & & \\ \vdots & \vdots & \vdots & \ddots & \\ R(i,0) & R(i,1) & R(i,2) & \cdots & R(i,i) \end{pmatrix}$$

Note that the most accurate approximation to the integral is always the last diagonal term of the array. The process is continued until the difference between two successive diagonal terms becomes sufficiently small.

The triangular matrix can be implemented using a one-dimensional array T. Note that after the first extrapolation, $R(0,0)$ is never used again, so it can be replaced with $R(1,1)$. As a result, we have the array

$$\begin{pmatrix} T(0) = R(1,1) \\ T(1) = R(1,0) \end{pmatrix}$$

In the next extrapolation round, $R(2,1)$ overwrites $R(1,0)$, and $R(2,2)$ replaces $R(1,1)$. The array now contains

$$\begin{pmatrix} T(0) = R(2,2) \\ T(1) = R(2,1) \\ T(2) = R(2,0) \end{pmatrix}$$

and so on. In this manner, $T(0)$ always contains the best current result. The extrapolation formula for the kth round is

$$T(i) = \frac{4^{k-i} T(i+1) - T(i)}{4^{k-i} - 1}, \qquad i = k-1, k-2, \cdots, 1$$

Implementation

Now we will implement the Romberg method for numerical integration. Add a new method, *Romberg*, to the *Intergration* class. Here is code listing of this method:

```
public static double Romberg(Function f, double a, double b,
                             int maxIterations, double tolerance)
{
    int n = (int)Math.Pow(2,maxIterations) + 1;
    double[] T = new double[n];
    double[] fn = new double[n - 2];
    double c1;
    double h = b - a;
    double fa = 0.5 * (f(a) + f(b));
    T[0] = h * fa;
    double result = T[0];
    int i, j;
    int nIterations = 0;
    int nSteps = 1;
    do
    {
        result = T[nIterations];
        nIterations++;
        h /= 2;
        nSteps *= 2;
        c1 = T[0];

        j = 0;
        i = nSteps - 1;
        do
        {
            j++;
            fn[i - 1] = f(a + i * h);
            if (i > 1)
                fn[i - 2] = fn[nSteps / 2 - j - 1];
            i -= 2;
        }
        while (i >= 1);

        T[0] = fa;
        for (i = 1; i < nSteps; i++)
        {
            T[0] += fn[i - 1];
        }
        T[0] *= h;
        for (i = 2; i < nIterations + 2; i++)
        {
            T[i - 1] = (Math.Pow(4,i-1)*T[i - 2] - c1) /
                       (Math.Pow(4, i - 1) - 1);
        }
    }
    while (nIterations < maxIterations &&
           Math.Abs(T[nIterations] - result) > tolerance) ;
    return T[nIterations];
}
```

The *Romberg* method takes as input parameters the delegate function *f*, the integration limits *a* and *b*, the maximum iterations, and the tolerance.

Testing the Romberg Method

Here, I will show you how to calculate the integration using methods implemented in the previous section. Add a new static method, *TestRomberg*, to the *Program.cs* file:

```
static void TestRomberg()
{
    double result;

    result = f1(1) - f1(0);
    Console.WriteLine("\n Analytic result = " + result.ToString());

    result = Integration.Romberg(f, 0, 1, 15, 1e-9);
    Console.WriteLine(" Result from Romberg method = " + result.ToString());
}
```

In the above, we use the same intgrand function *f* as in the Trapezoidal example. The function f_1 is the analytic integration of function *f*. Running this example generates the result shown in Figure 12-6.

```
Analytic result = 0.718281828459045
Result from Romberg method = 0.718281828577692
```

Figure 12-6 Results from the homberg method.

Gaussian Integration

In numerical integration, if we have the freedom to choose the points at which we evaluate the function values, a careful choice can lead to much more accuracy in evaluating the integral. The Gaussian integration method is specifically designed for this situation.

Gaussian formulas are good at estimating integrals of the form

$$I = \int_a^b w(x) f(x)\, dx$$

where *w(x)*, called the weighting function, can contains singularities, as long as they are integrable. Gaussian integration does not require us to evaluate the integrand at endpoints of the interval. This is very useful when it comes to evaluating various improper integrals, such as those with infinite limits.

Gaussian integration formulas have the same form as Newton-Cotes rules:

$$I = \sum_{i=0}^{n} w_i f(x_i)$$

The main difference lies in the way that the weights w_i and x_i are determined. In Newton-Cotes integration, the data points must be equally spaced in the interval (*a*, *b*); i.e., their locations are predetermined. On the other hand, in Gaussian integration, locations of the points and weights are chosen so that the above equation yields the exact integral if $f(x)$ is a polynomial of degree $2n + 1$ or less; that is,

$$I = \int_a^b w(x) P_m(x)\, dx = \sum_{i-0}^{n} w_i\, P_m(x_i), \qquad m \le 2n+1$$

One way of determining the weights and x_i is to set $P_i(x) = x^i$ in the above equation and solve the resulting $2n + 1$ equations

$$\int_a^b w(x) x^j\, dx = \sum_{i=0}^{n} w_i\, x_i^j, \qquad j = 0, 1, \cdots, 2n+1$$

for the unknown w_i and x_i.

Below, I will present several classic Gaussian integration methods, including Gauss-Legendre, Gauss-Laguerre, Gauss-Hermite, and Gauss-Chebyshev integrations. The weighting function for each type of Gauss integration is determined by a set of orthogonal polynomials.

Gauss-Legendre Integration

The Gauss-Legendre integration is used to numerically calculate the following integral

$$I = \int_{-1}^{1} f(x)\, dx = \sum_{i=0}^{n-1} w_i f(x_i)$$

This case corresponds to the Guass integration with the weighting function $w(x) = 1$.

With the help of a change in the variable, we can also use the Gauss-Legendre method to compute the integral defined in the interval $[a, b]$:

$$I = \int_a^b f(x)\, dx = \int_{-1}^{1} f\left(\frac{a+b}{2} + \frac{b-a}{2} x \right) \frac{b-a}{2}\, dx = \frac{b-a}{2} \sum_{i=0}^{n-1} w_i f\left(\frac{a+b}{2} + \frac{b-a}{2} x_i \right)$$

From the above equation, we can see that Gauss-Legendre integration is completely determined by a set of nodes x_i and weights w_i. The integration nodes x_i for the n points formula are found as roots of the nth-order Legendre polynomials $P_n(x)$, which occur symmetrically about 0. The recurrence formula for Legendre polynomials has already been discussed in Chapter 6.

After finding x_i, weights w_i are calculated by using the following formula:

$$w_i = \frac{2(1 - x_i^2)}{(n+1)^2 [P_{n+1}(x)]^2}$$

Implementation

The nodes and weights for the Gauss-Legendre integration have been computed with great precision and tabulated in literature. These data can be used without knowing the theory behind them, since all you need are the values of x_i and w_i. Here I will present a subroutine that allows you to compute the nodes and weights for any order of n.

In order to compute weights, we need to know the values of the Legendre polynomials. Add the *SpecialFunctions.cs* file from the project *SpecialFunctionsTest* in Chapter 6 to the current project. Add a new method, *LegendreNodesWeights*, to the *Integration* class. Here is the code listing of this method:

```
public static void LegendreNodesWeights(int n, out double[] x, out double[] w)
{
    double c, d, p1, p2, p3, dp;

    x = new double[n];
    w = new double[n];

    for (int i = 0; i < (n + 1) / 2; i++)
    {
        c = Math.Cos(Math.PI * (4 * i + 3) / (4 * n + 2));
        do
        {
            p2 = 0;
            p3 = 1;
            for (int j = 0; j < n; j++)
            {
                p1 = p2;
                p2 = p3;
                p3 = ((2 * j + 1) * c * p2 - j * p1) / (j + 1);
            }
            dp = n * (c * p3 - p2) / (c * c - 1);
            d = c;
            c -= p3 / dp;
        }
        while (Math.Abs(c - d) > 1e-12);
        x[i] = c;
        x[n - 1 - i] = -c;
        w[i] = 2 * (1 - x[i] * x[i]) / (n + 1) / (n + 1) /
                SpecialFunctions.Legendre(x[i], n + 1) /
                SpecialFunctions.Legendre(x[i], n + 1);
        w[n - 1 - i] = w[i];
    }
}
```

This method calculate nodes and weights, stored in *x* and *w* arrays, respectively, for Gauss-Legendre integration. Using this method, we can easily implement Gauss-Legendre integration. Add a new method, *GaussLegendre*, to the *Integration* class:

```
public static double GaussLegendre(Function f, double a, double b, int n)
{
    double[] x, w;
    LegendreNodesWeights(n, out x, out w);

    double sum = 0.0;
    for (int i = 0; i < n; i++)
```

```
    {
        sum += 0.5 * (b - a) * w[i] * f(0.5 * (a + b) + 0.5 * (b - a) * x[i]);
    }
    return sum;
}
```

Testing Gauss-Legendre Integration

Here I will show you how to compute the following integral using the Gauss-Legendre method:

$$I = \int_{1}^{2} \frac{1}{\sqrt{2\pi}} e^{-x^2/2} dx$$

Add a new static method, *TestGaussLegendre*, to the *Program.cs* file:

```
static void TestGaussLegendre()
{
    Console.WriteLine("\n Result from Gauss-Legendre method:\n");
    double result;
    for (int n = 1; n < 9; n++)
    {
        result = Integration.GaussLegendre(f2, 1, 2, n);
        Console.WriteLine(" n = {0}, result = {1}", n, result);
    }
}

static double f2(double x)
{
    return Math.Exp(-0.5 * x * x) / Math.Sqrt(2.0 * Math.PI);
}
```

Running this example creates the results shown in Figure 12-7. You can see that the result for $n = 3$ is already very accurate.

```
Result from Gauss-Legendre method:

n = 1, result = 0.129517595665892
n = 2, result = 0.136061775581924
n = 3, result = 0.135903762380603
n = 4, result = 0.13590512716461
n = 5, result = 0.135905121990046
n = 6, result = 0.135905121983062
n = 7, result = 0.135905121983279
n = 8, result = 0.135905121983278
```

Figure 12-7 Results from the Gauss-Legendre method.

Gauss-Laguerre Integration

Gauss-Laguerre integration is used to numerically calculate the following integral:

$$I = \int_{0}^{\infty} e^{-x} f(x) dx = \sum_{i=0}^{n-1} w_i f(x_i)$$

This corresponds to the Guass integration with the weighting function $w(x) = e^{-x}$.

We can see from the above equation that Gauss-Laguerre integration is completely determined by a set of nodes x_i and weights w_i. The integration nodes x_i for the n points formula are found as roots of nth-order Laguerre polynomials $L_n(x)$. The recurrence formula for Laguerre polynomials has been discussed in Chapter 6.

After finding x_i, weights w_i can be calculated by using the following formula:

$$w_i = \frac{x_i}{(n+1)^2 [L_{n+1}(x_i)]^2}$$

Implementation

The nodes and weights for the Gauss-Laguerre integration have been computed with great precision, and tabulated in the literature. These data can be used without knowing the theory behind them, since all you need are the values of x_i and w_i. Here I will present a subroutine that allows you to compute the nodes and weights for any order of n.

Add a new method, *LaguerreNodesWeights*, to the *Integration* class. Here is the code listing of this method:

```
public static void LaguerreNodesWeights(int n, out double[] x, out double[] w)
{
    double c = 0.0;
    double d, p1, p2, p3, dp;

    x = new double[n];
    w = new double[n];
    for (int i = 0; i < n; i++)
    {
        if (i == 0)
        {
            c = 3 / (1 + 2.4 * n);
        }
        else
        {
            if (i == 1)
            {
                c += 15 / (1 + 2.5 * n);
            }
            else
            {
                c+= (1 + 2.55 * (i - 1)) / (1.9 * (i - 1))  * (c - x[i - 2]);
            }
        }
        do
        {
            p2 = 0;
            p3 = 1;
            for (int j = 0; j < n; j++)
            {
                p1 = p2;
```

```
            p2 = p3;
            p3 = ((-c + 2 * j + 1) * p2 - j * p1) / (j + 1);
        }
        dp = (n * p3 - n * p2) / c;
        d = c;
        c = c - p3 / dp;
    }
    while (Math.Abs(c- d) > 1e-12);
    x[i] = c;
    w[i] = x[i] / (n + 1) / (n + 1) /
            SpecialFunctions.Laguerre(x[i], n + 1) /
            SpecialFunctions.Laguerre(x[i], n + 1);
    }
}
```

This method calculates nodes and weights for Gauss-Laguerre integration, which are stored in x and w arrays, respectively. With this method, we can easily implement Gauss-Laguerre integration. Add a new method, *GaussLaguerre*, to the *Integration* class:

```
public static double GaussLaguerre(Function f, int n)
{
    double[] x, w;
    LaguerreNodesWeights(n, out x, out w);

    double sum = 0.0;
    for (int i = 0; i < n; i++)
    {
        sum += w[i] * f(x[i]);
    }
    return sum;
}
```

Testing the Gauss-Laguerre Integration

Here I will show you how to compute the following integral using the Gauss-Laguerre method:

$$I = \int_0^\infty e^{-x} \sin x \, dx$$

Add a new static method, *TestGaussLaguerre*, to the *Program.cs* file:

```
static void TestGaussLaguerre()
{
    Console.WriteLine("\n Result from Gauss-Laguerre method:\n");
    double result;
    for (int n = 1; n < 9; n++)
    {
        result = Integration.GaussLaguerre(f3, n);
        Console.WriteLine(" n = {0}, result = {1}", n, result);
    }
}
static double f3(double x)
{
    return Math.Sin(x);
}
```

Note that the delegate function is simply a sine function, since the weighting function $w(x) = e^{-x}$ for the Gauss-Laguerre integration.

Running this example produces the results shown in Figure 12-8. The exact result of this integral is equal to 0.5. You can see that the result for $n = 6$ is already very accurate.

```
Result from Gauss-Laguerre method:
n = 1, result = 0.841470984807897
n = 2, result = 0.432459454679844
n = 3, result = 0.496029827480564
n = 4, result = 0.504879279460199
n = 5, result = 0.498903320956064
n = 6, result = 0.500049474797677
n = 7, result = 0.500038911994668
n = 8, result = 0.4999877537353
```

Figure 12-8 Results from the Gauss-Laguere method.

Gauss-Hermite Integration

Gauss-Hermite integration is used to numerically calculate the following integral:

$$I = \int_{-\infty}^{\infty} e^{-x^2} f(x)\,dx = \sum_{i=0}^{n-1} w_i f(x_i)$$

This corresponds to the Guass integration with the weighting function $w(x) = e^{-x^2}$.

You can see from the above equation that Gauss-Hermite integration is completely determined by a set of nodes x_i and weights w_i. The integration nodes x_i for the n points formula are found as roots of the nth-order Hermite polynomials $H_n(x)$. The recurrence formula for Hermite polynomials has already been discussed in Chapter 6.

After finding x_i, weights w_i are calculated by using the following formula:

$$w_i = \frac{2^{n+1} n! \sqrt{\pi}}{[H_{n+1}(x_i)]^2}$$

Implementation

The nodes and weights for the Gauss-Hermite integration have been computed with great precision and tabulated in the literature. These data can be used without knowing the theory behind them, since all you need are the values of x_i and w_i. Here I will present a subroutine that allows you to compute the nodes and weights for any order of n.

Add a new method, *HermiteNodesWeights*, to the *Integration* class. Here is the code listing of this method:

```
public static void HermiteNodesWeights(int n, out double[] x, out double[] w)
{
    double c = 0.0;
    double d, p1, p2, p3, dp;
```

```
x = new double[n];
w = new double[n];
for (int i = 0; i < (n + 1) / 2; i++)
{
    if (i == 0)
    {
        c = Math.Sqrt(2 * n + 1) -
            1.85575 * Math.Pow(2 * n + 1, -((double)(1) / (double)(6)));
    }
    else
    {
        if (i == 1)
        {
            c = c - 1.14 * Math.Pow(n, 0.426) / c;
        }
        else
        {
            if (i == 2)
            {
                c = 1.86 * c - 0.86 * x[0];
            }
            else
            {
                if (i == 3)
                {
                    c = 1.91 * c - 0.91 * x[1];
                }
                else
                {
                    c = 2 * c - x[i - 2];
                }
            }
        }
    }
    do
    {
        p2 = 0;
        p3 = Math.Pow(Math.PI, -0.25);
        for (int j = 0; j < n; j++)
        {
            p1 = p2;
            p2 = p3;
            p3 = p2 * c * Math.Sqrt((double)(2) / ((double)(j + 1))) -
                p1 * Math.Sqrt((double)(j) / ((double)(j + 1)));
        }
        dp = Math.Sqrt(2 * n) * p2;
        d = c;
        c -= p3 / dp;
    }
    while (Math.Abs(c - d) > 1e-12);
    x[i] = c;
    w[i] = Math.Pow(2, n + 1) *
        SpecialFunctions.Gamma(n + 1) * Math.Sqrt(Math.PI) /
        SpecialFunctions.Hermite(x[i], n + 1) /
```

```
                    SpecialFunctions.Hermite(x[i], n + 1);
            x[n - 1 - i] = -x[i];
            w[n - 1 - i] = w[i];
        }
    }
```

This method calculate nodes and weights for the Gauss-Hermite integration, which are stored in *x* and *w* arrays respectively. Using this method, we can easily implement Gauss-Hermite integration. Add a new method, *GaussHermite*, to the *Integration* class:

```
public static double GaussHermite(Function f, int n)
{
    double[] x, w;
    HermiteNodesWeights(n, out x, out w);

    double sum = 0.0;
    for (int i = 0; i < n; i++)
    {
        sum += w[i] * f(x[i]);
    }
    return sum;
}
```

Testing Gauss-Hermite Integration

Here I will show you how to compute the following Gaussian integral using the Gauss-Hermite method:

$$I = \int_{-\infty}^{\infty} e^{-x^2} x^2 \, dx$$

Add a new static method, *TestGaussHermite*, to the *Program.cs* file:

```
static void TestGaussHermite()
{
    Console.WriteLine("\n Result from Gauss-Hermit method:\n");
    double exact = Math.Sqrt(Math.PI) / 2;
    Console.WriteLine("\n Exact result = {0} \n", exact);
    double result;
    for (int n = 1; n < 9; n++)
    {
        result = Integration.GaussHermite(f4, n);
        Console.WriteLine(" n = {0}, result = {1}", n, result);
    }
}
```

Note that the delegate function is simply x^2, since the weighting function $w(x) = e^{-x^2}$ for the Gauss-Hermite integration.

Running this example produces the results shown in Figure 12-9. The exact result of this integral is equal to $\sqrt{\pi}/2$. You can see that the result for $n = 2$ is already very close to the exact result.

```
Result from Gauss-Hermit method:

Exact result = 0.886226925452758

n = 1, result = 0
n = 2, result = 0.886226925452759
n = 3, result = 0.886226925452759
n = 4, result = 0.886226925452758
n = 5, result = 0.88622692545276
n = 6, result = 0.88622692545276
n = 7, result = 0.88622692545276
n = 8, result = 0.88622692545276
```

Figure 12-9 Results from the Gauss-Hermite method.

Gauss-Chebyshev Integration

Gauss-Chebyshev integration is used to numerically calculate the following integral:

$$I = \int_{-1}^{1} \frac{1}{\sqrt{1-x^2}} f(x)\,dx = \sum_{i=0}^{n-1} w_i f(x_i)$$

This corresponds to the Guass integration with the weighting function $w(x) = 1/\sqrt{1-x^2}$.

By changing the variable, you can also use the Gauss-Chebyshev method to compute the integral defined in the interval $[a, b]$. But, here, I will consider only the simplest case, i.e., the interval is limited to $[-1, 1]$.

It is apparent from the above equation that the Gauss-Chebyshev integration is completely determined by a set of nodes x_i and weights w_i. For every n the nodes and weights can be expressed in closed form, so it is not necessary to implement a special method to calculate them:

$$x_i = \cos\left[\frac{(2i+1)\pi}{2n}\right], \quad i = 0, 1, 2, \cdots, n-1$$

$$w_i = \frac{\pi}{n}, \quad i = 0, 1, 2, \cdots, n-1$$

Implementation

With the closed form for nodes and weights, we can easily implement Gauss-Chebyshev integration. Add a new method, *GaussChebyshev*, to the *Integration* class:

```
public static double GaussChebyshev(Function f, int n)
{
    double x;
    double w = Math.PI / n;

    double sum = 0.0;
    for (int i = 0; i < n; i++)
    {
        x = Math.Cos(Math.PI * (i + 0.5) / n);
        sum += f(x);
    }
}
```

```
        return sum * w;
    }
```

Testing Gauss-Chebyshev Integration

Here I will show you how to compute the following Gaussian integral using the Gauss-Chebyshev method:

$$I = \int_{-1}^{1} (1-x^2)^{3/2} \, dx$$

Since the integrand in the above integral is smooth and free of singularities, we could use the *GaussLegendre* method to compute this integral. However, the integral can also be obtained by using the *GaussChebyshev* method we implemented in the previous section. We can rewrite the above integral in the following form:

$$I = \int_{-1}^{1} (1-x^2)^{3/2} \, dx = \int_{-1}^{1} \frac{1}{\sqrt{1-x^2}} (1-x^2)^2 \, dx$$

Add a new static method, *TestGaussChebyshev*, to the *Program.cs* file:

```
static void TestChebyshev()
{
    Console.WriteLine("\n Result from Gauss-Chebyshev method:\n");
    double result;
    for (int n = 1; n < 9; n++)
    {
        result = Integration.GaussChebyshev(f5, n);
        Console.WriteLine(" n = {0}, result = {1}", n, result);
    }
}

static double f5(double x)
{
    return (1 - x * x) * (1 - x * x);
}
```

Note that the delegate function is simply $(1-x^2)^2$, since the weighting function $w(x)=1/\sqrt{1-x^2}$ for the Gauss-Chebyshev integration.

Running this example produces results shown in Figure 12-10. The exact result of this integral is equal to $3\pi/8$. You can see that the result for $n = 3$ already gives the exact result, which is expected because the function $(1-x^2)^2$ is a polynomial of degree of four, meaning that the Gauss-Chebyshev integration is exact with three nodes.

```
Result from Gauss-Chebyshev method:

n = 1, result = 3.14159265358979
n = 2, result = 0.785398163397448
n = 3, result = 1.17809724509617
n = 4, result = 1.17809724509617
n = 5, result = 1.17809724509617
n = 6, result = 1.17809724509617
n = 7, result = 1.17809724509617
n = 8, result = 1.17809724509617
```

Figure 12-10 Results from the Gauss-Chebyshev method.

Chapter 13
Ordinary Differential Equations

In this chapter, I will present several numerical methods you can use to solve ordinary differential equations. These methods are especially useful when differential equations cannot be solved analytically. The general form of the first-order equations can be expressed by

$$y' = f(x, y) \tag{13.1}$$

and the higher-order equations can be written

$$y^{(n)} = f(x, y, y', y'', \cdots, y^{(n-1)}) \tag{13.2}$$

The above first- or higher-order equations have solutions which may be written

$$y = \varphi(x) \tag{13.3}$$

Our task in this chapter is to determine, from (13.1) or (13.2) and the necessary boundary conditions, the relationship (13.3) between x and y.

A common way of handling a second- or higher-order equation is to replace it with an equivalent system of first-order equations. Note that the higher-order equation (13.2) can always be transformed into a set of n first order equations. Using the notation

$$
\begin{aligned}
y_0 &= y \\
y_1 &= y' \\
y_2 &= y'' \\
&\vdots \\
y_{n-1} &= y^{(n-1)}
\end{aligned}
\tag{13.4}
$$

then the equivalent first-order equations become

$$y_0' = y_1$$
$$y_1' = y_2$$
$$y_2' = y_3 \qquad\qquad (13.5)$$
$$\vdots$$
$$y_n' = f(x, y_0, y_1, \cdots, y_{n-1})$$

For this reason, we will concentrate on methods for solving first-order equations.

Euler Method

The simplest method for the numerical integration of a first-order ordinary differential equation is Euler's method. Consider the problem

$$\frac{dy}{dx} = f(x, y) = g(x)$$

This method assume that for a small distance Δx along the x-axis from some initial point x_0, the function $g(x)$ is a constant equal to $g(x_0)$:

$$\left.\frac{dy}{dx}\right|_{x=x_0} = f(x_0, y_0) = g(x_0)$$

If we assume that $g(x) = g(x_0)$ for all values of x between x_0 and $x_1 = x + \Delta x$, then the change in y corresponding to the small change Δx in x is given approximately by

$$\frac{\Delta x}{\Delta y} = f(x_0, y_0)$$

If y_1 is used to denote $y_0 + \Delta y$, then the above equation becomes

$$y_1 = y_0 + \Delta x \, f(x_0, y_0) = y_0 + h f(x_0, y_0)$$

where $h = \Delta x$. y_1 can then be calculated from the above equation. This completes the solution process for one step along the x-axis. We can repeat the process by using the previous solution as the starting values for the current step, yielding a general solution:

$$y_2 = y_1 + h f(x_1, y_1)$$
$$y_2 = y_2 + h f(x_2, y_2)$$
$$\vdots$$
$$y_{n+1} = y_n + h f(x_n, y_n)$$

This is known as Euler's method. The calculation process, step by step, along the x-axis from the initial point x_0 to the required finishing point.

Euler's method can be explained geometrically, as shown in Figure 13-1. The slope of the curve of y against x is dy/dx, i.e. $f(x, y)$. Thus, at the starting point (x_n, y_n) of each interval the slope is exactly $f(x_n, y_n)$. This value is a good approximation, if h is small, of the average slope over the interval h from x_n to x_{n+1}.

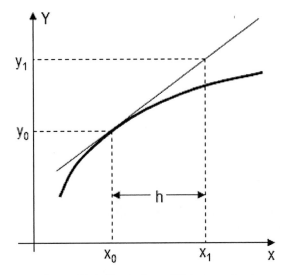

Figure 13-1 Illustration of Euler's method.

Implementation

Here we will implement the numerical solution for first-order ordinary differential equations by using Euler's method. Start with a new C# Console application and name it *ODETest*. Add a new class, *ODE*, to the project and change its namespace to *XuMath*. Add a new method, *Euler*, to the *ODE* class. The following is the code listing of this method:

```
using System;

namespace XuMath
{
    public class ODE
    {
        public delegate double Function(double x, double y);

        public static double Euler(Function f, double x0, double y0,
                            double h, double x)
        {
            double y;
            if (x < x0)
            {
                return double.NaN;
            }
            if (x== x0)
            {
                return y0;
            }

            do
            {
                y = y0 + h * f(x0, y0);
                y0 = y;
```

```
                x0 += h;
            }
            while (x0 < x);
            return y;
        }
    }
}
```

Here, we first define a delegate function $f(x,y)$, which allows the user to input any function. The method takes the delegate function f, the initial values x_0 and y_0, the increment h along the x axis, and the position x, at which you want to find the solution, as input parameters.

Testing Euler's Method

Here I will show you how to numerically solve the following first-order differential equation by using Euler's method, as implemented in the previous section:

$$\frac{dy}{dx} = x + y$$

With the initial condition $x_0 = 0$ and $y_0 = 1$. This equation has an analytical solution:

$$y = 2e^x - x - 1$$

which allows us to compare the calculated result to the exact solution.

Add a new static method, *TestEuler*, to the *Program.cs* file:

```
using System;
using XuMath;
namespace ODETest
{
    class Program
    {
        static void Main(string[] args)
        {
            TestEuler();
            Console.ReadLine();
        }

        static void TestEuler()
        {
            double h = 0.001;
            double x0 = 0;
            double y0 = 1.0;
            Console.WriteLine("\n x0 = {0}, h = {1}", x0, h);
            double result = y0;

            for (int i = 0; i < 11; i++)
            {
                double x = 0.1 * i;
                result = ODE.Euler(f, x0, result, h, x);
                double exact = 2.0 * Math.Exp(x) - x - 1.0;
                Console.WriteLine("\n x = {0}, y = {1},
                                exact = {2}", x, result, exact);
```

```
            x0 = x;
        }
    }

    static double f(double x, double y)
    {
        return x + y;
    }
  }
}
```

Running this example generates results shown in Figure 13-2. It is clear that Euler's method provides results with double-digit precision for $h = 0.001$, which means the method is accurate in the order of $O(h)$.

```
Results from the Euler's method with h = 0.001
x = 0.0, y = 1.000000000000e+000, exact = 1.000000000000e+000
x = 0.1, y = 1.110231395442e+000, exact = 1.110341836151e+000
x = 0.2, y = 1.242561410698e+000, exact = 1.242805516320e+000
x = 0.3, y = 1.399312957609e+000, exact = 1.399717615152e+000
x = 0.4, y = 1.583053122515e+000, exact = 1.583649395283e+000
x = 0.5, y = 1.796618832826e+000, exact = 1.797442541400e+000
x = 0.6, y = 2.043145221558e+000, exact = 2.044237600781e+000
x = 0.7, y = 2.326096973420e+000, exact = 2.327505414941e+000
x = 0.8, y = 2.649302965873e+000, exact = 2.651081856985e+000
x = 0.9, y = 3.016994551502e+000, exact = 3.019206222314e+000
x = 1.0, y = 3.433847864472e+000, exact = 3.436563656918e+000
```

Figure 13-2 Results from the Euler's method.

It is can be seen that Euler's method, with an overall error of order of $O(h)$, is not very accurate. Despite its simplicity, it is seldom used in practice.

Second-Order Runge-Kutta Method

The names of Runge and Kutta are traditionally associated with a class of methods for the numerical integration of ordinary differential equations. In this section, we will consider the second-order Runge-Kutta method.

From the discussion in the previous sections, we see that the main reason why Euler's method has a large truncation error per step is that in evolving the solution from x_n to x_{n+1}, the method only evaluates derivatives at the beginning of the interval, i.e., at x_n. The method is, therefore, very asymmetric in regards to the beginning and the end of the interval. We can construct a more symmetric integration method by making an Euler-like trial step to the midpoint of the interval, and then using the values of both x and y at the midpoint to make the real step across the interval.

To be more specific, we can introduce two parameters k_1 and k_2:

$$k_1 = h f(x_n, y_n)$$
$$k_2 = h f(x_n + h/2, y_n + k_1/2)$$
$$y_{n+1} = y_n + k_2$$

This symmetrization cancels out the first-order error, making the method second-order. The above relation is generally known as the second-order Runge-Kutta method. Euler's method can be regarded as a first-order Runge-Kutta method.

Implementation

Here we will implement the numerical solution for first-order ordinary differential equations by using the second-order Runge-Kutta method. Add a new method, *RungeKutta2*, to the *ODE* class. Here is the code listing of this method:

```
public static double RungeKutta2 (Function f, double x0, double y0,
                                  double h, double x)
{
    double y;
    if (x < x0)
    {
        return double.NaN;
    }
    if (x == x0)
    {
        return y0;
    }

    do
    {
        double k1 = h * f(x0, y0);
        double k2 = h * f(x0 + 0.5 * h, y0 + 0.5 * k1);
        y = y0 + k2;
        y0 = y;
        x0 += h;
    }
    while (x0 < x);

    return y;
}
```

This method takes the delegate function f, the initial values x_0 and y_0, the increment h along the x axis, and the position x, at which you want to find the solution, as input parameters.

Testing the Second-Order Runge-Kutta Method

Here I will show you how to use the second-order Runge-Kutta method to numerically solve the same first-order differential equation which we previously solved using Euler's method.

Add a new static method, *TestRungeKutta2*, to the *Program.cs* file:

```
static void TestRungeKutta2 ()
{
    double h = 0.01;
    double x0 = 0;
    double y0 = 1.0;
    Console.WriteLine("\n Results from the second-order Runge-Kutta method
                       with h = {0}\n", h);
    double result = y0;
```

```
for (int i = 0; i < 11; i++)
{
    double x = 0.1 * i;
    result = ODE.RungeKutta2(f, x0, result, h, x);
    double exact = 2.0 * Math.Exp(x) - x - 1.0;
    Console.WriteLine(" x = {0}, y = {1}, exact = {2}", x, result, exact);
    x0 = x;
}
}
```

Here the delegate function *f* and the initial conditions are the same as those in Euler's method. Running this example generates the results shown in Figure 13-3. Compared to the results from Euler's method, you can see that the second-order Runge-Kutta method provides much more accurate results.

```
Results from the second-order Runge-Kutta method with h = 0.001
x = 0.0, y = 1.0000000000000e+000, exact = 1.0000000000000e+000
x = 0.1, y = 1.110341799340e+000, exact = 1.110341836151e+000
x = 0.2, y = 1.242805434955e+000, exact = 1.242805516320e+000
x = 0.3, y = 1.399717480267e+000, exact = 1.399717615152e+000
x = 0.4, y = 1.583649196522e+000, exact = 1.583649395283e+000
x = 0.5, y = 1.797442266819e+000, exact = 1.797442541400e+000
x = 0.6, y = 2.044237236630e+000, exact = 2.044237600781e+000
x = 0.7, y = 2.327504945418e+000, exact = 2.327505414941e+000
x = 0.8, y = 2.651081263952e+000, exact = 2.651081856985e+000
x = 0.9, y = 3.019205484986e+000, exact = 3.019206222314e+000
x = 1.0, y = 3.436562751504e+000, exact = 3.436563656918e+000
```

Figure 13-3 Results from the second-order Runge-Kutta method.

Fourth-Order Runge-Kutta Method

Even though the second-order Runge-Kutta method provides more accurate results than Euler's method does, the second-order Runge-Kutta method is still not used often in numerical applications. Most programmers prefer integration formulas of the fourth order, which achieve great accuracy with less computational effort.

In most problems encountered in computational physics, the fourth-order Runge-Kutta integration method represents an appropriate compromise between the competing requirements of a low truncation error per step and a low computational cost per step.

The fourth-order Runge-Kutta method can be derived by three trial steps per interval. The standard form of this method can be expressed by the following equations:

$$k_1 = h f(x_n, y_n)$$
$$k_2 = h f(x_n + h/2, y_n + k_1/2)$$
$$k_3 = h f(x_n + h/2, y_n + k_2/2)$$
$$k_4 = h f(x_n + h, y_n + k_3)$$
$$y_{n+1} = y_n + (k_1 + 2k_2 + 2k_3 + k_4)/6$$

Implementation

Here we will implement the fourth-order Runge-Kutta method, which we can use to find numerical solutions for the first-order ordinary differential equation. Add a new method, *RungeKutta4*, to the *ODE* class. Here is the code listing of this method:

```
public static double RungeKutta4(Function f, double x0, double y0,
                                 double h, double x)
{
    double y;
    if (x < x0)
    {
        return double.NaN;
    }
    if (x == x0)
    {
        return y0;
    }

    do
    {
        double k1 = h * f(x0, y0);
        double k2 = h * f(x0 + 0.5 * h, y0 + 0.5 * k1);
        double k3 = h * f(x0 + 0.5 * h, y0 + 0.5 * k2);
        double k4 = h * f(x0 + h, y0 + k3);
        y = y0 + (k1 + 2 * k2 + 2 * k3 + k4) / 6;
        y0 = y;
        x0 += h;
    }
    while (x0 < x);

    return y;
}
```

This method takes the delegate function *f*, the initial values x_0 and y_0, the increment *h* along the *x* axis, and the position *x*, at which you want to find solution, as input parameters.

Testing the Fourth-Order Runge-Kutta Method

Here I will show you how to use the fourth-order Runge-Kutta method to numerically solve the same first-order differential equation that we previously solved using Euler's method.

Add a new static method, *TestRungeKutta4*, to the *Program.cs* file:

```
static void TestRungeKutta4()
{
    double h = 0.001;
    double x0 = 0;
    double y0 = 1.0;
    Console.WriteLine("\n Results from the fourth-order Runge-Kutta method
                      with h = {0}\n", h);
    double result = y0;
    for (int i = 0; i < 11; i++)
    {
        double x = 0.1 * i;
```

```
        result = ODE.RungeKutta4(f, x0, result, h, x);
        double exact = 2.0 * Math.Exp(x) - x - 1.0;
        Console.WriteLine(" x = {0}, y = {1}, exact = {2}", x, result, exact);
        x0 = x;
    }
}
```

where the delegate function f, the increment h, and the initial conditions are the same as those in Euler's method. Running this example generates the results shown in Figure 13-4. You can see that the fourth-order Runge-Kutta method provides results better than 12-digit accuracy!

```
Results from the fourth-order Runge-Kutta method with h = 0.001
x = 0.0, y = 1.0000000000000e+000, exact = 1.0000000000000e+000
x = 0.1, y = 1.110341836151e+000, exact = 1.110341836151e+000
x = 0.2, y = 1.242805516320e+000, exact = 1.242805516320e+000
x = 0.3, y = 1.399717615152e+000, exact = 1.399717615152e+000
x = 0.4, y = 1.583649395283e+000, exact = 1.583649395283e+000
x = 0.5, y = 1.797442541400e+000, exact = 1.797442541400e+000
x = 0.6, y = 2.044237600781e+000, exact = 2.044237600781e+000
x = 0.7, y = 2.327505414941e+000, exact = 2.327505414941e+000
x = 0.8, y = 2.651081856985e+000, exact = 2.651081856985e+000
x = 0.9, y = 3.019206222314e+000, exact = 3.019206222314e+000
x = 1.0, y = 3.436563656918e+000, exact = 3.436563656918e+000
```

Figure 13-4 Results from the fourth-order Runge-Kutta method.

Adaptive Runge-Kutta Method

The main disadvantage of the Runge-Kutta method is that it is hard to estimate the truncation error. In addition, a constant step size may not be appropriate for the entire range of integration. This is where adaptive methods come in. These methods estimate the truncation error at each integration step and automatically adjust the step size to keep the error within prescribed limits.

The Runge-Kutta-Fehlberg method is one way to try to resolve this problem. It determines whether the proper step size h is being used. At each step, two different approximations for the solution are made and compared. If the two answers are in close agreement, the approximation is accepted. If the two answers do not agree to a specified accuracy, the step size is reduced. If the answers agree by more significant digits than is required, the step size is increased.

In the following, I list the formulas used in the Runge-Kutta-Fehlberg method:

$$k_1 = h f(x_n, y_n)$$
$$k_2 = h f(x_n + h/4, y_n + k_1/4)$$
$$k_3 = h f(x_n + 3h/8, y_n + 3k_1/32 + 9k_2/32)$$
$$k_4 = h f(x_n + 12h/13, y_n + 1932k_1/2197 - 7200k_2/2197 + 7296k_3/2197)$$
$$k_5 = h f(x_n + h, y_n + 439k_1/216 - 8k_2 + 3680k_3/513 - 845k_4/4104)$$
$$k_6 = h f(x_n + h/2, y_n - 8k_1/27 + 2k_2 - 3544k_3/2565 + 1859k_4/4104 - 11k_5/40)$$

Then an approximation to the solution is made using a Runge-Kutta method of order 4:

$$y_{n+1} = y_n + \frac{25}{216}k_1 + \frac{1408}{2565}k_3 + \frac{2197}{4104}k_4 - \frac{1}{5}k_5$$

And a better value for the solution is determined using a Runge-Kutta method of order 5:

$$z_{n+1} = y_n + \frac{16}{135}k_1 + \frac{6656}{12825}k_3 + \frac{28561}{56430}k_4 - \frac{9}{50}k_5 + \frac{2}{55}k_6$$

The optimal step size $s \cdot h$ can be determined by multiplying the scaling factor s by the current step size h. The scaling factor s is given by

$$s = \left(\frac{\varepsilon}{2\,error}\right)^{1/4}$$

where ε is the specified error control tolerance parameter and $error$ is the truncation error in this method, given by

$$error = \left|\frac{1}{360}k_1 - \frac{128}{4275}k_3 - \frac{2197}{75240}k_4 + \frac{1}{50}k_5 + \frac{2}{55}k_6\right|\Big/h$$

Implementation

Here we will implement the Runge-Kutta-Fehlberg method, which we can use to find numerical solutions for first-order ordinary differential equations. Add a new method, *RungeKuttaFehlberg*, to the *ODE* class. The following is the code listing of this method:

```
public static double RungeKuttaFehlberg(Function f, double x0, double y0,
        double x, double h, double tolerance)
{
    double hmin = 0.0001;
    double hmax = 0.5;
    if (h > hmax)
        h = hmax;
    if (h < hmin)
        h = hmin;

    while (x0 < x)
    {
        double k1 = h * f(x0, y0);
        double k2 = h * f(x0 + 0.25 * h, y0 + 0.25 * k1);
        double k3 = h * f(x0 + 3 * h / 8, y0 + 3 * k1 / 32 + 9 * k2 / 32);
        double k4 = h * f(x0 + 12 * h / 13, y0 + 1932 * k1 / 2197 -
                    7200 * k2 / 2197 + 7296 * k3 / 2197);
        double k5 = h * f(x0 + h, y0 + 439 * k1 / 216 - 8 * k2 +
                    3680 * k3 / 513 - 845 * k4 / 4104);
        double k6 = h * f(x0 + 0.5 * h, y0 - 8 * k1 / 27 + 2 * k2 -
                    3544 * k3 / 2565 + 1859 * k4 / 4104 - 11 * k5 / 40);
        double error = Math.Abs(k1 / 360 - 128 * k3 / 4275 -
                    2197 * k4 / 75240 + k5 / 50 + 2 * k6 / 55) / h;
        double s = Math.Pow(0.5 * tolerance / error, 0.25);

        if (error < tolerance)
```

```
        {
            y0 += 25 * k1 / 216 + 1408 * k3 / 2565 +
                    2197 * k4 / 4104 - 0.2 * k5;
            x0 += h;
        }

        if (s < 0.1)
            s = 0.1;
        if (s > 4)
            s = 4;
        h *= s;
        if (h > hmax)
            h = hmax;
        if (h < hmin)
            h = hmin;
        if (h > x - x0)
            h = x - x0;
    }
    return y0;
}
```

This method takes the delegate function *f*, the initial values x_0 and y_0, the increment *h* along the *x* axis, the tolerance, and the position *x*, at which you want to find solution, as input parameters.

Testing the Adaptive Runge-Kutta Method

Here I will show you how to use the adaptive Runge-Kutta method, or the Runge-Kutta-Fehlberg method, to numerically solve the same first-order differential equation that we used in discussing the Euler's method.

Add a new static method, *TestRungeKuttaFehlberg*, to the *Program.cs* file:

```
static void TestRungeKuttaFehlberg()
{
    double h = 0.2;
    double x0 = 0;
    double y0 = 1.0;
    Console.WriteLine("\n Results from the fourth-order Runge-Kutta-Fehlberg
                        method with h = {0}\n", h);
    double result = y0;
    for (int i = 0; i < 11; i++)
    {
        double x = 0.1 * i;
        result = ODE.RungeKuttaFehlberg(f, x0, result, x, h, 1e-8);
        double exact = 2.0 * Math.Exp(x) - x - 1.0;
        Console.WriteLine(" x = {0}, y = {1}, exact = {2}", x, result, exact);
        x0 = x;
    }
}
```

You can see that the step size *h* is set to be a rather large value of 0.2. The Runge-Kutta-Fehlberg method will automatically refine the step size according to the accuracy requirements, which the tolerance parameter controls.

In this example, the delegate function f and the initial conditions are the same as those in Euler's method. Running this example generates the results shown in Figure 13-5. It is clear that the results from the adaptive Runge-Kutta method provide a 8-digit precision, as specified by the tolerance parameter, regardless of the initial step size.

```
Results from the fourth-order Runge-Kutta-Fehlberg method with h = 0.2
x = 0.0, y = 1.0000000000000e+000, exact = 1.0000000000000e+000
x = 0.1, y = 1.110341836579e+000, exact = 1.110341836151e+000
x = 0.2, y = 1.242805517225e+000, exact = 1.242805516320e+000
x = 0.3, y = 1.399717616595e+000, exact = 1.399717615152e+000
x = 0.4, y = 1.583649397339e+000, exact = 1.583649395283e+000
x = 0.5, y = 1.797442544166e+000, exact = 1.797442541400e+000
x = 0.6, y = 2.044237604363e+000, exact = 2.044237600781e+000
x = 0.7, y = 2.327505419413e+000, exact = 2.327505414941e+000
x = 0.8, y = 2.651081862429e+000, exact = 2.651081856985e+000
x = 0.9, y = 3.019206228821e+000, exact = 3.019206222314e+000
x = 1.0, y = 3.436563664589e+000, exact = 3.436563656918e+000
```

Figure 13-5 Results from the adaptive Runge-Kutta method.

Runge-Kutta Method for Systems

The methods discussed in previous sections apply to only a single first-order ordinary differential equation as described in Equation (13.1). However, most problems in physics and engineering governed by differential equations are either higher-order equations or coupled differential equation systems.

As mentioned at the beginning of this chapter, a higher-order differential equation can always be transformed into a coupled first-order system of equations. The trick is to expand higher-order derivatives into a series of first-order equations. For example, suppose you want to model a spring-mass system with damping, which can be described by the following second-order differential equations:

$$m\frac{d^2x}{dt^2} = -kx - b\frac{dx}{dt}$$

where k is the spring constant and b is the damping coefficient. Since the velocity $v = dx/dt$, the equation of motion for a spring-mass system can be rewritten in terms of two first-order differential equations:

$$\frac{dv}{dt} = -\frac{k}{m}x - \frac{b}{m}v$$

$$\frac{dx}{dt} = v$$

In the above equation, the derivative of v is a function of v and x, and the derivative of x is a function of v. Since the solution of v as a function of time depends on x and the solution of x as a function of time depends on v, the two equations are coupled and must be solved simultaneously.

Most of the differential equations in physics and engineering are higher-order equations. This means that you must expand them into a series of first-order differential equations before they can be solved using numerrical methods.

Here, we will extend the fourth-order Runge-Kutta method discussed previously to a system of ordinary differential equations. The method will be written as generally as possible, so that it can be used to solve any number of coupled first-order differential equations.

Implementation

In dealing with a system of coupled first-order differential equations, we need vector operations. Add the *VectorR.cs* and *MatrixR.cs* from Chapter 3 into the current project.

In the following, we will implement the fourth-order Runge-Kutta method, which can be used to find numerical solutions for a system of first-order ordinary differential equations. Add a new method, *MultiRungeKutta4,* to the *ODE* class. Here is the code listing of this method:

```
public delegate VectorR MultiFunction(double x, VectorR y);

public static VectorR MultiRungeKutta4(MultiFunction f, VectorR y0,
                                       double x0, double h, double x)
{
    if (x <= x0)
        return y0;

    int n = y0.GetSize();
    VectorR k1 = new VectorR(n);
    VectorR k2 = new VectorR(n);
    VectorR k3 = new VectorR(n);
    VectorR k4 = new VectorR(n);
    VectorR y = y0;

    do
    {
        if (h > x - x0)
            h = x - x0;
        k1 = h * f(x0, y);
        k2 = h * f(x0 + h / 2, y + k1 / 2);
        k3 = h * f(x0 + h / 2, y + k2 / 2);
        k4 = h * f(x0 + h, y + k3);
        y = y + (k1 + 2 * k2 + 2 * k3 + k4) / 6;
        x0 += h;
    }
    while (x0 < x);

    return y;
}
```

Notice that here, we define another delegate function called *MultiFunction,* which takes a double variable x and a *VectorR* object y as its input parameters. The function itself is also a *VectorR* object. Then we implement a static method, *MultiRungeKutta4,* that returns a *VectorR* object as a solution to the system of coupled differential equations. This method takes a *MultiFunction f,* the initial values of x_0 and y_0, the step size h, and the position x, at which you want to find the solutions, as input parameters. Note that the k-parameters in the method also become *VectorR* objects.

The *MultiRungeKutta4* method looks quite similar to the *RungeKutta4* method for a single first-order differential equation. It only takes a very short code listing. However, it is very powerful in the sense that it can be used to solve first-order ordinary differential equations with any number of coupled equations. To apply the method to a specific physics or engineering problem, you simply supply it with the vector function, initial values, and step size. The following sections will show you how to solve real-world physics problems using this method.

Testing the MultiRungeKutta4 Method

Here, we will use the *MultiRungeKutta4* method to solve a coupled spring system with three springs and two masses, as shown in Figure 13-6. This system is fixed at both ends. The parameters m_1 and m_2 represent masses; k_1, k_2, and k_3 are spring constants that define how stiff the springs are; and b_1, b_2, and b_3 are damping coefficients that characterize how quickly the springs' motion will stop.

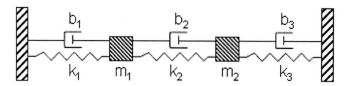

Figure 13-6 A spring-mass system.

The equations of motion for this system can be written in terms of two coupled second-order differential equations:

$$m_1 \frac{d^2 x_1}{dt^2} = -(k_1 + k_2)x_1 + k_2 x_2 - (b_1 + b_2)\frac{dx_1}{dt} + b_2 \frac{dx_2}{dt}$$

$$m_2 \frac{d^2 x_2}{dt^2} = -(k_2 + k_3)x_2 + k_2 x_1 - (b_2 + b_3)\frac{dx_2}{dt} + b_2 \frac{dx_1}{dt}$$

where x_1 and x_2 are the displacements of m_1 and m_2 respectively. There are no closed-form solutions to this set of coupled differential equations. In order to solve these equations numerically by using the Runge-Kutta method, you need to first convert them into a series of first-order differential equations. This can be easily done by introducing the velocity variables $v_1 = dx_1/dt$ and $v_2 = dx_2/dt$:

$$\frac{dx_1}{dt} = v_1$$

$$\frac{dx_2}{dt} = v_2$$

$$\frac{dv_1}{dt} = -\frac{1}{m_1}(k_1 + k_2)x_1 + \frac{k_2}{m_1}x_2 - \frac{1}{m_1}(b_1 + b_2)v_1 + \frac{b_2}{m_1}v_2$$

$$\frac{dv_2}{dt} = -\frac{1}{m_2}(k_2 + k_3)x_2 + \frac{k_2}{m_2}x_1 - \frac{1}{m_2}(b_2 + b_3)v_2 + \frac{b_2}{m_2}v_1$$

(13-6)

The initial conditions are described using the following set of parameters:

$$m_1 = m_2 = 0.2, \quad k_1 = k_3 = 10, \quad k_2 = 1, \quad b_1 = b_2 = b_3 = 0.01$$
$$x_1(0) = 1, \quad x_2(0) = 0, \quad v_1(0) = v_2(0) = 0$$

These coupled first-order differential equations can now be solved by using the Runge-Kutta method we implemented in the previous section.

Add a new method, *TestMultiRungeKutta4*, to the *Program.cs* file:

```
static double m1 = 0.2;
static double m2 = 0.2;
static double k1 = 10.0;
static double k2 = 1.0;
static double k3 = 10.0;
static double b1 = 0.01;
static double b2 = 0.01;
static double b3 = 0.01;
static double x10 = 1.0;
static double x20 = 0.0;
static double v10 = 0.0;
static double v20 = 0.0;

static void TestMultiRungeKutta4()
{
    double dt = 0.02;
    double t0 = 0.0;
    VectorR x0 = new VectorR(new double[] { x10, x20, v10, v20 });

    Console.WriteLine("\n Results for a coupled spring system with
                       h = {0}:\n", dt);
    Console.WriteLine(" t        x1          x2           v1          v2");

    VectorR x = x0;
    for (int i = 0; i < 21; i++)
    {
        double t = 0.1 * i;
        x = ODE.MultiRungeKutta4(f1, t0, x, dt, t);
        Console.WriteLine(" {0:n2}    {1,8:n5}    {2,8:n5}    {3,8:n5}    {4,8:n5}",
                          t, x[0], x[1], x[2], x[3]);
        t0 = t;
    }
}

static VectorR f1(double t, VectorR x)
{
    VectorR result = new VectorR(4);
    result[0] = x[2];
    result[1] = x[3];
    result[2] = -(k1 + k2) * x[0] / m1 + k2 * x[1] / m1 -
                 (b1 + b2) * x[2] / m1 + b2 * x[3] / m1;
    result[3] = -(k2 + k3) * x[1] / m2 + k2 * x[0] / m2 -
                 (b2 + b3) * x[3] / m2 + b2 * x[2] / m2;
    return result;
}
```

Here, we first define several static field members that can be used in computation. We also define a constant time increment *dt*, which corresponds to the step size *h*.

Pay special attention to how we create the vector delegate function using a static method. This function represents the functions on the right-hand side of equation (13-6). The Vector object *x* in the

MultirungeKutta4 method represents four dependent variables, x_1, x_2, v_1, and v_2, i.e., $x[0] = x_1, x[1] = x_2, x[2] = v_1, x[3] = v_2$. In this case, the result is also a *VectorR* object that gives solutions to x_1, x_2, v_1, and v_2.

Running this example generates results shown in Figure 13-7.

```
Results for a coupled spring system with h = 0.02:

t         x1          x2          v1          v2
0.00      1.00000     0.00000     0.00000     0.00000
0.10      0.73838     0.02226    -4.97972     0.39757
0.20      0.09477     0.06388    -7.28948     0.33973
0.30     -0.58892     0.07270    -5.75417    -0.22723
0.40     -0.95497     0.01416    -1.24140    -0.91551
0.50     -0.81825    -0.09504     3.83122    -1.15105
0.60     -0.26022    -0.18884     6.80049    -0.58931
0.70      0.41940    -0.19177     6.16505     0.58864
0.80      0.86475    -0.07268     2.34754     1.71560
0.90      0.85139     0.12345    -2.57712     2.01969
1.00      0.39987     0.29055    -6.01637     1.12803
1.10     -0.24241     0.31960    -6.23373    -0.62577
1.20     -0.73749     0.16757    -3.22914    -2.31070
1.30     -0.83581    -0.10538     1.32219    -2.89920
1.40     -0.50259    -0.35784     5.00718    -1.88682
1.50      0.07284    -0.44136     5.96217     0.33963
1.60      0.58466    -0.28746     3.81947     2.63368
1.70      0.77413     0.04332    -0.16966     3.68679
1.80      0.56091     0.38319    -3.86306     2.77654
1.90      0.07539     0.54317    -5.38431     0.23679
2.00     -0.42000     0.41876    -4.08025    -2.64589
```

Figure 13-7 Results from the MultiRungeKutta4 method.

Adaptive Runge-Kutta Method for Systems

As with solving a single first-order ordinary differential equation, the adaptive Runge-Kutta method is a more efficient and effective method for performing numerical integration, because it can estimate the truncation error at each integration step and automatically adjust the step size to keep the error within the prescribed tolerance.

Here, we will extend this adaptive Runge-Kutta method, i.e., the Runge-Kutta-Fehlberg method, to a system of first-order differential equations. One key parameter in the Runge-Kutta-Fehlberg method is the error function. For a system of equations, the error function becomes a vector. Its components $E_i(h)$ represents the errors in the dependent variables y_i. This brings up a question: what is the error measure *error(h)* that we wish to control in the computation? There is no single choice that works well in all problems. If we want to control the largest component of $E(h)$, that error measure would be

$$error(h) = \max_i |E_i(h)|$$

Here, we prefer to control some gross measure of the error, such as the root-main-square error defined by

$$error(h) = \sqrt{\frac{1}{n} \sum_{i=0}^{n-1} E_i^2 (h)}$$

where n is the number of first-order differential equations. Since the root-mean-square error is easier to handle, we will use it for our implementation.

Implementation

Here we will implement the adaptive Runge-Kutta method, which can be used to find numerical solutions for a system of first-order ordinary differential equations. Add a new method, *MultiRungeKuttaFehlberg*, to the *ODE* class. The following is the code listing of this method:

```
public static VectorR MultiRungeKuttaFehlberg(MultiFunction f, double x0,
            VectorR y0, double x, double h, double tolerance)
{
    if (x <= x0)
        return y0;

    double hmin = 0.0001;
    double hmax = 0.5;
    if (h > hmax)
        h = hmax;
    if (h < hmin)
        h = hmin;

    int n = y0.GetSize();
    VectorR k1 = new VectorR(n);
    VectorR k2 = new VectorR(n);
    VectorR k3 = new VectorR(n);
    VectorR k4 = new VectorR(n);
    VectorR k5 = new VectorR(n);
    VectorR k6 = new VectorR(n);

    while (x0 < x)
    {
        k1 = h * f(x0, y0);
        k2 = h * f(x0 + 0.25 * h, y0 + 0.25 * k1);
        k3 = h * f(x0 + 3 * h / 8, y0 + 3 * k1 / 32 + 9 * k2 / 32);
        k4 = h * f(x0 + 12 * h / 13, y0 + 1932 * k1 / 2197 -
                7200 * k2 / 2197 + 7296 * k3 / 2197);
        k5 = h * f(x0 + h, y0 + 439 * k1 / 216 - 8 * k2 +
                3680 * k3 / 513 - 845 * k4 / 4104);
        k6 = h * f(x0 + 0.5 * h, y0 - 8 * k1 / 27 + 2 * k2 -
                3544 * k3 / 2565 + 1859 * k4 / 4104 - 11 * k5 / 40);

        VectorR e1 = (k1 / 360 - 128 * k3 / 4275 - 2197 * k4 / 75240 +
                k5 / 50 + 2 * k6 / 55) / h;
        double error = Math.Sqrt(Math.Abs(VectorR.DotProduct(e1, e1)) / n);
        double s = Math.Pow(0.5 * tolerance / error, 0.25);

        if (error < tolerance)
        {
            y0 = y0 + 25 * k1 / 216 + 1408 * k3 / 2565 +
```

```
                    2197 * k4 / 4104 - 0.2 * k5;
            x0 += h;
        }

        if (s < 0.1)
            s = 0.1;
        if (s > 4)
            s = 4;
        h *= s;
        if (h > hmax)
            h = hmax;
        if (h < hmin)
            h = hmin;
        if (h > x - x0)
            h = x - x0;
    }
    return y0;
}
```

This method takes a vector MultiFunction f, the initial values of x_0 and y_0, the step size h, the tolerance, and the position x, at which you want to find solution, as input parameters. Note that the k-parameters in the method also become VectorR objects. The root-mean-square error is calculated by a vector dot product of the vector error function.

This method looks quite similar to the RungeKuttaFehlberg method for a single first-order differential equation. It only takes a very short code listing. However, it is very powerful in the sense that it can be used to solve first-order ordinary differential equations with any number of coupled equations. To apply the method to a specific physics or engineering problem, you simply need to supply the vector function, initial values, step size, and tolerance parameters. The following sections will show you how to solve real-world physics problems using this method.

Testing the MultiRungeKuttaFehlberg Method

Here I will show you how to use the adaptive Runge-Kutta method or the Runge-Kutta-Fehlberg method to numerically solve the same coupled spring system that we used in discussing the *MultiRungeKutta4* method.

Add a new static method, *TestMultiRungeKuttaFehlberg*, to the *Program.cs* file:

```
static void TestMultiRungeKuttaFehlberg()
{
    double dt = 0.5;
    double t0 = 0.0;
    VectorR x0 = new VectorR(new double[] { x10, x20, v10, v20 });
    Console.WriteLine("\n Results for a coupled spring system
                    with h = {0}:\n", dt);
    Console.WriteLine(" t        x1          x2          v1          v2");
    VectorR x = x0;
    for (int i = 0; i < 21; i++)
    {
        double t = 0.1 * i;
        x = ODE.MultiRungeKuttaFehlberg(f1, t0, x, t, dt, 1e-5);
        Console.WriteLine(" {0:n2}    {1,8:n5}    {2,8:n5}    {3,8:n5}    {4,8:n5}",
                    t, x[0], x[1], x[2], x[3]);
```

```
        t0 = t;
    }
}
```

The vector delegate function and the initial considitions used in this example are exactly the same as those used in the previous example. The only exception is that a much larger step size $h = dt = 0.5$ is used in this example, instead of the $dt = 0.02$ we used in the previous example. The Runge-Kutta-Fehlberg method will automatically refine the step size according to the accuracy requirement, which is controlled by the tolerance parameter.

Running this example produces results shown in Figure 13-8.

```
Results for a coupled spring system with h = 0.5:

t        x1         x2         v1         v2
0.00     1.00000    0.00000    0.00000    0.00000
0.10     0.73837    0.02227   -4.97974    0.39758
0.20     0.09477    0.06388   -7.28948    0.33973
0.30    -0.58893    0.07270   -5.75413   -0.22725
0.40    -0.95497    0.01416   -1.24131   -0.91554
0.50    -0.81824   -0.09504    3.83133   -1.15107
0.60    -0.26020   -0.18885    6.80053   -0.58929
0.70     0.41942   -0.19177    6.16498    0.58869
0.80     0.86476   -0.07267    2.34737    1.71567
0.90     0.85138    0.12346   -2.57731    2.01971
1.00     0.39984    0.29056   -6.01646    1.12799
1.10    -0.24244    0.31960   -6.23366   -0.62587
1.20    -0.73750    0.16756   -3.22893   -2.31081
1.30    -0.83581   -0.10540    1.32245   -2.89925
1.40    -0.50256   -0.35786    5.00734   -1.88676
1.50     0.07288   -0.44137    5.96212    0.33980
1.60     0.58469   -0.28745    3.81923    2.63386
1.70     0.77413    0.04335   -0.16998    3.68688
1.80     0.56088    0.38322   -3.86328    2.77645
1.90     0.07534    0.54318   -5.38430    0.23655
2.00    -0.42003    0.41875   -4.08002   -2.64616
```

Figure 13-8 Results from the MultiRungeKuttaFehlberg method.

Chapter 14
Boundary Value Problems

In the previous chapter, we discussed how to solve ordinary differential equations of orders higher than one, provided that the requisite number of initial conditions are available. However, problems originating in physics and engineering often require the solution of differential equations in which the data to be satisfied are located at two different values of the independent variable x. These are usually called the boundary conditions. This seemingly small difference from initial value problems has a major repercussion – it makes boundary value problems significantly more difficult to solve. In an initial value problem we can start at the point at which the initial values are given and process the solution forward as far as we need. However, this technique does not work for boundary value problems, because there are not enough starting conditions available at either end point to produce a unique solution.

Among the various techniques available for the solution of boundary problems, we will consider two methods in this chapter: solution by the shooting method and finite differences. The shooting method involves guessing the missing values. The resulting solution is unlikely to satisfy boundary conditions at the other end, but by inspecting the discrepancy we can estimate what changes we need to make to the initial conditions before integrating again. The finite difference method is to approximate the differential equations by finite differences at evenly spaced mesh points. As a result, a differential equation is transformed into a set of simultaneous algebraic equations.

Shooting Method

The shooting method for solving a two-point boundary value problem involves converting it to an initial value problem by determining sufficient additional conditions at one boundary. A second-order equation will require two initial conditions, and only one is provided; a third-order equation will require three, and only one or two are given; and so on. The missing initial conditions are determined in a way which causes the given conditions at the other boundary to be satisfied.

Let's consider the simplest two-point boundary value problem, i.e., the second-order differential equation with one condition specified at $x = a$ and another one at $x = b$. Here is an example of such a problem:

$$y'' = f(x, y, y'), \quad y(a) = \alpha, \quad y(b) = \beta \tag{14-1}$$

We now attempt to convert the about equation into the initial value problem

$$y'' = f(x, y, y'), \quad y(a) = \alpha, \quad y'(a) = u \tag{14-2}$$

The key step is how to find the correct value of u. This can be achieved by trial and error: guess u and solve the initial value problem by processing from $x = a$ to b. If the solution agrees with the prescribed boundary condition $y(b) = \beta$, we are done; otherwise we have to adjust u and try again.

In summary, the steps involved include:

- Split the second- (or higher-) order equation into two (or more) equivalent first-order equations as described in the previous chapter.

- Estimate values for the missing initial condition or conditions.

- Integrate the equations as an initial value problem.

- Compare the solution at the final boundary with the given final boundary conditions. If they do not agree, then go to next step.

- Adjust the estimated values of the missing initial conditions.

- Repeat the integration untill the process converges.

Clearly, this procedure is very tedious. More systematic methods become available if we realize that the determination of u is basically a root-finding problem. Because the solution of the initial value problem depends on the value of u, the computed value of $y(b)$ is a function of u:

$$y(b) = \theta(u)$$

Thus, u is a root of

$$r(u) = \theta(u) - \beta = 0 \tag{14-3}$$

where $r(u)$ is the boundary residual; i.e., the difference between the computed and specified boundary value at $x = b$. Equation (14-3) can be solved by one of the root-finding methods discussed in Chapter 5. We will use the false position method to find the root of u in Equation (14-3).

Here is the procedure we use to solve nonlinear boundary value problem:

- Specify the starting value u_1 and u_2 which must contain the root u of Equation (14-3).

- Apply the false position method to solve Eqation (14-3) for u. Note that each iteration requires evaluation of $\theta(u)$ by solving the differential equation as an initial value problem. Here we will use the fourth-order Runge-Kutta method to solve the initial value problem.

- Having determined the value of u, solve the differential equations once more and record the results.

Implementation

In this section, we will implement a shooting method for solving boundary value problems for a second-order differential equation. We hope that this method is general enough to handle boundary value problems of any second-order differential equation.

Start with a new C# Console application and name it *BoundaryValueTest*. Add a new class, *BoundaryValue*, to the project and change its namespace to *XuMath*. Since the shooting method requires the root finding method for a nonlinear equation as well as the method for solving the initial value problem, we also need to add the files, *NonlinearSystem.cs* from Chapter 5, *ODE.cs* from Chapter 13, and *VectorR.cs* from Chapter 3, to the current project. Here is the code listing of the *BoundaryValue* class:

```
using System;

namespace XuMath
{
    public class BoundaryValue
    {
        public double xa { get; set; }
        public double xb { get; set; }
        public double ya { get; set; }
        public double yb { get; set; }
        public double u1 { get; set; }
        public double u2 { get; set; }
        public double StepSize { get; set; }
        public double xOut { get; set; }
        public ODE.MultiFunction F1 { get; set; }

        public VectorR Shooting2()
        {
            double u =
                NonlinearSystem.FalsePosition(ResidualFunction, u1, u2, 1e-8);
            VectorR y0 = new VectorR(new double[] { ya, u });
            return ODE.MultiRungeKutta4(F1, xa, y0, StepSize, xOut);
        }

        private double ResidualFunction(double u)
        {
            VectorR y0 = new VectorR(new double[] { ya, u });
            VectorR y = ODE.MultiRungeKutta4(F1, xa, y0, StepSize, xb);
            return y[0] - yb;
        }
    }
}
```

In this method, we start off by creating several public properties using C# 3.0's new feature of automatic properties. These properties allow the user provide the boundary conditions, the search range of *u*, step size, and the vector delegate function to be used in the Runge-Kutta method. The *xOut* property specifies the position at which the final solution will be computed.

The method *ResidualFunction* returns the difference between the computed and specifed boundary value at x_b. Inside this method, we solve the initial value problem by using the fourth-order Runge-Kutta method.

Inside the *Shooting2* method, we first find the *u* value by solving the nonlinear equation using the *FalsePosition* method. Then we use this *u* value to construct an initial value problem, and solve this initial value problem using the *RungeKutta4* method. The result is returned by this method at *xOut* specified by the user.

Testing the Shooting2 Method

Here I will show you how to use the *Shooting2* method to numerically solve the boundary value problem of a second-order differential equation. As an example, we consider the equation

$$y''+2xy'-6y-2x = 0, \quad y(0) = 0, \quad y(1) = 2$$

This is a two-point boundary value problem. This equation happens to have an analytical solution:

$$y = x^3 + x$$

which will allow us to check on the accuracy of the numerical solution.

First we need to convert the second-order equation into a couple of first-order equations

$$y' = v, \quad v' = -2xv + 6y + 2x$$

Now, we can use the *Shooting2* method implemented in the previous section to solve the boundary value problem. Add a new static method, *TestShooting2*, to the *Program.cs* file:

```
using System;
using XuMath;

namespace BoundaryValueTest
{
    class Program
    {
        static void Main(string[] args)
        {
            TestShooting2();
            Console.ReadLine();
        }

        static void TestShooting2()
        {
            BoundaryValue bv = new BoundaryValue();
            bv.xa = 0.0;
            bv.xb = 1.0;
            bv.ya = 0.0;
            bv.yb = 2.0;
            bv.u1 = 0.0;
            bv.u2 = 2.0;
            bv.StepSize = 0.05;
            bv.F1 = f1;

            Console.WriteLine("\n Results from the shooting method:\n");
            Console.WriteLine(" x          y              y'        Exact y
                        Exact y'");
            for (int i = 0; i < 11; i++)
            {
                bv.xOut = 0.1 * i;
                VectorR y = bv.Shooting2();
                double yexact = bv.xOut * bv.xOut * bv.xOut + bv.xOut;
                double y1exact = 3 * bv.xOut * bv.xOut + 1.0;
                Console.WriteLine(" {0:n3}, {1,10:n6}, {2,10:n6} {3,10:n6},
                        {4,10:n6}", bv.xOut, y[0], y[1], yexact, y1exact);
```

```
        }
    }

    static VectorR f1(double x, VectorR y)
    {
        VectorR result = new VectorR(2);
        result[0] = y[1];
        result[1] = -2 * x * y[1] + 6 * y[0] + 2 * x;
        return result;
    }
}
}
```

Here, we set the step size = 0.05 and the *u* search range = [0, 2]. Running this example generates the results shown in Figure 14-1. You can see numerical results from the shooting method are very close to the analytical results.

x	y	y'	Exact y	Exact y'
0.000,	0.000000,	1.000002	0.000000,	1.000000
0.100,	0.101000,	1.030002	0.101000,	1.030000
0.200,	0.208000,	1.120002	0.208000,	1.120000
0.300,	0.327000,	1.270002	0.327000,	1.270000
0.400,	0.464000,	1.480001	0.464000,	1.480000
0.500,	0.625000,	1.750001	0.625000,	1.750000
0.600,	0.816000,	2.080000	0.816000,	2.080000
0.700,	1.043000,	2.470000	1.043000,	2.470000
0.800,	1.312000,	2.919999	1.312000,	2.920000
0.900,	1.629000,	3.429998	1.629000,	3.430000
1.000,	2.000000,	3.999997	2.000000,	4.000000

Figure 14-1 Results from the shooting method.

Following the similar procedure presented here, you can easily extend the shooting method to solve boundary value problems for higher-order differential equations.

Finite Difference for Linear Equation

The finite difference method is particularly suitable for linear equations; although, as will become apparent, it can also be used for nonlinear equations with a considerable increase in computational effort.

Algorithm

In this section, we consider the general second-order linear equation

$$P(x)y''+Q(x)y'+R(x)y+S(x) = 0$$

with boundary conditions

$$y(a) = \alpha, \quad \text{or} \quad y'(a) = \alpha$$

$$y(b) = \beta, \quad \text{or} \quad y'(b) = \beta$$

In the finite difference method, we divide the range of integration (a, b) into n equal subintervals of length h each, as shown in Figure 14-2. The values of the numerical solution at the mesh points are denoted by $y_i, i = 0, 1, 2, \cdots, n$. The purpose of the two points outside the interval (a, b) will be explained shortly. The finite difference method makes two approximations:

- The derivatives of y in the differential equation are replaced by the finite difference expressions. The first central difference approximation will be used: $y_i' = \dfrac{y_{i+1} - y_{i-1}}{2h}$, $y_i'' = \dfrac{y_{i-1} - 2y_i + y_{i+1}}{h^2}$.

- The differential equation is enforced only at the mesh points.

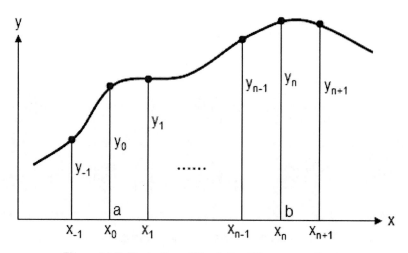

Figure 14-2 Illustration of the finite difference method.

Thus, the differential equations are replaced by $n + 1$ simultaneous algebraic equations, with the unknowns being $y_i, i = 0, 1, 2, \cdots, n$.

Since the truncation error in a first central difference approximation is in the order of $O(h^2)$, the finite difference method is not as accurate as the shooting method.

The finite difference version of our general second-order equations becomes

$$P_i \frac{y_{i-1} - 2y_i + y_{i+1}}{h^2} + Q_i \frac{y_{i+1} - y_{i-1}}{2h} + R_i y_i + S_i = 0$$

where P_i, Q_i, R_i, and S_i denote $P(x_i)$, $Q(x_i)$, $R(x_i)$, and $S(x_i)$, respectively. Rearraging the above equation, we obtain

$$\left(\frac{P_i}{h^2} - \frac{Q_i}{2h}\right) y_{i-1} + \left(R_i - \frac{2P_i}{h^2}\right) y_i + \left(\frac{P_i}{h^2} + \frac{Q_i}{2h}\right) y_{i+1} = -S_i, \quad i = 0, 1, \cdots, n$$

The boundary conditions under the finite difference approximation become

$$y_0 = \alpha \quad \text{or} \quad \frac{y_1 - y_{-1}}{2h} = \alpha$$

$$y_n = \beta \quad \text{or} \quad \frac{y_{n+1} - y_{n-1}}{2h} = \beta$$

Note that the presence of y_{-1} and y_{n+1}, which are both associated with points outside the solution domain (a, b). They can be eliminated by using the boundary conditions. But before we do that, let's rewrite the finite difference equation as

$$\left(\frac{P_0}{h^2} - \frac{Q_0}{2h}\right)y_{-1} + \left(R_0 - \frac{2P_0}{h^2}\right)y_0 + \left(\frac{P_0}{h^2} + \frac{Q_0}{2h}\right)y_1 = -S_0 \tag{a}$$

$$\left(\frac{P_i}{h^2} - \frac{Q_i}{2h}\right)y_{i-1} + \left(R_i - \frac{2P_i}{h^2}\right)y_i + \left(\frac{P_i}{h^2} + \frac{Q_i}{2h}\right)y_{i+1} = -S_i, \quad i = 1, 2, \cdots, n-1 \tag{b}$$

$$\left(\frac{P_n}{h^2} - \frac{Q_n}{2h}\right)y_{n-1} + \left(R_n - \frac{2P_n}{h^2}\right)y_n + \left(\frac{P_n}{h^2} + \frac{Q_n}{2h}\right)y_{n+1} = -S_i \tag{c}$$

The boundary conditions on y are easily dealt with. Equation (a) can be simply replaced by $y_0 - \alpha = 0$, and Equation (c) can be replaced by $y_m - \beta = 0$. If the derivative boundary conditions are specified, we obtain $y_{-1} = y_1 - 2h\alpha$ and $y_{m+1} = y_{m-1} + 2h\beta$, which are then substituted into Equations (a) and (c), respectively. Hence we end up with $n + 1$ equations in the unknowns $y_0, y_1, \cdots, y_m$:

$$y_0 = \alpha \qquad\qquad \text{if } y(a) = \alpha$$

$$\left(R_0 - \frac{2P_0}{h^2}\right)y_0 + \frac{2P_0}{h^2}y_1 = 2h\alpha\left(\frac{P_0}{h^2} - \frac{Q_0}{2h}\right) - S_0 \qquad\qquad \text{if } y'(a) = \alpha$$

$$\left(\frac{P_i}{h^2} - \frac{Q_i}{2h}\right)y_{i-1} + \left(R_i - \frac{2P_i}{h^2}\right)y_i + \left(\frac{P_i}{h^2} + \frac{Q_i}{2h}\right)y_{i+1} = -S_i \qquad\qquad i = 1, 2, \cdots, n-1$$

$$y_n = \beta \qquad\qquad \text{if } y(b) = \beta$$

$$\frac{2P_n}{h^2}y_{n-1} + \left(R_n - \frac{2P_n}{h^2}\right)y_n = -2h\beta\left(\frac{P_n}{h^2} + \frac{Q_n}{2h}\right) - S_n \qquad\qquad \text{if } y'(b) = \beta$$

The system described by the above $n + 1$ equations is tridiagonal. It may be written in a matrix form:

$$\begin{pmatrix} b_0 & c_0 & & & & \\ a_1 & b_1 & c_1 & & & \\ & a_2 & b_2 & c_2 & & \\ & & & \ddots & & \\ & & & a_{n-1} & b_{n-1} & c_{n-1} \\ & & & & a_n & b_n \end{pmatrix} \begin{pmatrix} y_0 \\ y_1 \\ y_2 \\ \vdots \\ y_{n-1} \\ y_n \end{pmatrix} = \begin{pmatrix} d_0 \\ d_1 \\ d_2 \\ \vdots \\ d_{n-1} \\ d_n \end{pmatrix}$$

where

$$\left.\begin{array}{l} a_0 = 0 \\ b_0 = 1 \\ c_0 = 0 \\ d_0 = \alpha \end{array}\right\} \quad \text{if } y(a) = \alpha$$

$$\left.\begin{array}{l} a_0 = 0 \\ b_0 = R_0 - \dfrac{2P_0}{h^2} \\ c_0 = \dfrac{2P_0}{h^2} \\ d_0 = 2h\alpha\left(\dfrac{P_0}{h^2} - \dfrac{Q_0}{2h}\right) - S_0 \end{array}\right\} \quad \text{if } y'(a) = \alpha$$

$$a_i = \dfrac{P_i}{h^2} - \dfrac{Q_i}{2h}, \quad i = 1,2,\cdots,n-1$$

$$b_i = R_i - \dfrac{2P_i}{h^2}, \quad i = 1,2,\cdots,n-1$$

$$c_i = \dfrac{P_i}{h^2} + \dfrac{Q_i}{2h}, \quad i = 1,2,\cdots,n-1$$

$$d_i = -S_i, \qquad i = 1,2,\cdots,n-1$$

$$\left.\begin{array}{l} a_n = 0 \\ b_n = 1 \\ c_n = 0 \\ d_n = \beta \end{array}\right\} \quad \text{if } y(b) = \beta$$

$$\left.\begin{array}{l} a_n = \dfrac{2P_n}{h^2} \\ b_n = R_n - \dfrac{2P_n}{h^2} \\ c_n = 0 \\ d_n = -2h\beta\left(\dfrac{P_n}{h^2} + \dfrac{Q_n}{2h}\right) - S_n \end{array}\right\} \quad \text{if } y'(b) = \beta$$

Since the coefficient matrix is tridiagonal, the system can be solved efficiently by the LU decomposition method described in Chapter 4.

Implementation

Here we will implement a finite difference method for solving the boundary value problem of a second-order linear differential equation. We hope that this method is general enough to handle boundary value problems with specified y values or derivatives at the boundary.

Since the method requires us to solve the linear system by using the LU decomposition method, we also need to add the files *LinearSystem.cs* from Chapter 4 to the current project. Add a new method, *FiniteDifferenceLinear2*, to the project. The following is the code listing of this method:

```
// Finite Difference method for the second-order linear differential equation:
// Note xa, xb, ya, and yb are the same as in the shooting method.

public double va { get; set; }              // va = y'(a)
public double vb { get; set; }              // vb = y'(b)
public int n { get; set; }

private int[] boundaryFlag = new int[] { 0, 0 };
public int[] BoundaryFlag
{
    get { return boundaryFlag; }
    set { boundaryFlag = value; }
}

public delegate VectorR MultiFunction(double x);

public VectorR FiniteDifferenceLinear2(MultiFunction f, out double[] x)
{
    double h = (xb - xa) / n;
    x = new double[n + 1];
    for (int i = 0; i < n + 1; i++)
        x[i] = xa + i * h;

    double[] a = new double[n + 1];
    double[] b = new double[n + 1];
    double[] c = new double[n + 1];
    double[] d = new double[n + 1];

    if (boundaryFlag[0] != 1)
    {
        a[0] = 0.0;
        b[0] = 1.0;
        c[0] = 0.0;
        d[0] = ya;
    }
    else
    {
        a[0] = 0.0;
        b[0] = f(x[0])[2] - 2 * f(x[0])[0] / h / h;
        c[0] = 2 * f(x[0])[0] / h / h;
        d[0] = 2 * h * va * (f(x[0])[0] / h / h -
                             f(x[0])[1] / 2 / h) - f(x[0])[3];
    }
    if (boundaryFlag[1] != 1)
    {
        a[n] = 0.0;
        b[n] = 1.0;
        c[n] = 0.0;
        d[n] = yb;
    }
    else
```

```
{
    a[n] = 2 * f(x[n])[0] / h / h;
    b[n] = f(x[n])[2] - 2 * f(x[n])[0] / h / h;
    c[n] = 0.0;
    d[n] = -2 * h * vb * (f(x[n])[0] / h / h +
                          f(x[n])[1] / 2 / h) - f(x[n])[3];
}

for (int i = 1; i < n; i++)
{
    a[i] = f(x[i])[0] / h / h - f(x[i])[1] / 2 / h;
    b[i] = f(x[i])[2] - 2 * f(x[i])[0] / h / h;
    c[i] = f(x[i])[0] / h / h + f(x[i])[1] / 2 / h;
    d[i] = -f(x[i])[3];
}

MatrixR A1 = new MatrixR(n + 1, n + 1);
A1[0, 0] = b[0];
for (int i = 1; i <= n; i++)
{
    A1[i, i] = b[i];
    A1[i, i - 1] = a[i];
    A1[i - 1, i] = c[i - 1];
}
VectorR b1 = new VectorR(d);

LinearSystem ls = new LinearSystem();
double d1 = ls.LUCrout(A1, b1);

return b1;
}
```

Here, we define a four-dimensional vector delegate function, which is used to store the coefficients [$P(x)$, $Q(x)$, $R(x)$, $S(x)$]. We also define several public properties by using the new C# feature of automatic properties. These properties allow the user to specify boundary conditions and the number of mesh intervals. The two-element integer array, *BoundaryFlag*, is defined using the conventional method with a default value [0, 0]. The first element 0 means the y value is specified at $x = a$; whereas first element 1 indicates the y' value is specified at $x = a$. The second element is used to specify the boundary conditions at the other end.

A major aspect of the code in this method is to construct the coefficient matrix. Then, the *LUCrout* method from the *LinearSystem* class is used to solve the linear system.

Testing FiniteDifferenceLinear2 Method

We can solve any second-order linear differential equation by using the method implemented in the previous section. Let's consider the following example:

$$xy'' - 2y' + 2 = 0, \quad y(0) = y(1) = 0$$

Thus,

$$P(x) = x, \quad Q(x) = -2, \quad R(x) = 0, \quad S(x) = 2$$

This differential equation has an analytical solution: $y = x - x^3$. Hence, we can examine the accuracy of the numerical results.

Add a new method, *TestFiniteDifferenceLinear2*, to the *Program.cs* file:

```
static void TestFiniteDifferenceLinear2()
{
    BoundaryValue bv = new BoundaryValue();
    bv.xa = 0.0;
    bv.xb = 1.0;
    bv.ya = 0.0;
    bv.yb = 0.0;
    bv.n = 8;
    double[] x;
    VectorR y = bv.FiniteDifferenceLinear2(f2, out x);

    Console.WriteLine("\n  x           y            Exact y");
    for (int i = 0; i < x.Length; i++)
    {
        double exact = x[i] - x[i] * x[i] * x[i];
        Console.WriteLine(" {0,8:n5}    {1,10:n6}    {2,10:n6}",
                          x[i], y[i], exact);
    }
}

static VectorR f2(double x)
{
    VectorR result = new VectorR(4);
    result[0] = x;
    result[1] = -2;
    result[2] = 0;
    result[3] = 2;
    return result;
}
```

Note that we did not specify the *BoundaryFlag* parameter for this example, because *y* values are specified at both ends. This is the default setting for the *BoundaryFlag*. Pay attention to how we define the vector delegate function.

Running this example produces the results shown in Figure 14-3. You can see that the numerical results are fairly good compared to the analytical solution, considering the small number of mesh intervals ($n = 8$). You can improve the accuracy by increasing the number of mesh points.

x	y	Exact y
0.00000	0.000000	0.000000
0.12500	0.125000	0.123047
0.25000	0.238095	0.234375
0.37500	0.327381	0.322266
0.50000	0.380952	0.375000
0.62500	0.386905	0.380859
0.75000	0.333333	0.328125
0.87500	0.208333	0.205078
1.00000	0.000000	0.000000

Figure 14-3 Results of the linear differential equation.

Next, we consider another example with mixed boundary conditions:

$$y'' + 4y - 8x = 0, \quad y(0) = 0, \quad y'(\pi/2) = 0$$

This equation also has an analytical solution:

$$y = 2x + \sin 2x$$

In this case, you simply modify the *TestFiniteDifferenceLinear2* method and provide a new delegate function *f3*:

```
static void TestFiniteDifferenceLinear2()
{
    BoundaryValue bv = new BoundaryValue();
    bv.BoundaryFlag = new int[] { 0, 1 };
    bv.xa = 0.0;
    bv.xb = 0.5 * Math.PI;
    bv.ya = 0.0;
    bv.vb = 0;
    bv.n = 10;
    double[] x;
    VectorR y = bv.FiniteDifferenceLinear2(f3, out x);

    Console.WriteLine("\n  x            y              Exact y");
    for (int i = 0; i < x.Length; i++)
    {
        double exact = 2 * x[i] + Math.Sin(2 * x[i]);
        Console.WriteLine(" {0,8:n5}    {1,10:n6}    {2,10:n6}",
                          x[i], y[i], exact);
    }
}

static VectorR f3(double x)
{
    VectorR result = new VectorR(4);
    result[0] = 1;
    result[1] = 0;
    result[2] = 4;
    result[3] = -8 * x;
    return result;
}
```

Here we set the *BoundaryFlag* parameter to [0, 1], because the y' value is specified at $x = b$. Running the example generates the results shown in Figure 14-4. Again, the agreement between numerical results and analytical solution is fairly good. More accurate results can be achieved by increasing n.

```
x              y              Exact y
0.00000        0.000000       0.000000
0.15708        0.628345       0.623176
0.31416        1.225682       1.216104
0.47124        1.764061       1.751495
0.62832        2.221353       2.207694
0.78540        2.583431       2.570796
0.94248        2.845567       2.836012
1.09956        3.012893       3.008132
1.25664        3.099904       3.101059
1.41372        3.129016       3.136450
1.57080        3.128363       3.141593
```

Figure 14-4 Results of the linear differential equation with mixed boundary conditions.

Finite Diference for Nonlinear Equations

If the differential equation to be solved is nonlinear, then it is apparent that its finite difference approximation will be nonlinear too. Hence, in order to solve such a problem by the finite difference method, we need to find solutions for a nonlinear system of equations.

Algorithm

We now consider the general second-order differential equation

$$y'' = f(x, y, y')$$

with the boundary conditions

$$y(a) = \alpha, \quad \text{or} \quad y'(a) = \alpha$$

$$y(b) = \beta, \quad \text{or} \quad y'(b) = \beta$$

Approximating the derivatives at the mesh points by finite difference, the problem then becomes

$$\frac{y_{i-1} - 2y_i + y_{i+1}}{h^2} = f\left(x_i, y_i, \frac{y_{i+1} - y_{i-1}}{2h}\right), \quad i = 0, 1, 2, \cdots, n$$

$$y_0 = \alpha \quad \text{or} \quad \frac{y_1 - y_{-1}}{2h} = \alpha$$

$$y_n = \beta \quad \text{or} \quad \frac{y_{n+1} - y_{n-1}}{2h} = \beta$$

Note that the presence of y_{-1} and y_{n+1}, which are associated with points outside the solution domain (a, b). These can be eliminated by using the boundary conditions. The final finite difference formulas for the second-order differential equation become

$$y_0 - \alpha = 0 \qquad\qquad\qquad \text{if } y(a) = \alpha$$

$$-2y_0 + 2y_1 - h^2 f(x_0, y_0, \alpha) - 2h\alpha = 0 \qquad \text{if } y'(a) = \alpha$$

$$y_{i-1} - 2y_i + y_{i+1} - h^2 f\left(x_i, y_i, \frac{y_{i+1} - y_{i-1}}{2h}\right) = 0, \quad i = 1, 2, \cdots, n-1$$

$$y_n - \beta = 0 \qquad\qquad\qquad \text{if } y(b) = \beta$$

$$2y_{n-1} - 2y_n - h^2 f(x_n, y_n, \beta) + 2h\beta = 0 \qquad \text{if } y'(b) = \beta$$

Implementation

Here we will implement a finite difference method for solve the boundary value problem of a second-order nonlinear differential equation. We hope that this method is general enough to handle boundary value problems with specified y or y' at the boundary.

Since the problem is nonlinear, the formulas for finite difference approximation presented in the previous section must be solved in a nonlinear manner. Here, we will use the *NewtonMultiEquations* method implemented in Chapter 5.

Add a new method, *FiniteDifferenceNonlinear2*, to the *BoundaryValue* class. The following is the corresponding code listing:

```
public delegate double FDFunction(double x, double y, double yprime);
public FDFunction fd { get; set; }

public VectorR FiniteDifferenceNonlinear2(out double[] x)
{
    double h = (xb - xa) / n;
    x = new double[n + 1];
    VectorR y = new VectorR(n + 1);
    for (int i = 0; i < n + 1; i++)
    {
        x[i] = xa + i * h;
        y[i] = 0.5 * x[i];
    }
    y = NonlinearSystem.NewtonMultiEquations(VF, y, 1e-8);
    return y;
}

private VectorR VF(VectorR y)
{
    double h = (xb - xa) / n;
    double[] x = new double[n + 1];
    for (int i = 0; i < n + 1; i++)
        x[i] = xa + i * h;

    VectorR result = new VectorR(n + 1);

    if (boundaryFlag[0] != 1)
    {
        result[0] = y[0] - ya;
```

```
    }
    else
    {
        result[0] = -2 * y[0] + 2 * y[1] -
                    h * h * fd(x[0], y[0], va) - 2 * h * va;
    }
    if (boundaryFlag[1] != 1)
    {
        result[n] = y[n] - yb;
    }
    else
    {
        result[n] = 2 * y[n - 1] - 2 * y[n] -
                    h * h * fd(x[n], y[n], vb) + 2 * h * yb;
    }
    for (int i = 1; i < n; i++)
    {
        result[i] = y[i - 1] - 2 * y[i] + y[i + 1] -
                    h * h * fd(x[i], y[i], (y[i + 1] - y[i - 1]) / 2 / h);
    }
    return result;
}
```

Here, we introduce a finite difference delegate function, *FDFunction*, which takes *x*, *y* and *y'* as its input parameters. This delegate function defines the differential equation $y'' = f(x, y, y')$. Using this function, we then define a public property, *fd*, so that it can be accessed using different methods. The private method, *VF*, defines a vector function, which is just the left-hand side of the difference equations. This vector function will be used by the nonlinear system solver – *NewtonMultiEquations*.

In order to use the *NewtonMultiEquations* method to solve nonlinear equations, we need to supply the initial guess values for the vector $y = [y_0, y_1, \cdots, y_n]$. Here we choose for the initial solution $y_i = 0.5x_i$. For different problems, you may need to change these initial values in order to obtain convergent solutions.

Testing FiniteDifferenceNonlinear2 Method

We can solve any second-order nonlinear differential equation using the method implemented in the previous section. Let's consider the example:

$$y'' = xyy' + \sin(2xy), \quad y(0) = 0, \quad y(1) = 1$$

Add a new method, *TestFiniteDifferenceNonlinear2*, to the *Program.cs* file:

```
static void TestfiniteDifferenceNonlinear2()
{
    BoundaryValue bv = new BoundaryValue();
    bv.xa = 0.0;
    bv.xb = 1.0;
    bv.ya = 0.0;
    bv.yb = 1.0;
    bv.n = 10;
    bv.fd = f4;
    double[] x;
```

```
        VectorR y = bv.FiniteDifferenceNonlinear2(out x);

        Console.WriteLine("\n   x        y");
        for (int i = 0; i < x.Length; i++)
        {
            Console.WriteLine(" {0,5:n2}    {1,10:n6}", x[i], y[i]);
        }
    }

    static double f4(double x, double y, double yprime)
    {
        return x * y * yprime + Math.Sin(2 * x * y);
    }
```

Running this example produces results shown in Figure 14-5.

x	y
0.00	0.000000
0.10	0.080138
0.20	0.160500
0.30	0.241764
0.40	0.325069
0.50	0.412053
0.60	0.504895
0.70	0.606374
0.80	0.719923
0.90	0.849612
1.00	1.000000

Figure 14-5 results of the nonlinear differential equation.

Finite Difference for Higher-Order Equation

So far, we have confined our discussion on the finite difference method to second-order differential equations. In fact, the finite difference method is a very general numerical method for solving various types of differential equations. It can be easily applied to higher-order differential equations.

Algorithm

In this section, we will consider a special type of higher-order differential equation – the fourth-order linear equation:

$$y^{(4)} = P(x)y + Q(x)$$

We assume that two boundary conditions are specified at each end of the solution domain $[a, b]$. We will consider two sets of boundary conditions because they are closely related to the beam theory:

Type I: $y(a) = y_a$, $y'(a) = v_a$ or Type II: $y'(a) = v_a$, $y'''(a) = g_a$

Type I: $y(b) = y_b$, $y'(b) = v_b$ or Type II: $y'(b) = v_b$, $y'''(b) = g_b$

Note that some other sets of boundary conditions are also possible. However, care must be taken when you specify boundary conditions, because some combinations of boundary conditions may not work for the fourth-order finite difference method. For example, the combination of $y(a)$ and $y'''(a)$ or $y'(a)$ and $y''(a)$ will not work.

Now we divide the solution domain into n intervals, each of length h. Replacing the derivatives of y by the central finite difference formulas discussed in Chapter 11 at the mesh points, we end up with the finite difference equations for the linear fourth-order differential equation:

$$\frac{y_{i-2} - 4y_{i-1} + 6y_i - 4y_{i+1} + y_{i+2}}{h^4} = P_i y_i + Q_i, \quad i = 0, 1, \cdots, n$$

You can see that there are four quantities, y_{-2}, y_{-1}, y_{n+1}, and y_{n+2} for $i = 1$ and $i = n$ that lie outside of $[a, b]$. These four quantities can be eliminated by applying the four boundary conditions – two at each end. Note that only the first two equations (for $i = 0, 1$) involve y_{-2}, y_{-1} and the last two equations (for $i = n-1, n$) contain y_{n+1} and y_{n+2}. We can eliminate these quantities by applying boundary conditions to these four equations. The resulting equations become

Type I boundary conditions: $y(a) = y_a$, $y'(a) = v_a$ and $y(b) = y_b$, $y'(b) = v_b$:

$$y_0 = y_a$$
$$-4y_0 + (7 - h^4 P_1)y_1 - 4y_2 + y_3 = h^4 Q_1 + 2hv_a$$
$$y_{n-3} - 4y_{n-2} + (7 - h^4 P_{n-1})y_{n-1} - 4y_n = h^4 Q_{n-1} - 2hv_b$$
$$y_n = y_b$$

Type II boundary conditions: $y'(a) = v_a$, $y'''(a) = g_a$ and $y'(b) = v_b$, $y'''(b) = g_b$:

$$(6 - h^4 P_0)y_0 - 8y_1 + 2y_2 = h^4 Q_0 + 2h^3 g_a - 4hv_a$$
$$-4y_0 + (7 - h^4 P_1)y_1 - 4y_2 + y_3 = h^4 Q_1 + 2hv_a$$
$$y_{n-3} - 4y_{n-2} + (7 - h^4 P_{n-1})y_{n-1} - 4y_n = h^4 Q_{n-1} - 2hv_b$$
$$2y_{n-2} - 8y_{n-1} + (6 - h^4 P_n)y_n = h^4 Q_n - 2h^3 g_b + 4hv_b$$

and the difference equations for $i = 2, 3, \cdots, n-2$ are given by

$$y_{i-2} - 4y_{i-1} + (6 - h^4 P_i)y_i - 4y_{i+1} + y_{i+2} = h^4 Q_i$$

The above difference equations are a linear system and can be expressed in a matrix form:

$$\begin{pmatrix} c_0 & d_0 & e_0 & & & & & & \\ b_1 & c_1 & d_1 & e_1 & & & & & \\ a_2 & b_2 & c_2 & d_2 & e_2 & & & & \\ & a_3 & b_3 & c_3 & d_3 & e_3 & & & \\ & & & \vdots & & & & & \\ & & a_{n-3} & b_{n-3} & c_{n-3} & d_{n-3} & e_{n-3} & & \\ & & & a_{n-2} & b_{n-2} & c_{n-2} & d_{n-2} & e_{n-2} \\ & & & & a_{n-1} & b_{n-1} & c_{n-1} & d_{n-1} \\ & & & & & a_n & b_n & c_n \end{pmatrix} \begin{pmatrix} y_0 \\ y_1 \\ y_2 \\ y_3 \\ \vdots \\ y_{n-3} \\ y_{n-2} \\ y_{n-1} \\ y_n \end{pmatrix} = \begin{pmatrix} t_0 \\ t_1 \\ t_2 \\ t_3 \\ \vdots \\ t_{n-3} \\ t_{n-2} \\ t_{n-1} \\ t_n \end{pmatrix}$$

where the constants are defined by

$$\left. \begin{aligned} c_0 &= 1 \\ d_0 &= 0 \\ e_0 &= 0 \\ t_0 &= y_a \end{aligned} \right\} \text{ if type I}$$

$$\left. \begin{aligned} c_0 &= 6 - h^4 P_0 \\ d_0 &= -8 \\ e_0 &= 2 \\ t_0 &= h^4 Q_0 + 2h^3 g_a - 4h v_a \end{aligned} \right\} \text{ if type II}$$

$$b_1 = -4$$
$$c_1 = 7 - h^4 P_1$$
$$d_1 = -4$$
$$e_1 = 1$$
$$t_1 = h^4 Q_1 + 2h v_a$$
$$a_i = 1, \qquad i = 2,3,\cdots n-2$$
$$b_i = -4, \qquad i = 2,3,\cdots,n-2$$
$$c_i = 6 - h^4 P_i, \quad i = 2,3,\cdots,n-2$$
$$d_i = -4, \qquad i = 2,3,\cdots,n-2$$
$$e_i = 1, \qquad i = 2,3,\cdots,n-2$$
$$t_i = h^4 Q_i, \qquad i = 2,3,\cdots,n-2$$
$$a_{n-1} = 1$$
$$b_{n-1} = -4$$
$$c_{n-1} = 7 - h^4 P_{n-1}$$
$$d_{n-1} = -4$$
$$t_{n-1} = h^4 Q_{n-1} - 2h v_b$$

$$\left.\begin{array}{c} a_n = 0 \\ b_n = 0 \\ c_{n=1} \\ t_n = y_b \end{array}\right\} \quad \text{if Type I}$$

$$\left.\begin{array}{c} a_n = 2 \\ b_n = -8 \\ c_n = 6 - h^4 P_n \\ t_n = h^4 Q_n - 2h^3 g_b + 4h v_b \end{array}\right\} \quad \text{if Type II}$$

This system can be solved efficiently by the *LU decomposition* method discussed in Chapter 4.

Implementation

Here we will implement a finite difference method for solving the boundary value problem of a linear fourth-order differential equation, based on the formulas developed in the previous section. Add a new method, *FiniteDifferenceLinear4*, to the project. Here is the code listing of this method:

```
public double ga { get; set; }
public double gb { get; set; }

public VectorR FiniteDifferenceLinear4(MultiFunction f, out double[] x)
{
    double h = (xb - xa) / n;
    double h3 = h * h * h;
    double h4 = h * h * h * h;
    x = new double[n + 1];
    for (int i = 0; i < n + 1; i++)
        x[i] = xa + i * h;

    double[] a = new double[n + 1];
    double[] b = new double[n + 1];
    double[] c = new double[n + 1];
    double[] d = new double[n + 1];
    double[] e = new double[n + 1];
    double[] t = new double[n + 1];

    if (boundaryFlag[0] != 1)
    {
        c[0] = 1.0;
        d[0] = 0.0;
        e[0] = 0.0;
        t[0] = ya;
    }
    else
    {
        c[0] = 6 - h4 * f(x[0])[0];
        d[0] = -8.0;
        e[0] = 2.0;
        t[0] = h4 * f(x[0])[1] + 2 * h3 * ga - 4 * h * va;
    }
```

```
b[1] = -4.0;
c[1] = 7 - h4 * f(x[1])[0];
d[1] = -4.0;
e[1] = 1.0;
t[1] = h4 * f(x[1])[1] + 2 * h * va;

for (int i = 2; i < n - 1; i++)
{
    a[i] = 1.0;
    b[i] = -4.0;
    c[i] = 6 - h4 * f(x[i])[0];
    d[i] = -4.0;
    e[i] = 1.0;
    t[i] = h4 * f(x[i])[1];
}

a[n - 1] = 1.0;
b[n - 1] = -4.0;
c[n - 1] = 7 - h4 * f(x[n - 1])[0];
d[n - 1] = -4.0;
t[n - 1] = h4 * f(x[n - 1])[1] - 2 * h * vb;

if (boundaryFlag[1] != 1)
{
    a[n] = 0.0;
    b[n] = 0.0;
    c[n] = 1.0;
    t[n] = yb;
}
else
{
    a[n] = 2.0;
    b[n] = -8.0;
    c[n] = 6 - h4 * f(x[n])[0];
    t[n] = h4 * f(x[n])[1] - 2 * h3 * gb + 4 * h * vb;
}

MatrixR A1 = new MatrixR(n + 1, n + 1);
A1[0, 0] = c[0];
A1[0, 1] = d[0];
A1[1, 0] = b[1];
A1[1, 1] = c[1];
for (int i = 2; i <= n; i++)
{
    A1[i, i - 2] = a[i];
    A1[i, i - 1] = b[i];
    A1[i, i] = c[i];
    A1[i - 1, i] = d[i - 1];
    A1[i - 2, i] = e[i - 2];
}

VectorR b1 = new VectorR(t);
LinearSystem ls = new LinearSystem();
double d1 = ls.LUCrout(A1, b1);
```

```
        return b1;
    }
```

The method takes the two-dimensional vector delegate function as its input that is used to store the coefficients [$P(x)$, $Q(x)$]. We also define two additional public properties g_a and g_b using the new C# feature of automatic properties, which allow the user to specify the y''' values at the boundary. The two-element integer array, *BoundaryFlag*, is also used in this method. The element = 0 (or 1) means that Type I (Type II) boundary conditions are used.

In this method, it's important to construct the coefficient matrix. Then, the *LUCrout* method from the *LinearSystem* class can be used to solve the linear system.

Testing FiniteDifferenceLinear4 Method

We can solve the fourth-order linear differential equation using the method implemented in the previous section. Let's consider the following example:

$$y^{(4)} = y\sin x + x, \quad y(0) = y'(0) = 0, \quad y'(1) = 0, \quad y'''(1) = -5$$

Thus,

$$P(x) = \sin x, \quad Q(x) = x$$

Add a new method, *TestFiniteDifferenceLinear4*, to the *Program.cs* file:

```
static void TestFiniteDifferenceLinear4()
{
    BoundaryValue bv = new BoundaryValue();
    bv.xa = 0.0;
    bv.xb = 1.0;
    bv.ya = 0.0;
    bv.va = 0.0;
    bv.vb = 0.0;
    bv.gb = -5;
    bv.n = 10;
    bv.BoundaryFlag = new int[] { 0, 1 };

    double[] x;
    VectorR y = bv.FiniteDifferenceLinear4(f5, out x);

    Console.WriteLine("\n  x              y");
    for (int i = 0; i < x.Length; i++)
    {
        Console.WriteLine(" {0,8:n5}    {1,10:n6}", x[i], y[i]);
    }
}

static VectorR f5(double x)
{
    VectorR result = new VectorR(2);
    result[0] = Math.Sin(x);
    result[1] = x;
    return result;
```

  }

Note that we specifically set the *BoundaryFlag* = [0, 1] for this example, because this is a mixed boundary value problem.

Running this example produces the results shown in Figure 14-6. You can improve its accuracy by increasing the number of mesh points.

By following the procedure presented in this chapter, you should be able to implement the finite difference method for any general higher-order differential equation.

x	y
0.00000	0.000000
0.10000	0.013877
0.20000	0.049863
0.30000	0.102318
0.40000	0.165629
0.50000	0.234210
0.60000	0.302527
0.70000	0.365102
0.80000	0.416539
0.90000	0.451532
1.00000	0.464886

Figure 14-6 Results of a fourth-order linear differential equation.

Chapter 15
Eigenvalue Problems

An n-dimensional vector $\mathbf{x}$ is called an eigenvector of a square $n \times n$ matrix $\mathbf{A}$ if and only if it satisfies the linear equation

$$\mathbf{A}\mathbf{x} = \lambda\mathbf{x} \tag{15-1}$$

Here λ is a scalar, and is refered to as the eigenvalue corresponding to $\mathbf{x}$. The above equation is usually called the eigenvalue equation or the eigenvalue problem.

Note that most vectors $\mathbf{x}$ will not satisfy the eigenvalue equation. A typical vector $\mathbf{x}$ changes direction when acted on by a matrix $\mathbf{A}$, so that $\mathbf{A}\mathbf{x}$ is not a multiple of $\mathbf{x}$. This means that only certain special vectors $\mathbf{x}$ are eigenvectors, and only certain special scalars λ are eigenvalues.

In general, the eigenvalues of a matrix $\mathbf{A}$ are determined by the equation

$$\det(\mathbf{A} - \lambda\mathbf{I}) = 0 \tag{15-2}$$

Equation (15-2) has the roots λ_i, $i = 1, 2, \cdots, n$, called the eigenvalues of the matrix $\mathbf{A}$. The solution $\mathbf{x}_i$ of $(\mathbf{A} - \lambda_i\mathbf{I})\mathbf{x} = 0$ are known as the eigenvectors.

Eigenvalue problems that originate from physics and engineering often end up with a symmetric or Hermitian matrix. In this chapter, I will implement several solvers for eigenvalue problems of real symmetric matrices.

Jacobi Method

The first algorithm solving the eigenvalue problem for a symmetric $n \times n$ matrix that we consider in this section is the Jacobi method. This algorithm uses orthogonal transformations to reduce the matrix to a diagonal form. It is used to compute all eigenvalues and eigenvectors of a real symmetric matrix. This method is usually used for smaller matrices because its computational intensity increases very rapidly with the size of the matrix. For matrices of order up to 10×10, the Jacobi method is competitive with more sophisticated methods. If speed is not a major consideration, it is even acceptable for matrices up to order 20×20. The advantage of this method is its robustness – when the Jacobi method is used, the solution is usually guaranteed for all real symmetric matrices.

Algorithm

A typical eigenvector x_i of a real symmetric matrix has the following orthogonality properties:

$$x_i^T x_i = 1 \text{ and } x_i^T x_j = 0 \text{ if } i \neq j$$

Therefore, we should have the following relations:

$$x_i^T A x_i = \lambda_i \text{ and } x_i^T A x_j = 0 \text{ if } i \neq j$$

if we apply the transformation $x = Tx^*$, where T is a nonsingular matrix. Substituting this into the eigenvalue equation (15-1) and multiplying each side by T^{-1}, we obtain

$$T^{-1} A T x^* = \lambda T^{-1} T x^*$$

or

$$A^* x^* = \lambda x^* \tag{15-3}$$

where $A^* = T^{-1} A T$. Because λ is untouched by the transformation, the eigenvalues of A should be the same as those of A^*. Matrices that have the same eigenvalues are considered to be similar, and the transformation between them is called a similarity transformation.

We can use the similarity transformations to change an eigenvalue problem to a form that is easier to solve. Suppose that we can find a T matrix that diagonalizes A^*. Equation (15-3) then becomes

$$\begin{pmatrix} A_{11}^* - \lambda & 0 & \cdots & 0 \\ 0 & A_{22}^* - \lambda & \cdots & 0 \\ \vdots & \vdots & \ddots & \vdots \\ 0 & 0 & \cdots & A_{nn}^* - \lambda \end{pmatrix} \begin{pmatrix} x_1^* \\ x_2^* \\ \vdots \\ x_n^* \end{pmatrix} = \begin{pmatrix} 0 \\ 0 \\ 0 \\ 0 \end{pmatrix}$$

which has the solutions

$$\lambda_i = A_{ii}^*, \quad i = 1, 2, \cdots n$$

$$x_1^* = \begin{pmatrix} 1 \\ 0 \\ \vdots \\ 0 \end{pmatrix} \quad x_2^* = \begin{pmatrix} 0 \\ 1 \\ \vdots \\ 0 \end{pmatrix} \quad \cdots \quad x_n^* = \begin{pmatrix} 0 \\ 0 \\ \vdots \\ 1 \end{pmatrix}$$

We can write the above eigenvectors in terms of a matrix form:

$$X^* = [x_1^* \; x_2^* \cdots x_n^*] = I$$

Then the original eigenvectors of A become

$$X = TX^* = TI = T \tag{15-4}$$

Thus the transformation matrix T contains the eigenvectors of A, and the eigenvalues of A are the diagonal terms of A^*.

The Jacobi method is a simple iterative algorithm in which the eigenvectors are computed from the following series of matrix multiplications (or transformation according to Equation (15-4)):

$$X = T = R^{(0)} R^{(1)} \cdots R^{(k)} \cdots R^{(n-1)} R^{(n)} \tag{15-5}$$

the starting transformation $R^{(0)}$ is set to a unit matrix. The iterative orthogonal transformation matrix $R^{(k)}$, with four nonzero terms in the I and J rows and columns, is of the following orthogonal form:

$$R^{(k)} = \begin{pmatrix}
1 & 0 & 0 & 0 & 0 & 0 & 0 & 0 \\
0 & 1 & 0 & 0 & 0 & 0 & 0 & 0 \\
0 & 0 & R_{II} & 0 & 0 & R_{IJ} & 0 & 0 \\
0 & 0 & 0 & 1 & 0 & 0 & 0 & 0 \\
0 & 0 & 0 & 0 & 1 & 0 & 0 & 0 \\
0 & 0 & R_{JI} & 0 & 0 & R_{JJ} & 0 & 0 \\
0 & 0 & 0 & 0 & 0 & 0 & 1 & 0 \\
0 & 0 & 0 & 0 & 0 & 0 & 0 & 1
\end{pmatrix} \tag{15-6}$$

The four nonzero terms are functions of an unknown rotation angle θ and are defined by

$$R_{II} = R_{JJ} = \cos\theta \quad \text{and} \quad R_{JI} = R_{IJ} = \sin\theta$$

Therefore, $R^{(k)T} R^{(k)} = I$, which is independent of the angle θ. The typical iteration involves the following matrix operation:

$$A^{(k)} = R^{(k)T} A^{(k-1)} R^{(k)}$$

The angle is selected to force the terms I, J and J, I in the matrix $A^{(k)}$ to be zero. This is satisfied if the angle is computed from

$$\tan 2\theta = \frac{-2 A_{IJ}^{(k-1)}}{A_{II}^{(k-1)} - A_{JJ}^{(k-1)}} \tag{15-7}$$

which could be solved for θ. However, the procedure can be further improved by the algorithm described below.

Introducing a new variable:

$$\phi = \cot 2\theta = -\frac{A_{II}^{(k-1)} - A_{JJ}^{(k-1)}}{2 A_{IJ}^{(k-1)}}$$

and using the relation

$$\tan 2\theta = \frac{2t}{1 - t^2}$$

where $t = \tan\theta$, we can then write Equation (15-7) as

$$t^2 + 2\phi t - 1 = 0$$

which has the solutions:

$$t = -\phi \pm \sqrt{1 + \phi^2}$$

It has been found that the solution $|t| \le 1$ results in a more stable transformation. Thus, we choose the plus sign if $\phi > 0$ and the minus sign if $\phi < 0$, which corresponds to the relation:

$$t = \text{sgn}(\phi)\left(\sqrt{1 + \phi^2} - |\phi| \right)$$

Having computed t, we can use the relation $t = \tan\theta$ to obtain:

$$\sin\theta = \frac{t}{\sqrt{1 + t^2}}, \quad \cos\theta = \frac{1}{\sqrt{1 + t^2}}$$

We can use these formulas to compute $\mathbf{A}^{(k)}$ and $\mathbf{R}^{(k)}$.

The Jacobi method is an iterative procedure that repeatedly applies Jacobi rotations until the off-diagonal terms have been virtually vanished.

Implementation

Start with a new C# Console application and name it *EigenvalueTest*. Add a new class, *Eigenvalue*, to the current project and change its namespace to *XuMath*. Since this project requires us to perform matrix and vector operations, we need to add the *VectorR.cs* and *MatrixR.cs* from Chapter 3 to the project. Add a new method, *Jacobi*, to the *Eigenvalue class*. Here is the code listing of this class:

```
using System;

namespace XuMath
{
    public class Eigenvalue
    {
        public static void Jacobi(MatrixR A, double tolerance,
                out MatrixR x, out VectorR lambda)
        {
            MatrixR AA = A.Clone();
            int n = A.GetCols();
            int maxTransform = 5 * n * n;
            MatrixR matrix = new MatrixR(n, n);
            MatrixR R = matrix.Identity();
            MatrixR R1 = R;
            MatrixR A1 = A;
            lambda = new VectorR(n);
            x = R;

            double maxTerm = 0.0;
            int I, J;

            do
            {
                maxTerm = MaxTerm(A, out I, out J);
                Transformation(A, R, I, J, out A1, out R1);
                A = A1;
                R = R1;
```

```
        }
    while (maxTerm > tolerance);

    x = R;
    for (int i = 0; i < n; i++)
        lambda[i] = A[i, i];

    for (int i = 0; i < n - 1; i++)
    {
        int index = i;
        double d = lambda[i];
        for (int j = i + 1; j < n; j++)
        {
            if (lambda[j] > d)
            {
                index = j;
                d = lambda[j];
            }
        }
        if (index != i)
        {
            lambda = lambda.GetSwap(i, index);
            x = x.GetColSwap(i, index);
        }
    }
}

private static double MaxTerm(MatrixR A, out int I, out int J)
{
    int n = A.GetCols();
    double result = 0.0;
    I = 0;
    J = 1;

    for (int i = 0; i < n-1; i++)
    {
        for (int j = i + 1; j < n; j++)
        {
            if (Math.Abs(A[i, j]) > result)
            {
                result = Math.Abs(A[i, j]);
                I = i;
                J = j;
            }
        }
    }
    return result;
}

private static void Transformation(MatrixR A, MatrixR R,
        int I, int J, out MatrixR A1, out MatrixR R1)
{
    int n = A.GetCols();
    double t = 0.0;
    double da = A[J, J] - A[I, I];
```

```
            if (Math.Abs(A[I, J]) < Math.Abs(da) * 1e-30)
            {
                t = A[I, J] / da;
            }
            else
            {
                double phi = da / (2.0 * A[I, J]);
                t = 1.0 / (Math.Abs(phi) + Math.Sqrt(1.0 + phi * phi));
                if (phi < 0.0)
                    t = -t;
            }

            double c = 1.0 / Math.Sqrt(Math.Abs(t * t + 1.0));
            double s = t * c;
            double tau = s / (1.0 + c);
            double temp = A[I, J];
            A[I, J] = 0.0;
            A[I, I] -= t * temp;
            A[J, J] += t * temp;

            for (int i = 0; i < I; i++)
            {
                temp = A[i, I];
                A[i, I] = temp - s * (A[i, J] + tau * temp);
                A[i, J] += s * (temp - tau * A[i, J]);
            }
            for (int i = I + 1; i < J; i++)
            {
                temp = A[I, i];
                A[I, i] = temp - s * (A[i, J] + tau * A[I, i]);
                A[i, J] += s * (temp - tau * A[i, J]);
            }
            for (int i = J + 1; i < n; i++)
            {
                temp = A[I, i];
                A[I, i] = temp - s * (A[J, i] + tau * temp);
                A[J, i] += s * (temp - tau * A[J, i]);
            }

            for (int i = 0; i < n; i++)
            {
                temp = R[i, I];
                R[i, I] = temp - s * (R[i, J] + tau * R[i, I]);
                R[i, J] += s * (temp - tau * R[i, J]);
            }

            A1 = A;
            R1 = R;
        }
    }
}
```

The Jacobi method computes all eigenvalues and eigenvectors of a real symmetric matrix. The principal diagonal of the matrix is replaced by the eigenvalues, and the columns of the transformation matrix become the normalized eigenvectors.

The above code also contains two private methods, *MaxItem* and *Transformation*. The *MaxTerm* method is used to find the largest off-diagonal term of the matrix, and the *Transformation* method is used to rotate the matrix with a specific angle such that the largest off-diagonal term of the matrix is equal to zero.

Testing the Jacobi Method

We can find all eigenvalues and eigenvectors of a real symmetric matrix by using the Jacobi method implemented in the previous section. For example, we want to compute the eigenvalues and eigenvectors of the following matrix:

$$\begin{pmatrix} 4 & 3 & 6 \\ 3 & 7 & -1 \\ 6 & -1 & 9 \end{pmatrix}$$

Add a new static method, *TestJacobi*, to the *Program.cs* file:

```
using System;
using XuMath;

namespace EigenvalueTest
{
    class Program
    {
        static void Main(string[] args)
        {
            TestJacobi();

            Console.ReadLine();
        }

        static void TestJacobi()
        {
            MatrixR A =
                new MatrixR(new double[,] { { 4,3,6}, {3,7,-1}, {6,-1,9} });
            MatrixR x;
            VectorR lambda;
            Eigenvalue.Jacobi(A, 1e-10, out x, out lambda);
            Console.WriteLine("\n x = \n {0}", x);
            Console.WriteLine("\n lambda = \n {0}", lambda);
        }
    }
}
```

Running this example produces results shown in Figure 15-1.

```
x =
(0.577262787554022, 0.206639019358934, 0.789979740109668
0.151677039485513, 0.923478466319118, -0.35239409464093
0.802347648964132, -0.323245785623368, -0.501747359014934)

lambda =
(13.127761438716, 8.02131546982304, -1.14907690853908)
```

Figure 15-1 Eigenvectors and eigenvalues of a symmetric matrix using the Jacobi method.

Power Iterration

The power iteration is an eigenvalue algorithm. For any given real symmetric matrix $\mathbf{A}$, the algorithm will generate an eigenvalue λ and a nonzero eigenvector $\mathbf{x}$, such that $\mathbf{A}\mathbf{x} = \lambda\mathbf{x}$.

The power iteration is probably the simplest method for finding the eigenvalue and eigenvector of a real symmetric matrix. It does not compute matrix decomposition, and hence can be used when $\mathbf{A}$ is a very large and sparse matrix. However, it will find only a single eigenvalue (the one with the largest absolute value) and it may converge slowly.

Algorithm

Suppose $\mathbf{A}$ is an $n \times n$ real symmetric matrix and v_0 is any vector of length n. Perform the following operations:

- Multiply v_0 by $\mathbf{A}$.
- Normalize the new vector $\mathbf{A}v_0$ to unit length.
- Repeat the above process.

The above operations generate a series of vectors, $v_0, v_1, v_2, \cdots$. It turns out that these vectors converge to the eigenvector that corresponds to the eigenvalue with the largest absolute value. This can be easily be confirmed. Let's first find a formula for v_k. At each stage of this process, the vector is multiplied by $\mathbf{A}$ and then by some number. Thus v_k must be a multiple of $\mathbf{A}^k v_0$. Since the resulting vector has a unit length, that number must be $1/|\mathbf{A}^k v_0|$. Therefore

$$\mathbf{v}_k = \frac{\mathbf{A}^k \mathbf{v}_0}{|\mathbf{A}^k \mathbf{v}_0|}$$

We know that $\mathbf{A}$ has a basis of eigenvectors $\mathbf{x}_1, \mathbf{x}_2, \cdots, \mathbf{x}_n$. We can order these eigenvectors such that $|\lambda_1| > |\lambda_2| \geq |\lambda_3| \geq \cdots \geq |\lambda_n|$. Note that here, we assume $|\lambda_1| \neq |\lambda_2|$, otherwise the power iteration runs into difficulty. We can expand the initial vector v_0 in this basis:

$$\mathbf{v}_0 = c_1 \mathbf{x}_1 + c_2 \mathbf{x}_2 + \cdots + c_n \mathbf{x}_n$$

Here we require $c_1 \neq 0$ for this method to work. Since $\mathbf{A}^k \mathbf{x}_i = \lambda_i^k \mathbf{x}_i$ we have

$$\mathbf{A}^k \mathbf{v}_0 = c_1 \lambda_1^k \mathbf{x}_1 + c_2 \lambda_2^k \mathbf{x}_2 + \cdots + c_n \lambda_n^k \mathbf{x}_n = \lambda_1^k \left[c_1 \mathbf{x}_1 + c_2 \left(\frac{\lambda_2}{\lambda_1} \right)^k \mathbf{x}_2 + \cdots + c_n \left(\frac{\lambda_n}{\lambda_1} \right)^k \mathbf{x}_n \right]$$

Theoretically, $\mathbf{A}^k \mathbf{v}_0 \to c_1 \lambda_1^k \mathbf{x}_1$ for a large number of k. which is because $| \lambda_i / \lambda_1 | < 1$ for every $i > 1$ so the powers tend to zero. Thus, for a large k, we have:

$$\mathbf{v}_k = \frac{\mathbf{A}^k \mathbf{v}_0}{| \mathbf{A}^k \mathbf{v}_0 |} \approx \left(\frac{\lambda_1}{| \lambda_1 |} \right)^k \frac{c_1 \mathbf{x}_1}{| c_1 \mathbf{x}_1 |} = (\pm)^k \pm \frac{\mathbf{x}_1}{| \mathbf{x}_1 |}$$

This confirms that $\mathbf{v}_k$ converges, except for a possible sign flip at each stage, to a normalized eigenvector corresponding to the largest eigenvalue λ_1. The sign flip is present exactly when $\lambda_1 < 0$. Knowing $\mathbf{x}_1$ or a multiple of it we can find $| \lambda_1 |$ with

$$| \lambda_1 | = \frac{| \mathbf{A}\mathbf{x}_1 |}{| \mathbf{x}_1 |}$$

This determines λ_1 since we can determine its sign by the presence or absence of the sign flip. This gives us a method for finding the largest eigenvalue (in absolute value) and the corresponding eigenvector.

Implementation

The power iteration method can easily be implemented in C#. Add a new public static method, *Power*, to the *Eigenvalue* class:

```
public static void Power(MatrixR A, double tolerance,
                         out VectorR x, out double lambda)
{
    int n = A.GetCols();
    x = new VectorR(n);
    lambda = 0.0;
    double delta = 0.0;
    Random random = new Random();
    for (int i = 0; i < n; i++)
    {
        x[i] = random.NextDouble();
    }

    do
    {
        VectorR temp = x;
        x = MatrixR.Transform(A, x);
        x.Normalize();
        if (VectorR.DotProduct(temp, x) < 0)
            x = -x;

        VectorR dx = temp - x;
        delta = dx.GetNorm();
    }
    while (delta > tolerance);
```

```
        lambda = VectorR.DotProduct(x, MatrixR.Transform(A,x));
    }
```

This method takes the matrix **A** and the tolerance parameter as input. Note that the initial vector is filled with the random numbers. The outputs will give you the largest eigenvalue (in absolute value) and its corresponding eigenvector.

Testing Power Method

I will show you how to use the power method implemented in the previous section to find the largest eigenvalue and its corresponding eigenvector. Here, we consider the same matrix we used to examine the Jacobi method, so that we can check the results.

Add a method, *TestPower*, to the *Program.cs* file:

```
    static void TestPower()
    {
        MatrixR A =
            new MatrixR(new double[,] { { 4,3,6}, {3,7,-1}, {6,-1,9} });
        VectorR x;
        double lambda;
        Eigenvalue.Power(A, 1e-5, out x, out lambda);
        Console.WriteLine("\n lambda = {0} \n x= {1}", lambda, x);
    }
```

Running this example generates the results:

$$\lambda = 13.12776, \quad \mathbf{x} = (0.57726, 0.15169, 0.80234)$$

By comparing these results to the results from the Jacobi method, we find that the power method indeed gives the largest eigenvalue and corresponding eigenvector.

Inverse Iteration

The inverse iteration is an iteration algorithm based on the power method.

Algorithm

We know that if λ is an eigenvalue of matrix **A**, then $1/\lambda$ is an eigenvalue of $\mathbf{A}^{-1}$. This suggests a method to compute the smallest eigenvalue of **A**. Arrange eigenvalues of $\mathbf{A}^{-1}$ such that

$$| \lambda_n^{-1} | > | \lambda_{n-1}^{-1} | \geq \cdots \geq | \lambda_1^{-1} |$$

Then apply the power method to $\mathbf{A}^{-1}$. You can explicitly compute the inverse matrix of **A**, which is efficient because it has to be done only once. Alternately, you can solve

$$\mathbf{A}\mathbf{x}^{(k+1)} = \mathbf{x}^{(k)}$$

for $\mathbf{x}^{(k+1)}$ by using an efficient linear algebra solver. You can consider using LU factorization, since it only has to be done once.

So far, the power and inverse iteration methods only compute the eigenvalue with the largest or smallest absolute value, and its corresponding eigenvector. By employing the power and inverse method, we find that we can actually calculate the eigenvalue closest to any number s. Note that the eigenvalue of $(\mathbf{A} - s\mathbf{I})^{-1}$ is exactly $(\lambda_i - s)^{-1}$. Moreover, the eigenvectors of $\mathbf{A}$ and $(\mathbf{A} - s\mathbf{I})^{-1}$ are the same. This can be confirmed by the following derivation.

If we have

$$\mathbf{A}\mathbf{x} = \lambda\mathbf{x}$$

Then

$$(\mathbf{A} - s\mathbf{I})\mathbf{x} = (\lambda - s)\mathbf{x}$$

Multiply both sides of the above equation by $(\mathbf{A} - s\mathbf{I})^{-1}$ and divide by $\lambda - s$ we obtain

$$(\lambda - s)^{-1}x = (\mathbf{A} - s\mathbf{I})^{-1}\mathbf{x}$$

Now start with an arbitrary vector $\mathbf{v}_0$ and define

$$\mathbf{v}_{k+1} = \frac{(\mathbf{A} - s\mathbf{I})^{-1}\mathbf{v}_k}{|(\mathbf{A} - s\mathbf{I})^{-1}\mathbf{v}_k|}$$

Then $\mathbf{v}_k$ will converge to the eigenvector $\mathbf{x}_i$ of $(\mathbf{A} - s\mathbf{I})^{-1}$ for which $|\lambda_i - s|^{-1}$ is the largest. Since the eigenvectors of $\mathbf{A}$ and $(\mathbf{A} - s\mathbf{I})$ are the same, $\mathbf{x}_i$ is also an eigenvector of $\mathbf{A}$. And since $|\lambda_i - s|^{-1}$ is the largest when λ_i is the closest to s, we have computed the eigenvector $\mathbf{x}_i$ of $\mathbf{A}$ for which λ_i is closest to s. We can now compute λ_i by comparing $\mathbf{A}\mathbf{x}_i$ with $\mathbf{x}_i$.

Implementation

Here we will implement the inverse method by explicitly computing the inverse matrix. Of course, you can code a more efficient inverse iteration method by not directly calculating the inverse matrix. For example, you don't need to know the whole matrix $(\mathbf{A} - s\mathbf{I})^{-1}$, just the vector $(\mathbf{A} - s\mathbf{I})^{-1}\mathbf{v}_k$. This vector is the solution $\mathbf{y}$ of the linear equation $(\mathbf{A} - s\mathbf{I})\mathbf{y} = \mathbf{x}_k$.

Add a new public static method, *Inverse*, to the *Eigenvalue* class:

```
public static void Inverse(MatrixR A, double s, double tolerance,
                        out VectorR x, out double lambda)
{
    int n = A.GetCols();
    x = new VectorR(n);
    lambda = 0.0;
    double delta = 0.0;
    MatrixR identity = new MatrixR(n, n);
    A = A - s * (identity.Identity());
    LinearSystem ls = new LinearSystem();
    A = ls.LUInverse(A);

    Random random = new Random();
```

```
    for (int i = 0; i < n; i++)
    {
        x[i] = random.NextDouble();
    }
    do
    {
        VectorR temp = x;
        x = MatrixR.Transform(A, x);
        x.Normalize();
        if (VectorR.DotProduct(temp, x) < 0)
            x = -x;
        VectorR dx = temp - x;
        delta = dx.GetNorm();
    }
    while (delta > tolerance);
    lambda = s + 1.0 / (VectorR.DotProduct(x, MatrixR.Transform(A, x)));
}
```

This method takes as input the matrix **A**, the shift number s, and the tolerance parameter. Note that the initial vector is filled with random numbers. The outputs will give you the eigenvalue closest to s, and corresponding eigenvector.

Testing the Inverse Method

I will now show you how to use the inverse method implemented in the previous section to find the eigenvalue closest to the specified value s and the corresponding eigenvector. Here, we consider the same matrix we used in examining the Jacobi method, so that we can easily check the results.

Add a method, *TestInverse*, to the *Program.cs* file:

```
static void TestInverse()
{
    MatrixR A =
        new MatrixR(new double[,] { { 4, 3, 6 }, { 3, 7, -1 }, { 6, -1, 9 } });
    VectorR x;
    double lambda;
    Eigenvalue.Inverse(A, 7, 1e-5, out x, out lambda);
    Console.WriteLine("\n lambda = {0} \n x= {1}", lambda, x);
}
```

Here, we want to calculate the eigenvalue closest to 7. Running this example generates the following results:

$$\lambda = 8.02132, \quad \mathbf{x} = (0.20664, 0.92348, -0.32325)$$

Compared to the results from the Jacobi method, we find that the inverse iteration method indeed gives the eigenvalue closest to 7 and the corresponding eigenvector.

Rayleigh Method

In this section, we consider the Rayleigh method, which is a variant of the power iteration method used to compute the dominant eigenvalue of a symmetric matrix.

Algorithm

Note that a symmetric $n \times n$ matrix $\mathbf{A}$ has n linearly independent eigenvectors, which can be orthonormal. Given an arbitrary starting vector $\mathbf{v}_1$, which can be written as

$$\mathbf{v}_1 = c_1 \mathbf{x}_1 + c_2 \mathbf{x}_2 + \cdots + c_n \mathbf{x}_n$$

where $\{\mathbf{x}_i\}$ are the n orthonormal eigenvectors and c_i are scalar constants. Applying matrix $\mathbf{A}$ to $\mathbf{v}_1$:

$$\mathbf{A}\mathbf{v}_1 = c_1 \lambda_1 \mathbf{x}_1 + c_2 \lambda_2 \mathbf{x}_2 + \cdots + c_n \lambda_n \mathbf{x}_n$$

where λ_i is the eigenvalue with corresponding eigenvector $\mathbf{x}_i$. Let $\mathbf{v}_k$ be the result of applying $\mathbf{A}$ to $\mathbf{v}_1$ k times. Then $\mathbf{v}_k = \mathbf{A}\mathbf{v}_{k-1}$. If λ_1 is the dominant eigenvalue, as k becomes large, we have

$$\mathbf{v}_k = \lambda_1^k \left[c_1 \mathbf{x}_1 + \left(\frac{\lambda_2}{\lambda_1}\right)^k c_2 \mathbf{x}_2 + \cdots + \left(\frac{\lambda_n}{\lambda_1}\right)^k c_n \mathbf{x}_n \right] \approx \lambda_1^k c_1 \mathbf{x}_1 \approx \lambda_1 \mathbf{v}_{k-1}$$

Thus, at each step if $\mathbf{v}_{k-1}$ is normalized, then we have

$$\lambda_1 = \mathbf{v}_k \cdot \mathbf{v}_{k-1}$$

In order to use the Rayleigh method to compute the dominant eigenvalue and the corresponding eigenvector, we can process as following:

- Start with an initial vector $\mathbf{x}_0$.

- Normalize the current $\mathbf{x}_k$.

- Compute $\mathbf{x}_{k+1} = \mathbf{A}\mathbf{x}_k$.

- Calculate the quantity $\lambda = \mathbf{x}_{k+1} \cdot \mathbf{x}_k$.

- Repeat the above steps until the relative change in λ is less than the tolerance.

Implementation

The Rayleigh method can be easily implemented in C#. Add a new public static method, *Rayleigh*, to the *Eigenvalue* class:

```
public static void Rayleigh(MatrixR A, double tolerance,
                            out VectorR x, out double lambda)
{
    int n = A.GetCols();
    double delta = 0.0;
    Random random = new Random();
    x = new VectorR(n);
    for (int i = 0; i < n; i++)
    {
        x[i] = random.NextDouble();
    }
    x.Normalize();
    lambda = VectorR.DotProduct(x, MatrixR.Transform(A, x));
    double temp = lambda;
    do
```

```
        {
            temp = lambda;
            VectorR x0 = x;
            x0.Normalize();
            lambda = VectorR.DotProduct(x, MatrixR.Transform(A, x0));
            delta = Math.Abs((temp - lambda) / lambda);
        }
        while (delta > tolerance);
        x.Normalize();
    }
```

This method takes the matrix **A** and the tolerance parameter as input parameters. Note that the initial vector is filled with random numbers. The outputs from this method give you the dominant eigenvalue and the corresponding eigenvector.

Testing the Rayleigh Method

I will now show you how to use the *Rayleigh* method we implemented in the previous section to find the dominant eigenvalue and the corresponding eigenvector. Here, we consider the same matrix that we used to examine the Jacobi method, so that we can easily check the results.

Add a new static method, *TestRayleigh*, to the *Program.cs* file:

```
static void Testrayleigh()
{
    MatrixR A =
        new MatrixR(new double[,] { { 4, 3, 6 }, { 3, 7, -1 }, { 6, -1, 9 } });
    VectorR x;
    double lambda;
    Eigenvalue.Rayleigh(A, 1e-8, out x, out lambda);
    Console.WriteLine("\n lambda = {0} \n x= {1}", lambda, x);
}
```

Running this example generates the results:

$$\lambda = 13.12776, \quad \mathbf{x} = (0.57728, 0.15174, 0.80233)$$

Compared to the results from the Jacobi method, we find that the Rayleigh method does indeed give the dominant eigenvalue and corresponding eigenvector.

Rayleigh-Quotient Method

The Rayleigh-quotient method is a variant of the inverse iteration for computing one of the eigenvalues and the corresponding eigenvector. Note that this method changes the shift in each iteration, so it does not necessarily find the dominant eigenvalue.

Algorithm

For a given symmetric matrix **A** and nonzero vector **v**, the Rayleigh quotient is defined as

$$R(\mathbf{A}, \mathbf{v}) = \frac{\mathbf{v}^T \mathbf{A} \mathbf{v}}{\mathbf{v}^T \mathbf{v}}$$

where $\mathbf{v}^T$ is the transpose of $\mathbf{v}$. Note that $R(\mathbf{A}, c\mathbf{v}) = R(\mathbf{A}, \mathbf{v})$ for any real scalar constant c.

We can show that the Rayleigh quotient reaches its smallest eigenvalue of $\mathbf{A}$, λ_{min}, when $\mathbf{v}$ is the $\mathbf{x}_{min}$ (the corresponding eigenvector). Similarly,

$$R(\mathbf{A}, \mathbf{v}) \leq \lambda_{max} \quad \text{and} \quad R(\mathbf{A}, \mathbf{x}_{max}) = \lambda_{max}$$

In order to use the Rayleigh-quotient method to compute the eigenvalue and the corresponding eigenvector, we can follow the process below:

- For a given $n \times n$ symmetric $\mathbf{A}$ and an initial vector $\mathbf{x}_0$, normalize $\mathbf{x}_0$.

- Calculate the quantity $\lambda = \mathbf{x}_0^T \mathbf{A} \mathbf{x}_0$.

- Solve the equation $(\mathbf{A} - \lambda \mathbf{I})\mathbf{x} = \mathbf{x}_0$ for $\mathbf{x}$.

- Set $\mathbf{x}_0 = \mathbf{x}$.

- Repeat the above steps until the relative change in λ is less than the tolerance.

The Rayleigh-quotient method has a better convergence property than that of the Rayleigh method and is usually used after the Rayleigh method has obtained a close estimate of the dominant eigenvalue and corresponding eigenvector.

Implementation

Add a public static method, *RayleighQuotient*, to the *Eigenvalue* class:

```
public static void RayleighQuotient(MatrixR A, double tolerance, int flag,
                        out VectorR x, out double lambda)
{
    int n = A.GetCols();
    double delta = 0.0;
    Random random = new Random();
    x = new VectorR(n);
    if (flag != 2)
    {
        for (int i = 0; i < n; i++)
        {
            x[i] = random.NextDouble();
        }
        x.Normalize();
        lambda = VectorR.DotProduct(x, MatrixR.Transform(A, x));
    }
    else
    {
        lambda = 0.0;
        Rayleigh(A, 1e-2, out x, out lambda);
    }

    double temp = lambda;
    MatrixR identity = new MatrixR(n, n);
    LinearSystem ls = new LinearSystem();
```

```
do
{
    temp = lambda;
    double d = ls.LUCrout(A - lambda * identity.Identity(), x);
    x.Normalize();
    lambda = VectorR.DotProduct(x, MatrixR.Transform(A, x));
    delta = Math.Abs((temp - lambda) / lambda);
}
while (delta > tolerance);
}
```

Note that in adition to the matrix **A** and the tolerance, this method also takes as input an integer flag parameter. This flag parameter takes two values, 1 and 2, corresponding to the case that the initial vector is filled with random numbers and the case that initial vector and eigenvalue are approximate solutions of the Rayleigh method, respectively. For the former case, the RayleighQuotient method will give you one of the eigenvalues (not necessarily the dominant eigenvalue) and the corresponding eigenvector. For the latter case, the method will give you the dominant eigenvalue and the corresponding eigenvector, since the initial vector is already the approximate dominant eigenvector. The tolerance parameter used in calling the Rayleigh method is set to 1e–2, which is just a rough estimate of the eigenvalue and the corresponding eigenvector.

Testing Rayleigh-Quotient Method

In this section, I will show you how to use the Rayleigh-quotient method we implemented in the previous section to find the eigenvalue and the corresponding eigenvector. Here, we consider the same matrix we used to examine the Jacobi method, so that we can check the results.

Add a new static method, *TestRayleighQuotient*, to the *Program.cs* file:

```
static void TestrayleighQuotient()
{
    MatrixR A =
        new MatrixR(new double[,] { { 4, 3, 6 }, { 3, 7, -1 }, { 6, -1, 9 } });
    VectorR x;
    double lambda;
    Eigenvalue.RayleighQuotient(A, 1e-8, 1, out x, out lambda);
    Console.WriteLine("\n Results for an initial vector filled
                        with random numbers:");
    Console.WriteLine(" lambda = {0} \n x= {1}", lambda, x);

    x = new VectorR(3);
    lambda = 0.0;
    Eigenvalue.RayleighQuotient(A, 1e-8, 2, out x, out lambda);
    Console.WriteLine("\n\n Results for an  initial vector
                        generated from the Rayleigh method:");
    Console.WriteLine(" lambda = {0} \n x= {1}", lambda, x);
}
```

Here, we compute the eigenvalue and the corresponding eigenvector with two different initial vectors: one is an arbitrary vector filled with random numbers (corresponding to flag = 1); the other is the vector generated by the Rayleigh method (corresponding to flag = 2).

Running this example produces the results shown in Figure 15-2. You can see from the figure that the Rayleigh-quotient method with an initial vector filled with random numbers finds the eigenvalue which is not the dominant eigenvalue. Since the initial vector is filled with random numbers, different runs may give you different eigenvalues and corresponding eigenvectors. On the other hand, if the intial vector is estimated from the Rayleigh method, the Rayleigh-quotient method always give you the dominant eigenvalue and the corresponding eigenvector.

```
Results for an initial vector filled with random numbers:
lambda = 8.02131546982304
x= (-0.206639019358934, -0.923478466319118, 0.323245785623368)


Results for an  initial vector generated from the Rayleigh method:
lambda = 13.127761438716
x= (0.577262787554022, 0.151677039485512, 0.802347648964132)
```

Figure 15-2 Results from the Rayleigh-quotient method.

Matrix Tridiagonalization

As mententioned before, similarity transformations can be used to transform an eigenvalue problem into a form that is easier to solve. If we can find a way to transform the original square matrix into a tridiagonal form, we can much more efficiently find the eigenvalues and eigenvectors.

The householder algorithm is such an efficient means that can be used to reduce an $n \times n$ real symmetric matrix $\mathbf{A}$ to a tridiagonal real symmetric matrix $\mathbf{T}$ by $n - 2$ orthogonal transformations. Each transformation annihilates the required part of a whole column and whole corresponding row. Using this tridiagonal matrix, we can easily find eigenvalues and eigenvectors of a square matrix.

Algorithm

The key step of the Householder method is to construct a householder matrix $\mathbf{P}$, which has the form:

$$\mathbf{P} = \mathbf{I} - \frac{\mathbf{u} \cdot \mathbf{u}^T}{H}, \qquad H = \frac{1}{2} |\mathbf{u}|^2$$

Here $\mathbf{u}$ can be any vector. It can be easily shown that the matrix $\mathbf{P}$ is symmetric and orthogonal.

Suppose $\mathbf{x}$ is an arbitrary vector composed of the first column of matrix $\mathbf{A}$. Choose

$$\mathbf{u} = \mathbf{x} \mp |\mathbf{x}| \mathbf{e}_1, \quad \mathbf{e}_1 = [1, 0, \cdots, 0]^T$$

We now consider the transformation $\mathbf{Px}$:

$$\mathbf{P} \cdot \mathbf{x} = \mathbf{x} - \frac{1}{H} \mathbf{u} \cdot (\mathbf{x} \mp |\mathbf{x}| \mathbf{e}_1)^T \cdot \mathbf{x} = \mathbf{x} - \frac{2\mathbf{u} \cdot (|\mathbf{x}|^2 \mp |\mathbf{x}| x_1)}{2|\mathbf{x}|^2 \mp 2|\mathbf{x}| x_1} = \mathbf{x} - \mathbf{u} = \pm |\mathbf{x}| \mathbf{e}_1$$

This result shows that the householder matrix $\mathbf{P}$ acts on a given vector $\mathbf{x}$ to zero all its elements except the first one.

To reduce a real symmetric matrix A to tridigonal form, we choose the vector $\mathbf{x}$ as the first column of A , with the first element omitted, i.e., the lower $n - 1$ elements of the first column. Then the lower $n - 2$ elements will vanish:

$$
\mathbf{P_1 \cdot A} =
\begin{pmatrix}
1 & | & 0 & 0 & \cdots & 0 \\
-- & | & -- & -- & -- & -- \\
0 & | & & & & \\
0 & | & & & & \\
\vdots & | & & {}^{(n-1)}\mathbf{P_1} & & \\
0 & |
\end{pmatrix}
\cdot
\begin{pmatrix}
a_{00} & | & a_{01} & a_{02} & \cdots & a_{0n-1} \\
-- & | & -- & -- & -- & -- \\
a_{10} & | & & & & \\
a_{20} & | & & & & \\
\vdots & | & & & \text{irrelevant} & \\
a_{n-10} & |
\end{pmatrix}
$$

$$
=
\begin{pmatrix}
a_{00} & | & a_{01} & a_{02} & \cdots & a_{0n-1} \\
-- & | & -- & -- & -- & -- \\
k & | & & & & \\
0 & | & & & & \\
\vdots & | & & & \text{irrelevant} & \\
0 & |
\end{pmatrix}
$$

Here we write the matrices in partitioned form, with ${}^{(n-1)}\mathbf{P}$ denoting a householder matrix with dimensions $(n-1) \times (n-1)$. The quantity k is simply plus or minus the magnitude of the vector $[a_{10}, a_{20}, \cdots, a_{n-10}]^T$.

The complete orthogonal transformation becomes:

$$
\mathbf{A' = P \cdot A \cdot P} =
\begin{pmatrix}
a_{00} & | & k & 0 & \cdots & 0 \\
-- & | & -- & -- & -- & -- \\
k & | & & & & \\
0 & | & & & & \\
\vdots & | & & & \text{irrelevant} & \\
0 & |
\end{pmatrix}
$$

Here we have used the fact that $\mathbf{P}^T = \mathbf{P}$.

Now choose the vector $\mathbf{x}$ for the second householder matrix to be the bottom $n-2$ elements of the second column, and from it construct:

$$
\mathbf{P_2} =
\begin{pmatrix}
1 & 0 & | & 0 & \cdots & 0 \\
0 & 1 & | & 0 & \cdots & 0 \\
-- & -- & | & -- & -- & -- \\
0 & 0 & | & & & \\
\vdots & \vdots & | & & {}^{(n-2)}\mathbf{P_2} & \\
0 & 0 & |
\end{pmatrix}
$$

The identity block in the upper-left corner insures that the tridiagonalization achieved in the first step will not be spoiled by this transformation, while the $(n-2)$-dimensional Householder matrix

$^{(n-2)}\mathbf{P}_2$ creates one additional row and column in the tridigonal output. It is apparent that it takes a total of $(n-2)$ transformations to obtain the final tridiagonal matrix.

In practice, instead of performing the matrix multiplication $\mathbf{A}' = \mathbf{P} \cdot \mathbf{A} \cdot \mathbf{P}$, we compute the vector

$$\mathbf{v} = \frac{\mathbf{A} \cdot \mathbf{u}}{H}$$

Then

$$\mathbf{A} \cdot \mathbf{P} = \mathbf{A} \cdot \left(\mathbf{I} - \frac{\mathbf{u} \cdot \mathbf{u}^T}{H} \right) = \mathbf{A} - \mathbf{v} \cdot \mathbf{u}^T$$

$$\mathbf{A}' = \mathbf{P} \cdot \mathbf{A} \cdot \mathbf{P} = \mathbf{A} - \mathbf{v} \cdot \mathbf{u}^T - \mathbf{u} \cdot \mathbf{v}^T + 2g\mathbf{u} \cdot \mathbf{u}^T$$

where

$$g = \frac{\mathbf{u}^T \cdot \mathbf{v}}{2H}$$

Introducing a new vector $\mathbf{w} = \mathbf{v} - g\mathbf{u}$, we can easily verify that the transformation can be written as

$$\mathbf{A}' = \mathbf{P} \cdot \mathbf{A} \cdot \mathbf{P} = \mathbf{A} - \mathbf{w} \cdot \mathbf{u}^T - \mathbf{u} \cdot \mathbf{w}^T$$

If the eigenvectors of the final tridiagonal matrix $\mathbf{T}$ are found, then the eigenvectors of the original square matrix $\mathbf{A}$ can be obtained by applying the accumulated transformation

$$\mathbf{V} = \mathbf{P}_1 \cdot \mathbf{P}_2 \cdots \mathbf{P}_{n-2}$$

to those eigenvectors. We can construct the accumulated transformation matrix by initializing $\mathbf{V}$ to an $n \times n$ identity matrix and then applying the transformation $\mathbf{P} \leftarrow \mathbf{P} \cdot \mathbf{P}_i$ with $i = 1, 2, \cdots, n-2$.

We don't need to store the final tridiagonal $n \times n$ matrix $\mathbf{T}$; instead, we can introduce two arrays $\alpha[n]$ and $\beta[n-1]$ to represent the diagonal and off-diagonal elements of $\mathbf{T}$. We can then easily construct a tridiagonal matrix using the α and β arrays:

$$\mathbf{T} = \begin{pmatrix} \alpha_0 & \beta_0 & 0 & 0 & \cdots & 0 & 0 \\ \beta_0 & \alpha_1 & \beta_1 & 0 & \cdots & 0 & 0 \\ 0 & \beta_1 & \alpha_2 & \beta_2 & \cdots & 0 & 0 \\ \vdots & \vdots & \vdots & \vdots & \vdots & \vdots & \vdots \\ 0 & 0 & 0 & 0 & \cdots & \alpha_{n-2} & \beta_{n-2} \\ 0 & 0 & 0 & 0 & \cdots & \beta_{n-2} & \alpha_{n-1} \end{pmatrix}$$

Implementation

In this section, we will implement the householder algorithm, which can be used to transform the original real symmetric square matrix into a real symmetric tridiagonal matrix.

Add a new public static method, *Tridiagonalize*, to the *Eigenvalue* class:

```
public static double[] Alpha { get; set; }
public static double[] Beta { get; set; }

public static MatrixR Tridiagonalize(MatrixR A)
{
    int n = A.GetCols();
    MatrixR A1 = new MatrixR(n, n);
    A1 = A.Clone();
    double h, g, unorm;
    for (int i = 0; i < n - 2; i++)
    {
        VectorR u = new VectorR(n - i - 1);
        for (int j = i + 1; j < n; j++)
        {
            u[j - i - 1] = A[i, j];
        }
        unorm = u.GetNorm();
        if (u[0] < 0.0)
            unorm = -unorm;
        u[0] += unorm;

        for (int j = i + 1; j < n; j++)
        {
            A[j, i] = u[j - i - 1];
        }

        h = VectorR.DotProduct(u, u) * 0.5;
        VectorR v = new VectorR(n - i - 1);
        MatrixR a1 = new MatrixR(n - i - 1, n - i - 1);
        for (int j = i + 1; j < n; j++)
        {
            for (int k = i + 1; k < n; k++)
                a1[j - i - 1, k - i - 1] = A[j, k];
        }
        v = MatrixR.Transform(a1, u) / h;
        g = VectorR.DotProduct(u, v) / (2.0 * h);
        v -= g * u;

        for (int j = i + 1; j < n; j++)
        {
            for (int k = i + 1; k < n; k++)
                A[j, k] = A[j, k] - v[j - i - 1] * u[k - i - 1] -
                          u[j - i - 1] * v[k - i - 1];
        }
        A[i, i + 1] = -unorm;
    }
    Alpha = new double[n];
    Beta = new double[n - 1];
    Alpha[0] = A[0, 0];
    for (int i = 1; i < n; i++)
    {
        Alpha[i] = A[i, i];
        Beta[i - 1] = A[i - 1, i];
    }
}
```

```
    MatrixR V = new MatrixR(n, n);
    V = V.Identity();

    for (int i = 0; i < n - 2; i++)
    {
        VectorR u = new VectorR(n - i - 1);
        for (int j = i + 1; j < n; j++)
            u[j - i - 1] = A.GetColVector(i)[j];
        h = VectorR.DotProduct(u, u) * 0.5;
        VectorR v = new VectorR(n - 1);
        MatrixR v1 = new MatrixR(n - 1, n - i - 1);
        for (int j = 1; j < n; j++)
        {
            for (int k = i + 1; k < n; k++)
                v1[j - 1, k - i - 1] = V[j, k];
        }

        v = MatrixR.Transform(v1, u) / h;
        for (int j = 1; j < n; j++)
        {
            for (int k = i + 1; k < n; k++)
            {
                V[j, k] -= v[j - 1] * u[k - i - 1];
            }
        }
    }
    return V;
}
```

Here, we first define two public static propertis, *Alpha* and *Beta* double arrays, which are then used to construct the final tridiagonal matrix. This method takes the original real symmetric matrix **A** as input, and returns the transformed matrix **V**. The tridiagonal matrix generated by this method is stored in the public properties, *Alpha* and *Beta*.

Note that only the upper triangular portion of the matrix **A** is reduced to triangular form. The part below the principal diagonal is used to store the vectors **u**.

We also implement a utility method, *SetTridiagonalMatrix*, which allows you to construct the final tridiagonal matrix **T** using the computed *Alpha* and *Beta* arrays:

```
public static MatrixR SetTridiagonalMatrix()
{
    int n = Alpha.GetLength(0);
    MatrixR t = new MatrixR(n, n);
    t[0, 0] = Alpha[0];
    for (int i = 1; i < n; i++)
    {
        t[i, i] = Alpha[i];
        t[i - 1, i] = Beta[i - 1];
        t[i, i - 1] = Beta[i - 1];
    }
    return t;
}
```

Testing the Tridiagonalize Method

Now we can use the householder algorithm implemented in the previous section to reduce the following real symmetric matrix into symmetric tridiagonal form:

$$A = \begin{pmatrix} 5 & 1 & 2 & 2 & 4 \\ 1 & 1 & 2 & 1 & 0 \\ 2 & 2 & 0 & 2 & 1 \\ 2 & 1 & 2 & 1 & 2 \\ 4 & 0 & 1 & 2 & 4 \end{pmatrix}$$

Add a new method, *TestTridiagonalize*, to the *Program.cs* file:

```
static void TestTridiagonalize()
{
    MatrixR A = new MatrixR(new double[,]{{ 5, 1, 2, 2, 4 },
                                          { 1, 1, 2, 1, 0},
                                          { 2, 2, 0, 2, 1},
                                          { 2, 1, 2, 1, 2},
                                          { 4, 0, 1, 2, 4}});
    MatrixR V = Eigenvalue.Tridiagonalize(A);
    MatrixR T = Eigenvalue.SetTridiagonalMatrix();
    Console.WriteLine("\n Tridiagonal matrix T = \n{0}", T);
    Console.WriteLine("\n Transform matrix V = \n{0}", V);
}
```

Here, we first call the *Tridiagonalize* method to compute the *Alpha* and *Beta* arrays and the transformation matrix **V**. To display the results easily, we then call the *SetTridiagonalMatrix* method to construct the final tridiagonal matrix **T** by using the computed *Alpha* and *Beta* arrays.

Running this example produces the results shown in Figure 15-3.

```
Tridiagonal matrix T =
<5, -5, 0, 0, 0
 -5, 5.8, -0.824621125123532, 0, 0
 0, -0.824621125123532, -0.882352941176469, -1.57787470393402, 0
 0, 0, -1.57787470393402, 1.37321351489228, 1.27901542088873
 0, 0, 0, 1.27901542088873, -0.290860573715811>

Transform matrix V =
<1, 0, 0, 0, 0
 0, -0.2, 0.2910427500436, -0.27342353322451, -0.89472548255098
 0, -0.4, -0.388057000058133, -0.803633718022304, 0.208769279261896
 0, -0.4, 0.824621125123532, -0.098374604546384, 0.387714375772092
 0, -0.8, -0.2910427500436, 0.519360044590471, -0.0745604568792485>
```

Figure 15-3 Tridiagonal and transformation matrices calculated from the Householder method.

Eigenvalues of Symmetric Tridiagonal Matrices

Once the original symmetric $n \times n$ matrix has been reduced to tridiagonal form, we can determine its eigenvalues by finding the roots of the characteristic polynomial $P_n(\lambda)$:

$$P_n(\lambda) = |\mathbf{A} - \lambda\mathbf{I}| = \begin{vmatrix} \alpha_0 - \lambda & \beta_0 & 0 & \cdots & 0 & 0 \\ \beta_0 & \alpha_1 - \lambda & \beta_1 & \cdots & 0 & 0 \\ 0 & \beta_1 & \alpha_2 - \lambda & \cdots & 0 & 0 \\ \vdots & \vdots & \vdots & \ddots & \vdots & \vdots \\ 0 & 0 & 0 & \cdots & \alpha_{n-2} - \lambda & \beta_{n-2} \\ 0 & 0 & 0 & \cdots & \beta_{n-2} & \alpha_{n-1} - \lambda \end{vmatrix}$$

This root-finding process can be achieved with only $3(n-1)$ multiplications by using the following sequence of operations:

$$P_0(\lambda) = 1$$
$$P_1(\lambda) = \alpha_0 - \lambda$$
$$P_i(\lambda) = (\alpha_{i-1} - \lambda)P_{i-1}(\lambda) - \beta_{i-2}^2 P_{i-2}(\lambda), \quad i = 2, 3, \cdots, n$$

The polynomials $P_0(\lambda), P_1(\lambda), \cdots, P_n(\lambda)$ form a Sturm sequence. The Sturm sequence possesses an important property: the number of sign changes in the sequence $P_0(a), P_1(a), \cdots, P_n(a)$ is equal to the number of roots of $P_n(\lambda)$ that are smaller than a. This property makes it possible to bracket the eigenvalues of a tridiagonal matrix.

In order to determine the global bounds on the eigenvalues of a symmetric matrix $\mathbf{A}$, we can apply Gerschgorin's theorem. The global bounds enclose all the eigenvalues. This theorem states that if λ is an eigenvalue of $\mathbf{A}$, then

$$a_i - r_i \leq \lambda \leq a_i + r_i, \quad i = 1, 2, \cdots, n$$

where

$$a_i = A_{ii}, \quad r_i = \sum_{j=0, j \neq i}^{n-1} |A_{ij}|$$

It follows that the limits on the smallest and the largest eigenvalues are given by

$$\lambda_{\min} \geq \min_i (a_i - r_i), \quad \lambda_{\max} \leq \max_i (a_i + r_i)$$

The Sturm sequence property and Gerschgorin's theorem provide us with convenient tools for bracketing each eigenvalue of a symmetric tridiagonal matrix.

Once the desired eigenvalues are bracketed, they can be found by determining the roots of $P_n(\lambda) = 0$ with a nonlinear system solver, such as the false position method described in Chapter 5.

After the eigenvalues have been computed, the best means of computing the corresponding eigenvectors becomes the inverse power method with eigenvalue shifting. This method has been discussed before, but the algorithm we used was developed for a general real square symmetric method.

Here we will present a version of the method implemented specifically for real symmetric tridiagonal matrices.

Before implementing the method for computing eigenvalues and eigenvectors of a tridiagonal matrix, we need first to develop an efficient LU decomposition and LU solver for linear systems described by a tridiagonal matrix.

LU Decomposition of Tridiagonal Matrices

In Chapter 4, we discussed LU decomposition for general real matrices. Here we present the LU decomposition for tridiagonal matrices.

Consider the solution of $\mathbf{Tx} = \mathbf{b}$ by the LU decomposition, where $\mathbf{T}$ is the general (asymmetric) $n \times n$ tridiagonal matrix:

$$\mathbf{T} = \begin{pmatrix} \alpha_0 & \beta_0 & 0 & 0 & \cdots & 0 & 0 \\ \gamma_0 & \alpha_1 & \beta_1 & 0 & \cdots & 0 & 0 \\ 0 & \gamma_1 & \alpha_2 & \beta_2 & \cdots & 0 & 0 \\ \vdots & \vdots & \vdots & \vdots & \vdots & \vdots & \vdots \\ 0 & 0 & 0 & 0 & \cdots & \alpha_{n-2} & \beta_{n-2} \\ 0 & 0 & 0 & 0 & \cdots & \gamma_{n-2} & \alpha_{n-1} \end{pmatrix}$$

Following a procedure similar to that presented in Chapter 4, we implement an LU decomposition method and an LU solver for the tridiagonal matrix.

Add two public methods, *LUDecomposition* and *LUSolver*, to the *Eigenvalue* class:

```
public static void LUDecomposition(double[] gamma, double[] alpha,
    double[] beta, out double[] gamma1, out double[] alpha1, out double[] beta1)
{
    int n = alpha.GetLength(0);
    for (int i = 1; i < n; i++)
    {
        double c = gamma[i - 1] / alpha[i - 1];
        alpha[i] -= c * beta[i - 1];
        gamma[i - 1] = c;
    }
    gamma1 = gamma;
    alpha1 = alpha;
    beta1 = beta;
}

public static VectorR LUSolver(double[] gamma, double[] alpha,
                    double[] beta, VectorR b)
{
    int n = alpha.GetLength(0);
    for (int i = 1; i < n; i++)
    {
        b[i] -= gamma[i - 1] * b[i - 1];
    }
    b[n - 1] /= alpha[n - 1];
    for (int i = n - 2; i > -1; i--)
```

```
        {
            b[i] = (b[i] - beta[i] * b[i + 1]) / alpha[i];
        }

        return b;
    }
```

The *LUDecomposition* method is used to construct the coefficient matrix. The coefficients after the LU decomposition are stored in new arrays, *gamma1*, *alpha1*, and *beta1*, which will be used by the *LUSolver* method.

Within the *LUSolver* method, the vector **b** will be changed during forward substitution. The solution is stored in the vector **b**, which is returned by the solver.

Implementation

In this section, we implement the methods in C#, which can be used to compute eigenvalues and eigenvectors of a tridiagonal matrix. We first need to implement several utility methods. Add the following methods to the *Eigenvalue* class:

```
public static void SetAlphaBeta(MatrixR T)
{
    int n = T.GetCols();
    Alpha = new double[n];
    Beta = new double[n - 1];
    Alpha[0] = T[0, 0];
    for (int i = 1; i < n; i++)
    {
        Alpha[i] = T[i, i];
        Beta[i - 1] = T[i - 1, i];
    }
}

public static double[] SturmSequence(double lambda)
{
    int n = Alpha.GetLength(0);
    double[] p = new double[n + 1];
    p[0] = 1;
    p[1] = Alpha[0] - lambda;
    for (int i = 2; i < n + 1; i++)
    {
        p[i] = (Alpha[i - 1] - lambda) * p[i - 1] -
                Beta[i - 2] * Beta[i - 2] * p[i - 2];
    }
    return p;
}

public static int NumberOfEigenvalues(double[] p)
{
    int n = p.GetLength(0);
    int sign1 = 1;
    int sign = 0;
    int nEigenvalues = 0;
    for (int i = 1; i < n; i++)
    {
        {
```

```
            if (p[i] > 0.0)
                sign = 1;
            else if (p[i] < 0.0)
                sign = -1;
            else
                sign = -sign1;
            if (sign1 * sign < 0)
                nEigenvalues++;
            sign1 = sign;
        }
        return nEigenvalues;
    }

    public static double[] EigenvalueBound()
    {
        int n = Alpha.GetLength(0);
        double min = Alpha[0] - Math.Abs(Beta[0]);
        double max = Alpha[0] + Math.Abs(Beta[0]);
        double lambda;
        for (int i = 1; i < n-1; i++)
        {
            lambda = Alpha[i] - Math.Abs(Beta[i]) - Math.Abs(Beta[i - 1]);
            if (lambda < min)
                min = lambda;
            lambda = Alpha[i] + Math.Abs(Beta[i]) + Math.Abs(Beta[i - 1]);
            if (lambda > max)
                max = lambda;
        }
        lambda = Alpha[n - 1] - Math.Abs(Beta[n - 2]);
        if (lambda < min)
            min = lambda;
            lambda = Alpha[n - 1] + Math.Abs(Beta[n - 2]);
        if (lambda > max)
            max = lambda;
        return new double[] { min, max };
    }

    public static double[] EigenvalueRange(int nEigenvalues)
    {
        int n = nEigenvalues;
        double[] r = new double[n + 1];
        double lambda, h;
        int nlambda;
        double[] bound = EigenvalueBound();
        r[0] = bound[0];

        for (int i = n; i > 0; i--)
        {
            lambda = 0.5 * (bound[0] + bound[1]);
            h = 0.5 * (bound[1] - bound[0]);
            for (int j = 0; j < 1000; j++)
            {
                double[] p = SturmSequence(lambda);
                nlambda = NumberOfEigenvalues(p);
                h *= 0.5;
```

```
            if (nlambda < i)
                lambda += h;
            else if (nlambda > i)
                lambda -= h;
            else
                break;
        }
        bound[1] = lambda;
        r[i] = lambda;
    }
    return r;
}
```

The *SetAlphaBeta* method allows us to set public properties, *Alpha* and *Beta* arrays, from a symmetric tridiagonal matrix. This method is useful when the user provides the tridiagonal matrix rather than the *Alpha* and *Beta* arrays.

The *SturmSequence* method returns the Sturm sequence associated with the characteristic polynomial of a tridiagonal matrix. The Sturm sequence is needed to bracket the eigenvalues of a tridiagonal matrix.

The *NumberOfEigenvalues* method returns the number of eigenvalues in a tridiagonal matrix that are smaller than a specified value.

Based on Gerschgorin's theorem, the method, *EigenvalueBound*, returns the lower and upper global limits on the eigenvalues of a symmetric diagonal matrix.

Combined with the *SturmSequence* and *EigenvalueBound* methods, the *EigenvalueRange* method brackets the *nEigenvalues* ($\leq n$, specified by the user), the smallest eigenvalues of a symmetric tridiagonal martrix. It returns the sequence with each interval containing exactly one eigenvalue. The algorithm first finds the global limits on the eigenvalues by using the *EigenvalueBound* method. Then, in conjunction with the *SturmSequence* method it determines the order of the returned sequence.

Once the desired eigenvalues have been bracketed, they can be computed by finding the roots of $P_n(\lambda) = 0$ using the *FalsePosition* method. Add a new public static method, *TridiagonalEigenvalues*, to the *Eigenvalue* class:

```
public static double[] TridiagonalEigenvalues(int nEigenvalues)
{
    double[] lambda = new double[nEigenvalues];
    double[] range = EigenvalueRange(nEigenvalues);
    for (int i = 0; i < nEigenvalues; i++)
    {
        lambda[i] =
            NonlinearSystem.FalsePosition(f, range[i], range[i + 1], 1e-8);
    }
    return lambda;
}

private static double f(double x)
{
    double[] p = SturmSequence(x);
    return p[p.GetLength(0) - 1];
}
```

This method computes the *nEigenvalues* – the smallest eigenvalues of a symmetric tridiagonal matrix, using the *FalsePosition* method.

Next, we can calculate the eigenvectors by using the technique similar to the inverse power method. Add a new method, *TridiagonalEigenvectors*, to the *Eigenvalue* class:

```
public static VectorR TridiagonalEigenvector(double s,
                        double tolerance, out double lambda)
{
        int n = Alpha.GetLength(0);
        double[] gamma = (double[])Beta.Clone();
        double[] beta = (double[])Beta.Clone();
        double[] alpha = new double[n];
        for (int i = 0; i < n; i++)
        {
            alpha[i] = Alpha[i] - s;
        }
        double[] gamma1, alpha1, beta1;
        LUDecomposition(gamma, alpha, beta, out gamma1, out alpha1, out beta1);
        VectorR x = new VectorR(n);
        Random random = new Random();
        for (int i = 0; i < n; i++)
            x[i] = random.NextDouble();
        x.Normalize();
        VectorR x1 = new VectorR(n); ;
        double sign;
        do
        { .
            x1 = x.Clone();
            LUSolver(gamma1, alpha1, beta1, x);
            x.Normalize();
            if (VectorR.DotProduct(x1, x) < 0.0)
            {
                sign = -1.0;
                x = -x;
            }
            else
            {
                sign = 1.0;
            }
        }
        while ((x - x1).GetNorm() > tolerance);
        lambda = s + sign / x.GetNorm();
        return x;
}
```

In this method, the LU decomposition and LU solver are used to solve the linear system described by the tridiagonal matrix. This method executes much faster than the inverse power method because it takes advantage of the tridiagonal nature of the matrix.

The methods implemented in this section, combined with the householder technique, are the most efficient techniques for finding all the eigenvalues and eigenvectors of a large real symmetric matrix.

Examples

Here, we will show you how to compute eigenvalues and eigenvectors using the method we developed in this section. First, we want to compute all the eigenvalues and eigenvectors of the following symmetric matrix:

$$\mathbf{A} = \begin{pmatrix} 5 & 1 & 2 & 2 & 4 \\ 1 & 1 & 2 & 1 & 0 \\ 2 & 2 & 0 & 2 & 1 \\ 2 & 1 & 2 & 1 & 2 \\ 4 & 0 & 1 & 2 & 4 \end{pmatrix}$$

Add a new static method, *TestTridiagonalEigenvalues*, to the *Program.cs* file:

```
static void TestTridiagonalEigenvalues()
{
    MatrixR A = new MatrixR(new double[,]{{ 5, 1, 2, 2, 4 },
                                          { 1, 1, 2, 1, 0},
                                          { 2, 2, 0, 2, 1},
                                          { 2, 1, 2, 1, 2},
                                          { 4, 0, 1, 2, 4}});
    int nn = 5;
    MatrixR xx = new MatrixR(A.GetCols(), nn);
    MatrixR V = Eigenvalue.Tridiagonalize(A);
    double[] lambda = Eigenvalue.TridiagonalEigenvalues(nn);
    for (int i = 0; i < nn; i++)
    {
        double s = lambda[i] * 1.001;
        double lam;
        VectorR x = Eigenvalue.TridiagonalEigenvector(s, 1e-8, out lam);
        for (int j = 0; j < A.GetCols(); j++)
        xx[j, i] = x[j];
    }
    xx = V * xx;

    Console.WriteLine("\n Results from the tridiagonalization method:");
    Console.WriteLine("\n Eigenvalues: \n ({0,10:n6}  {1,10:n6}  {2,10:n6}
                    {3,10:n6}  {4,10:n6})", lambda[0], lambda[1], lambda[2],
                    lambda[3], lambda[4]);
    Console.WriteLine("\n Eigenvectors:");
    for (int i = 0; i < 5; i++)
    {
        Console.WriteLine(" ({0,10:n6}  {1,10:n6}  {2,10:n6}  {3,10:n6}
                    {4,10:n6})", xx[i,0], xx[i,1], xx[i,2], xx[i,3], xx[i,4]);
    }

    A = new MatrixR(new double[,]{{ 5, 1, 2, 2, 4 },
                                  { 1, 1, 2, 1, 0},
                                  { 2, 2, 0, 2, 1},
                                  { 2, 1, 2, 1, 2},
                                  { 4, 0, 1, 2, 4}});

    MatrixR xm;
```

```
VectorR lamb;
Eigenvalue.Jacobi(A, 1e-8, out xm, out lamb);

Console.WriteLine("\n\n Results from the Jacobi method:");
Console.WriteLine("\n Eigenvalues: \n ({0,10:n6}   {1,10:n6}   {2,10:n6}
                    {3,10:n6}   {4,10:n6})", lamb[4], lamb[3], lamb[2],
                    lamb[1], lamb[0]);
Console.WriteLine("\n Eigenvectors:");
for (int i = 0; i < 5; i++)
{
    Console.WriteLine(" ({0,10:n6}   {1,10:n6}   {2,10:n6}   {3,10:n6}
                        {4,10:n6})", xm[i, 4], xm[i, 3], xm[i, 2],
                        xm[i, 1], xm[i, 0]);
}
}
```

Here we first call the *Tridiagonalize* method to reduce the original square matrix to tridiagonal form and compute the corresponding transformation matrix **V**. Then we call the *TridiagonalEigenvalues* method to compute all of the five eigenvalues. Next, the *TridiagonalEigenvector* method is used to calculate the corresponding eigenvectors. Note that the *TridiagonalEigenvector* method only returns a single eigenvector for each corresponding eigenvalue (slight shift from the exact eigenvalue). Therefore a for-loop is used to compute all of the eigenvectors. The resulting eigenvectors are stored in the matrix *xx*.

For comparison's sake, we also solve the same problem using the Jacobi method, which gives us all of the eigenvalues and corresponding eigenvectors.

Running this example produces the results shown in Figure 15-4.

```
Results from the tridiagonalization method:

Eigenvalues:
( -2.043362    -0.577672     0.519702      2.652048     10.449283)

Eigenvectors:
(  0.105359     0.173821     0.703475      0.090258      0.675027)
(  0.381641    -0.605547     0.019381     -0.677839      0.166799)
( -0.828741     0.074549    -0.103553     -0.467130      0.280532)
(  0.390652     0.632436    -0.505677     -0.275822      0.340042)
( -0.061886    -0.444484    -0.488175      0.487955      0.567617)


Results from the Jacobi method:

Eigenvalues:
( -2.043362    -0.577672     0.519702      2.652048     10.449283)

Eigenvectors:
( -0.105359     0.173821    -0.703475     -0.090258      0.675027)
( -0.381641    -0.605547    -0.019381      0.677839      0.166799)
(  0.828741     0.074549     0.103553      0.467130      0.280532)
( -0.390652     0.632436     0.505677      0.275822      0.340042)
(  0.061886    -0.444484     0.488175     -0.487955      0.567617)
```

Figure 15-4 Eigenvalues and corresponding eigenvectors from different methods.

You can see that the results from the tridiagonalization-based method are almost identical to those obtained from the Jacobi method, except for a 180-degree phase difference in some eigenvectors.

The real power of the tridiagonalization-based method is its ability to find eigenvalues and eigenvectors for large matrices. Now we want to use the *TridiagonalEigenvalues* method to determine the 20 smallest eigenvalues of a 200×200 tridiagonal matrix:

$$
\mathbf{T} = \begin{pmatrix}
3 & -1.5 & 0 & 0 & \cdots & 0 & 0 \\
-1.5 & 3 & -1.5 & 0 & \cdots & 0 & 0 \\
0 & -1.5 & 3 & -1.5 & \cdots & 0 & 0 \\
\vdots & \vdots & \vdots & \vdots & \vdots & \vdots & \vdots \\
0 & 0 & 0 & 0 & \cdots & 3 & -1.5 \\
0 & 0 & 0 & 0 & \cdots & -1.5 & 3
\end{pmatrix}
$$

Add a new static method, *TestLargeMatrix*, to the *Program.cs* file:

```
static void TestLargeMatrix()
{
    int n = 200;
    Eigenvalue.Alpha = new double[n];
    Eigenvalue.Beta = new double[n-1];
    Eigenvalue.Alpha[0] = 3.0;
    for (int i = 0; i < n-1; i++)
    {
        Eigenvalue.Alpha[i+1] = 3.0;
        Eigenvalue.Beta[i] = -1.5;
    }

    double[] lambda = Eigenvalue.TridiagonalEigenvalues(20);
    Console.WriteLine("\n 20 smallest eigenvalues for a 200 x 200
                    tridiagonal matrix:\n");
    Console.WriteLine(" {0,8:n6}   {1,8:n6}   {2,8:n6}   {3,8:n6}   {4,8:n6}",
            lambda[0], lambda[1], lambda[2], lambda[3], lambda[4]);
    Console.WriteLine(" {0,8:n6}   {1,8:n6}   {2,8:n6}   {3,8:n6}   {4,8:n6}",
            lambda[5], lambda[6], lambda[7], lambda[8], lambda[9]);
    Console.WriteLine(" {0,8:n6}   {1,8:n6}   {2,8:n6}   {3,8:n6}   {4,8:n6}",
            lambda[10], lambda[11], lambda[12], lambda[13], lambda[14]);
    Console.WriteLine(" {0,8:n6}   {1,8:n6}   {2,8:n6}   {3,8:n6}   {4,8:n6}",
            lambda[15], lambda[16], lambda[17], lambda[18], lambda[19]);
}
```

Here we directly specify the *Alpha* and *Beta* arrays. Running this example generates results shown in Figure 15-5.

```
20 smallest eigenvalues of a 200 x 200 tridiagonal matrix:

0.000366   0.001466   0.003297   0.005861   0.009156
0.013182   0.017937   0.023421   0.029632   0.036569
0.044230   0.052612   0.061715   0.071535   0.082071
0.093320   0.105279   0.117944   0.131314   0.145385
```

Figure 15-5 First 20 Eigenvalues of a large tridiagonal matrix.

Index

Also Available from Dr. Jack Xu:

Programming expert Jack Xu provides you with everything you need to add advanced graphics to your applications in this in-depth introduction to graphics programming with the Windows Presentation Foundation (WPF). From basic 2D shapes to complex interactive 3D models, Dr. Xu clearly explains every step it takes to build a variety of WPF graphics applications using code examples. You'll learn how to use WPF to create impressive graphics effects and high-fidelity user interfaces. This book includes:

- An overview of WPF graphics capabilities and the mathematical basics of computer graphics.
- Step-by-step procedures to create a variety of 2D and 3D custom geometries and shapes with complete ready-to-run XAML and C# code for each application.
- Powerful 2D chart applications and user controls that can be directly used in your WPF applications or can be easily modified to create your own sophisticated chart packages.
- Detailed procedures of how to create various 3D surfaces in WPF using rectangular meshes.
- An introduction to building physics-based models, games, and fractals.
- Advanced color, lighting, and shading effects for 3D graphics objects.
- Direct interaction with graphics models, including animation, transformations, hit-testing, and mouse events.

For details, see www.drxudotnet.com

Also Available from Dr. Jack Xu:

For .NET developers, creating professional charts and graphics in your C# applications is now easier than ever before. Practical C# Charts and Graphics is the perfect guide to learning all the basics for creating advanced chart and graphics applications in C#. The book clearly explains practical chart and graphics methods and their underlying algorithms. The book contains:

- Overview of GDI+ graphics capabilities and mathematical basics of computer charting and graphics.
- Step-by-step procedures to create a variety of 2D and 3D charts and graphics with complete ready-to-run C# code for each application.
- Powerful 2D and 3D chart packages and user controls that can be directly used in your C# applications or can be easily modified to create your own sophisticated charts and graphics packages.
- Detailed procedures to create C# spreadsheet-like chart and graphics applications.
- Introduction for how to use Microsoft's Excel charts in your C# applications.

For details, see www.drxudotnet.com

Also Available from Dr. Jack Xu:

Programming Expert Jack Xu provides everything you need to create advanced graphics and user interfaces in your Web applications using Silverlight 2. From simple user interfaces and 2D shapes to complex custom user controls and 3D graphics objects, Dr. Xu uses code examples to clearly explain every step it takes to build a variety of Silverlight applications. You'll learn how to use Silverlight to develop impressive graphics effects and high-fidelity user interfaces. This book includes:

- A complete, in-depth instruction on practical Silverlight programming. After reading this book and running the example programs, you will be able to add various sophisticated graphics and interactive user interfaces to your Web applications.
- About 100 ready-to-run example programs that allow you to explore the UI and graphics programming techniques described in the book. These examples can be used to better understand how graphics algorithms work. You can modify the code examples or add new features to them to form the basis of your own projects. Some of the example code listings provided in this book already are sophisticated graphics and user interfaces that can be used directly in your own real-world Silverlight applications.
- Many classes in the sample code listings that you will find useful in your Silverlight programming. These classes contain matrix manipulation, coordinate transformation, color maps, 3D model library, custom controls, and other useful utility classes. You can extract these classes and plug them into your own Silverlight applications.

For details, see www.drxudotnet.co

Printed in the United States
144887LV00003B/88/P

9 780979 372537